Numerical Methods for Fractional Calculus

CHAPMAN & HALL/CRC
Numerical Analysis and Scientific Computing

Aims and scope:
Scientific computing and numerical analysis provide invaluable tools for the sciences and engineering. This series aims to capture new developments and summarize state-of-the-art methods over the whole spectrum of these fields. It will include a broad range of textbooks, monographs, and handbooks. Volumes in theory, including discretisation techniques, numerical algorithms, multiscale techniques, parallel and distributed algorithms, as well as applications of these methods in multi-disciplinary fields, are welcome. The inclusion of concrete real-world examples is highly encouraged. This series is meant to appeal to students and researchers in mathematics, engineering, and computational science.

Proposals for the series should be submitted to one of the series editors above or directly to:
CRC Press, Taylor & Francis Group
3 Park Square, Milton Park
Abingdon, Oxfordshire OX14 4RN
UK

Published Titles

Classical and Modern Numerical Analysis: Theory, Methods and Practice
Azmy S. Ackleh, Edward James Allen, Ralph Baker Kearfott, and Padmanabhan Seshaiyer

Cloud Computing: Data-Intensive Computing and Scheduling *Frédéric Magoulès, Jie Pan, and Fei Teng*

Computational Fluid Dynamics *Frédéric Magoulès*

A Concise Introduction to Image Processing using C++ *Meiqing Wang and Choi-Hong Lai*

Coupled Systems: Theory, Models, and Applications in Engineering *Juergen Geiser*

Decomposition Methods for Differential Equations: Theory and Applications *Juergen Geiser*

Designing Scientific Applications on GPUs *Raphaël Couturier*

Desktop Grid Computing *Christophe Cérin and Gilles Fedak*

Discrete Dynamical Systems and Chaotic Machines: Theory and Applications
Jacques M. Bahi and Christophe Guyeux

Discrete Variational Derivative Method: A Structure-Preserving Numerical Method for Partial Differential Equations *Daisuke Furihata and Takayasu Matsuo*

Grid Resource Management: Toward Virtual and Services Compliant Grid Computing
Frédéric Magoulès, Thi-Mai-Huong Nguyen, and Lei Yu

Fundamentals of Grid Computing: Theory, Algorithms and Technologies *Frédéric Magoulès*

Handbook of Sinc Numerical Methods *Frank Stenger*

Introduction to Grid Computing *Frédéric Magoulès, Jie Pan, Kiat-An Tan, and Abhinit Kumar*

Iterative Splitting Methods for Differential Equations *Juergen Geiser*

Mathematical Objects in C++: Computational Tools in a Unified Object-Oriented Approach
Yair Shapira

Numerical Linear Approximation in C *Nabih N. Abdelmalek and William A. Malek*

Numerical Methods and Optimization: An Introduction *Sergiy Butenko and Panos M. Pardalos*

Numerical Methods for Fractional Calculus *Changpin Li and Fanhai Zeng*

Numerical Techniques for Direct and Large-Eddy Simulations *Xi Jiang and Choi-Hong Lai*

Parallel Algorithms *Henri Casanova, Arnaud Legrand, and Yves Robert*

Parallel Iterative Algorithms: From Sequential to Grid Computing *Jacques M. Bahi, Sylvain Contassot-Vivier, and Raphaël Couturier*

Particle Swarm Optimisation: Classical and Quantum Perspectives *Jun Sun, Choi-Hong Lai, and Xiao-Jun Wu*

XML in Scientific Computing *C. Pozrikidis*

Numerical Methods
for
Fractional Calculus

Changpin Li

Department of Mathematics
Shanghai University
China

Fanhai Zeng

Department of Mathematics
Shanghai University
China

CRC Press
Taylor & Francis Group
Boca Raton London New York

CRC Press is an imprint of the
Taylor & Francis Group, an **informa** business
A CHAPMAN & HALL BOOK

CRC Press
Taylor & Francis Group
6000 Broken Sound Parkway NW, Suite 300
Boca Raton, FL 33487-2742

First issued in paperback 2020

ISBN-13: 978-1-4822-5380-1 (hbk)
ISBN-13: 978-0-367-65879-3 (pbk)

Visit the Taylor & Francis Web site at
http://www.taylorandfrancis.com

and the CRC Press Web site at
http://www.crcpress.com

Contents

Foreword

Fractional calculus is often regarded as a branch of mathematical analysis which deals with integro-differential equations where the integrals are of the convolution type and exhibit (weakly singular) kernels of the power-law type. It has a history of at least three hundred years, since it can be dated back to a letter from G.W. Leibniz to G. L'Hôpital, dated 30 September 1695, in which the meaning of the one-half order derivative was first discussed and some remarks about its possibility were made. Subsequent mention of fractional derivatives was made by L. Euler (1730), J.L. Lagrange (1772), P.S. Laplace (1812), S.F. Lacroix (1819), J.B.J. Fourier (1822), N.H. Abel (1823), J. Liouville (1832), B. Riemann (1847), H.L. Green (1859), H. Holmgren (1865), A.K. Grünwald (1867), A.V. Letnikov (1868), N.Ya. Sonin (1869), H. Laurent (1884), P.A. Nekrassov (1888), A. Krug (1890), O. Heaviside (1892), S. Pincherle (1902), H. Weyl (1919), P. Lévy (1923), A. Marchaud (1927), H.T. Davis (1936), A. Zygmund (1945), M. Riesz (1949), and W. Feller (1952), to cite some relevant contributors up to the middle of the last century. For the entire history of fractional calculus, refer to K.B. Oldham and J. Spanier's book– *The Fractional Calculus: Theory and Applications of Differentiation and Integration to Arbitrary Order* (Academic Press, New York, renewed 2002). Some complementary materials can be found from J.A. Tenreiro Machado, V. Kiryakova and F. Mainardi's posters (poster depicting the recent history of fractional calculus, *Fractional Calculus and Applied Analysis* 13(3), 329-334, 2010; poster depicting the old history of fractional calculus, *Fractional Calculus and Applied Analysis* 13(4), 447-454, 2010) and a brief introduction also by them (Recent history of fractional calculus, *Communications in Nonlinear Science and Numerical Simulation* 16, 1140-1153, 2011).

Roughly speaking, fractional calculus underwent two stages: from its beginning to the 1970s, and after 1970s. In the first stage, fractional calculus was studied mainly by mathematicians as an abstract area containing only pure mathematical manipulations of little or no use. In the second stage, the paradigm began to shift from pure mathematical research to application in various fields, such as long-memory processes and materials, anomalous diffusion, long-range interactions, long-term behaviors, power laws, allometric scaling laws, and so on.

Due to applications of fractional calculus, various kinds of numerical methods have independently appeared in periodicals. This book aims to collect and sort out these studies, including the authors' work. Loosely speaking, the present book contains (1) numerical methods for fractional integrals and fractional derivatives, (2) finite difference methods for fractional ordinary/partial differential equations, and (3) finite element methods for fractional partial differential equations. Due to the rapid development of fractional numerical methods, more and more publications are emerging. However, very recent publications are not included or introduced since this book is designed for beginners.

Last but not least, we thank Professors Vo Anh, Kevin Burrage, Guanrong Chen, Wen Chen, YangQuan Chen, Qiang Du, Jinqiao Duan, Roberto Garrappa, Benyu Guo, Haiyan Hu, George Em Karniadakis, Virginia Kiryakova, Jürgen Kurths, Fawang Liu, Francesco Mainardi, Igor Podlubny, Zhongci Shi, Yifa Tang, Ian Turner, Blas M. Vinagre, Hong Wang, Xiaohua Xia, Dingyu Xue, and Weiqiu Zhu for their strong support, unselfish cooperation, and for providing suggestions for revision. We greatly appreciate Sunil Nair and Sarfraz Khan for sparing no pains to inform us, replying to us and explaining various details regarding this book. The first author particularly thanks his PhD students Fanhai Zeng (who is also the second author of this book) and An Chen for collecting the materials and difficult typesetting. He also thanks his PhD students Jianxiong Cao, Hengfei Ding, Peng Guo, Yutian Ma, Fengrong Zhang, Zhengang Zhao, and Yunying Zheng for their careful reading and for providing correction suggestions. CL acknowledges the financial support from National Natural Science Foundation of China (10872119, 11372170) and the Key Program of Shanghai Municipal Education Commission (12ZZ084).

Changpin Li and Fanhai Zeng
April 2015

Preface

Fractional calculus (which includes fractional integration and fractional differentiation) is as old as its familiar counterpart, classical calculus (or integer order calculus). For quite a long time it developed slowly. However, in the past few decades, fractional calculus has attracted increasing interest due to its applications in science and engineering. Fractional derivatives have provided excellent tools to describe various materials and processes with memory and hereditary properties, etc.; and fractional differential equation models in these applied fields are thus established.

There are several analytical methods used to solve very special (mostly linear) fractional differential equations (FDEs), such as the Fourier transform method, the Laplace transform method, the Mellin transform method and the Green function method. Hence, developing efficient and reliable numerical methods for solving general FDEs is of particular usefulness in application. The book mainly focuses on investigating numerical methods for fractional integrals, fractional derivatives, and fractional differential equations.

There are five chapters in this book. In Chapter 1, the basic definitions and properties of fractional integrals and derivatives are introduced, including the most frequently used Riemann–Liouville integral, the Riemann–Liouville derivative, the Caputo derivative, and some other fractional derivatives. Furthermore, important and/or complicated properties are studied. Also, the corresponding physical meaning of fractional calculus and the definite conditions for fractional differential equations are included.

In Chapter 2, numerical methods for fractional integrals and fractional derivatives are displayed in detail. We first derive the numerical schemes based on polynomial interpolation, Gauss interpolation and linear multistep methods for the fractional integrals (or Riemann–Liouville integrals). Then we investigate the Grünwald–Letnikov approximation, L1, L2 and L2C methods for the Riemann–Liouville derivatives. The natural generalization of the above methods for the Caputo derivatives and the Riesz derivatives are also introduced. These discretized schemes are useful for the discussions in the subsequent chapters.

In the next chapter, the finite difference methods for fractional ordinary differential equations are investigated. These finite difference methods mainly include the fractional Euler method, the fractional Adams method, the high order method, the fractional linear multistep method, and their various variants. The stability, convergence, and error estimates of these methods are also carefully studied.

Next, the finite difference methods for fractional partial differential equations are presented in Chapter 4. The fractional partial differential equations in this chapter include (1) the time-fractional differential equations (with Riemann–Liouville derivative or Caputo derivative) in one spatial dimension, (2) the space-fractional differential equations (with one-sided Riemann–Liouville derivative, or two-sided Riemann–Liouville derivative, or the Riesz derivative) in one spatial dimension, (3) the time-space fractional differential equation in one spatial dimension, and (4) the fractional differential equations in two spatial dimensions. The derived numerical methods mainly consist of the Euler method, the Crank–Nicolson method and the fractional linear multistep methods. The stability, convergence, and error estimates are studied. Many numerical examples are also displayed, which support the theoretical analysis.

Generally speaking, the fractional finite difference methods are convenient to implement but the smooth conditions of the solutions often need to be assumed. If the solutions have good smoothness (and the domains are regular), then spectral methods are possibly the best solvers. However, fractional calculus seems to be a useful tool to deal with nonsmooth problems. So the finite element method is often regarded as one of main methods for solving fractional differential equations. In the last chapter, the finite element methods for fractional partial differential equations are presented and analyzed. We first introduce the basic framework of the finite element methods for fractional differential equations. Then we establish the fully discrete schemes for time-space fractional equations, where the time-fractional derivatives are discretized by the finite difference methods, and the space-fractional derivatives are approximated by the finite element methods. The stability, convergence, and error estimates for the established methods are also studied. Additional material is available from the CRC Web site: http://www.crcpress.com/product/isbn/9781482253801

Due to broad applications of fractional calculus, seeking numerical algorithms with high accuracy, rapid convergence, and less storage is becoming more and more important. This book is just a primer in this respect. We hope it can offer fresh stimuli for the fractional calculus community to further promote and develop the cutting-edge research on numerical fractional calculus.

List of Figures

List of Tables

Chapter 1

Introduction to Fractional Calculus

In this chapter, we first introduce fractional calculus (i.e., fractional integration and fractional differentiation). Generally speaking, the fractional integral mainly means (fractional) Riemann–Liouville integral. The fractional derivatives consist of at least six kinds of definitions, but they are not equivalent. Here, we present the most frequently used fractional integral and derivatives, i.e., the Riemann–Liouville integral, the Riemann–Liouville derivative, the Caputo derivative, etc. Then we study their important properties, some of which are easily confused. Besides, we further introduce the definite conditions of fractional differential equations which are often misused.

1.1 Fractional Integrals and Derivatives

Fractional calculus is not a new topic, in reality it has almost the same history as that of classical calculus. It can be dated back to the Leibniz's letter to L'Hôpital, see [72, 115, 118], dated 30 September 1695, in which the meaning of the one-half order derivative was first discussed with some remarks about its possibility. Nowadays, fractional calculus is undergoing rapid development, with more and more convincing applications in the real world, see [74, 75, 80, 85] and references therein. Maybe one notices that another word "fractal" sometimes takes the place of "fractional calculus" in some situations. However, this may be not proper. As far as we know, *fractal* [108] is in the realm of geometry, while fractional calculus belongs to analysis. Although some studies displayed the relations between them, they are different.

It is known that calculus means integration and differentiation. Fractional calculus, as its name suggests, refers to fractional integration and fractional differentiation. Fractional integration often means *Riemann–Liouville integral*. But for fractional differentiation, there are several kinds of fractional derivatives. In the following, some definitions are introduced [68, 124, 134].

Definition 1 *The left fractional integral (or the left Riemann–Liouville integral) and right fractional integral (or the right Riemann–Liouville integral) with order $\alpha > 0$ of the given function $f(t)$, $t \in (a,b)$ are defined as*

$$\mathrm{D}_{a,t}^{-\alpha} f(t) = {}_{RL}\mathrm{D}_{a,t}^{-\alpha} f(t) = \frac{1}{\Gamma(\alpha)} \int_a^t (t-s)^{\alpha-1} f(s)\,\mathrm{d}s, \tag{1.1}$$

and

$$\mathrm{D}_{t,b}^{-\alpha} f(t) = {}_{RL}\mathrm{D}_{t,b}^{-\alpha} f(t) = \frac{1}{\Gamma(\alpha)} \int_t^b (s-t)^{\alpha-1} f(s)\,\mathrm{d}s, \tag{1.2}$$

respectively, where $\Gamma(\cdot)$ is the Euler's gamma function.

Definition 2 *The left and right Grünwald–Letnikov derivatives with order $\alpha > 0$ of the given function $f(t)$, $t \in (a,b)$ are defined as*

$$_{GL}\mathrm{D}_{a,t}^{\alpha} f(t) = \lim_{\substack{h\to 0\\ Nh=t-a}} h^{-\alpha} \sum_{j=0}^{N} (-1)^j \binom{\alpha}{j} f(t-jh), \tag{1.3}$$

and

$$_{GL}\mathrm{D}_{t,b}^{\alpha} f(t) = \lim_{\substack{h\to 0\\ Nh=b-t}} h^{-\alpha} \sum_{j=0}^{N} (-1)^j \binom{\alpha}{j} f(t+jh), \tag{1.4}$$

respectively.

Definition 3 *The left and right Riemann–Liouville derivatives with order $\alpha > 0$ of the given function $f(t)$, $t \in (a,b)$ are defined as*

$$\begin{aligned} {}_{RL}\mathrm{D}_{a,t}^{\alpha} f(t) &= \frac{\mathrm{d}^m}{\mathrm{d}t^m}\left[\mathrm{D}_{a,t}^{-(m-\alpha)} f(t)\right] \\ &= \frac{1}{\Gamma(m-\alpha)} \frac{\mathrm{d}^m}{\mathrm{d}t^m} \int_a^t (t-s)^{m-\alpha-1} f(s)\,\mathrm{d}s, \end{aligned} \tag{1.5}$$

and

$$\begin{aligned} {}_{RL}\mathrm{D}_{t,b}^{\alpha} f(t) &= (-1)^m \frac{\mathrm{d}^m}{\mathrm{d}t^m}\left[\mathrm{D}_{t,b}^{-(m-\alpha)} f(t)\right] \\ &= \frac{(-1)^m}{\Gamma(m-\alpha)} \frac{\mathrm{d}^m}{\mathrm{d}t^m} \int_t^b (s-t)^{m-\alpha-1} f(s)\,\mathrm{d}s, \end{aligned} \tag{1.6}$$

respectively, where m is a positive integer satisfying $m-1 \le \alpha < m$.

Definition 4 *The left and right Caputo derivatives with order $\alpha > 0$ of the given function $f(t)$, $t \in (a,b)$ are defined as*

$$\begin{aligned} {}_{C}\mathrm{D}_{a,t}^{\alpha} f(t) &= \mathrm{D}_{a,t}^{-(m-\alpha)}\left[f^{(m)}(t)\right] \\ &= \frac{1}{\Gamma(m-\alpha)} \int_a^t (t-s)^{m-\alpha-1} f^{(m)}(s)\,\mathrm{d}s, \end{aligned} \tag{1.7}$$

and

$$_{C}\mathrm{D}_{t,b}^{\alpha} f(t) = \frac{(-1)^m}{\Gamma(m-\alpha)} \int_t^b (s-t)^{m-\alpha-1} f^{(m)}(s)\,\mathrm{d}s, \tag{1.8}$$

respectively, where m is a positive integer satisfying $m-1 < \alpha \le m$.

Definition 5 *The Riesz derivative with order $\alpha > 0$ of the given function $f(t)$, $t \in (a,b)$ is defined as*

$$ {}_{RZ}\mathrm{D}_t^\alpha f(t) = c_\alpha \left({}_{RL}\mathrm{D}_{a,t}^\alpha f(t) + {}_{RL}\mathrm{D}_{t,b}^\alpha f(t) \right), \tag{1.9} $$

where $c_\alpha = -\frac{1}{2\cos(\alpha\pi/2)}, \alpha \neq 2k+1, k = 0, 1, \cdots$. ${}_{RZ}\mathrm{D}_t^\alpha f(t)$ *is sometimes expressed as* $\frac{\partial^\alpha f(t)}{\partial |t|^\alpha}$.

In the above definitions, the initial value a is often set to zero. When we say the fractional integral (or Riemann–Liouville integral), the Grünwald–Letnikov derivative, the Riemann–Liouville derivative, and the Caputo derivative, we often mean the left fractional integral (or the left Riemann–Liouville integral), the left Grünwald–Letnikov derivative, the left Riemann–Liouville derivative, and the left Caputo derivative, respectively if no confusion is caused.

Generally speaking, the above definitions of the Grünwald–Letnikov derivative, the Riemann–Liouville derivative, and the Caputo derivative are not equivalent. If $f(t)$ is suitably smooth, i.e. $f \in C^m[a,b]$, then the Grünwald–Letnikov derivative of $f(t)$ and the Riemann–Liouville derivative of $f(t)$ are equivalent, that is

$$ {}_{RL}\mathrm{D}_{a,t}^\alpha f(t) = {}_{GL}\mathrm{D}_{a,t}^\alpha f(t), \quad {}_{RL}\mathrm{D}_{t,b}^\alpha f(t) = {}_{GL}\mathrm{D}_{t,b}^\alpha f(t). \tag{1.10} $$

The Riemann–Liouville derivative and Caputo derivative of $f(t)$ have following relation [124]

$$ {}_{RL}\mathrm{D}_{a,t}^\alpha f(t) = {}_{C}\mathrm{D}_{a,t}^\alpha f(t) + \sum_{k=0}^{m-1} \frac{f^{(k)}(a)(t-a)^{k-\alpha}}{\Gamma(k+1-\alpha)}, \tag{1.11} $$

where $m-1 < \alpha < m$, m is a positive integer, $f \in C^{m-1}[a,t]$ and $f^{(m)}$ is integrable on $[a,t]$. In fact, (1.11) can be obtained by repeatedly performing integration by parts. Furthermore, if $f \in C^m[a,t]$, then from (1.11) or the Taylor series expansion, we have

$$ {}_{RL}\mathrm{D}_{a,t}^\alpha [f(t) - \phi(t)] = {}_{C}\mathrm{D}_{a,t}^\alpha f(t), \tag{1.12} $$

where $\phi(t) = \sum\limits_{k=0}^{m-1} \frac{f^{(k)}(a)}{\Gamma(k+1)}(t-a)^k$. On the other hand, it is easy to find that

$$ {}_{RL}\mathrm{D}_{a,t}^\alpha f(t) = {}_{C}\mathrm{D}_{a,t}^\alpha f(t) \tag{1.13} $$

if $f^{(k)}(a) = 0$ $(k = 0, 1, 2, \cdots, m-1, m-1 < \alpha < m)$, or $a = -\infty$.

For the continuous function $f(t)$, one has

$$ \lim_{\alpha \to 0^+} \mathrm{D}_{a,t}^{-\alpha} f(t) = f(t), \quad \alpha > 0. $$

Suppose that $f(t)$ is suitably smooth, $m-1 < \alpha < m$, m is a positive integer. Then one has

$$ \begin{aligned} &\lim_{\alpha \to m^-} {}_{RL}\mathrm{D}_{a,t}^\alpha f(t) = f^{(m)}(t), \qquad \lim_{\alpha \to (m-1)^+} {}_{RL}\mathrm{D}_{a,t}^\alpha f(t) = f^{(m-1)}(t); \\ &\lim_{\alpha \to m^-} {}_{C}\mathrm{D}_{a,t}^\alpha f(t) = f^{(m)}(t), \qquad \lim_{\alpha \to (m-1)^+} {}_{C}\mathrm{D}_{a,t}^\alpha f(t) = f^{(m-1)}(t) - f^{(m-1)}(0). \end{aligned} \tag{1.14} $$

Obviously, the Riemann–Liouville derivative is reduced to the classical derivative when the fractional order α approaches an integer for the fixed t, but it is not the case for the Caputo derivative if the homogeneous initial conditions are not satisfied.

Because of the relations (1.11) and (1.10), we mainly focus on introducing the properties of the Riemann–Liouville derivative operator below. Next, we introduce some further properties for the fractional integral operator and the Riemann–Liouville derivative operator.

Proposition 1.1.1 ([131]) *The left and right Riemann–Liouville fractional integral operators satisfy the following semi-group properties*

$$\mathrm{D}_{a,t}^{-\alpha}\mathrm{D}_{a,t}^{-\beta}f(t)=\mathrm{D}_{a,t}^{-\beta}\mathrm{D}_{a,t}^{-\alpha}f(t)=\mathrm{D}_{a,t}^{-\alpha-\beta}f(t), \tag{1.15}$$

$$\mathrm{D}_{t,b}^{-\alpha}\mathrm{D}_{t,b}^{-\beta}f(t)=\mathrm{D}_{t,b}^{-\beta}\mathrm{D}_{t,b}^{-\alpha}f(t)=\mathrm{D}_{t,b}^{-\alpha-\beta}f(t), \tag{1.16}$$

where $\alpha,\beta>0$. *If* $f(t)$ *is continuous on* $[a,b]$, *then*

$$\lim_{t\to a}\mathrm{D}_{a,t}^{-\alpha}f(t)=\lim_{t\to b}\mathrm{D}_{t,b}^{-\alpha}f(t)=0,\quad \forall\alpha>0. \tag{1.17}$$

Proposition 1.1.2 ([124, 131]) *The left and right Riemann–Liouville fractional derivative operators satisfy the following properties*

$${}_{RL}\mathrm{D}_{a,t}^{\alpha}\mathrm{D}_{a,t}^{-\alpha}f(t)=f(t), \tag{1.18}$$

$${}_{RL}\mathrm{D}_{t,b}^{\alpha}\mathrm{D}_{t,b}^{-\alpha}f(t)=f(t), \tag{1.19}$$

where $\alpha>0$.

Proposition 1.1.3 ([124, 131]) *The left and right Riemann–Liouville fractional derivative operators satisfy the following properties*

$$\mathrm{D}_{a,t}^{-\alpha}\left({}_{RL}\mathrm{D}_{a,t}^{\alpha}f(t)\right)=f(t)-\sum_{j=1}^{m}\left[{}_{RL}\mathrm{D}_{a,t}^{\alpha-j}f(t)\right]_{t=a}\frac{(t-a)^{\alpha-j}}{\Gamma(\alpha-j+1)}, \tag{1.20}$$

$$\mathrm{D}_{t,b}^{-\alpha}\left({}_{RL}\mathrm{D}_{t,b}^{\alpha}f(t)\right)=f(t)-\sum_{j=1}^{m}\left[{}_{RL}\mathrm{D}_{t,b}^{\alpha-j}f(t)\right]_{t=b}\frac{(b-t)^{\alpha-j}}{\Gamma(\alpha-j+1)}, \tag{1.21}$$

where $m-1\le\alpha<m$, *m is a positive integer. Furthermore,*

$$\mathrm{D}_{a,t}^{-\alpha}\,{}_{RL}\mathrm{D}_{a,t}^{\alpha}f(t)=f(t),\quad \mathrm{D}_{t,b}^{-\alpha}\,{}_{RL}\mathrm{D}_{t,b}^{\alpha}f(t)=f(t)$$

when

$$\left[{}_{RL}\mathrm{D}_{a,t}^{\alpha-j}f(t)\right]_{t=a}=0,\quad \left[{}_{RL}\mathrm{D}_{t,b}^{\alpha-j}f(t)\right]_{t=b}=0,\quad j=1,2,\cdots,m. \tag{1.22}$$

If $f(t)$ *has a sufficient number of continuous derivatives, then the conditions* (1.22) *are equivalent to*

$$f^{(j)}(a)=0,\quad f^{(j)}(b)=0,\quad j=0,1,\cdots,m-1. \tag{1.23}$$

In effect, (1.22) and (1.23) are generally not equivalent. (1.23) is often chosen to take the place of (1.22) in numerically studying Riemann–Liouville type differential equations, mainly for convenience.

Note that

$$\lim_{t\to a} {}_{C}\mathrm{D}^{\alpha}_{a,t}f(t) = \lim_{t\to b} {}_{C}\mathrm{D}^{\alpha}_{t,b}f(t) = 0, \quad \alpha > 0, \tag{1.24}$$

if $f(t)$ is sufficiently smooth and $f^{(m)}$ is bounded [164]. So the equivalence of (1.22) and (1.23) can be derived easily from (1.11) and (1.24).

Next, we present the more general cases of (1.18)–(1.21).

For any $m-1 \le \alpha < m, n-1 \le \beta < n$, m, n are positive integers, one has [124]

$$ {}_{RL}\mathrm{D}^{\alpha}_{a,t}\left(\mathrm{D}^{-\beta}_{a,t}f(t)\right) = {}_{RL}\mathrm{D}^{\alpha-\beta}_{a,t}f(t), \quad {}_{RL}\mathrm{D}^{\alpha}_{t,b}\left(\mathrm{D}^{-\beta}_{t,b}f(t)\right) = {}_{RL}\mathrm{D}^{\alpha-\beta}_{t,b}f(t) \tag{1.25}$$

and

$$\begin{cases} \mathrm{D}^{-\beta}_{a,t}\left({}_{RL}\mathrm{D}^{\alpha}_{a,t}f(t)\right) = {}_{RL}\mathrm{D}^{\alpha-\beta}_{a,t}f(t) - \sum_{j=1}^{m}\left[{}_{RL}\mathrm{D}^{\alpha-j}_{a,t}f(t)\right]_{t=a}\dfrac{(t-a)^{\beta-j}}{\Gamma(1+\beta-j)}, \\ \mathrm{D}^{-\beta}_{t,b}\left({}_{RL}\mathrm{D}^{\alpha}_{t,b}f(t)\right) = {}_{RL}\mathrm{D}^{\alpha-\beta}_{t,b}f(t) - \sum_{j=1}^{m}\dfrac{(b-t)^{\beta-j}}{\Gamma(1+\beta-j)}. \end{cases} \tag{1.26}$$

Next, let us consider the composition of two Riemann–Liouville derivative operators: ${}_{RL}\mathrm{D}^{\alpha}_{a,t}$ $(m-1 \le \alpha < m)$ and ${}_{RL}\mathrm{D}^{\beta}_{a,t}$ $(n-1 \le \beta < n)$, where m, n are positive integers.

Proposition 1.1.4 ([90, 91, 124]) *If $m-1 \le \alpha < m, n-1 \le \beta < n$, where m, n are positive integers, ${}_{RL}\mathrm{D}^{\alpha+\beta}_{a,t}f(t)$, ${}_{RL}\mathrm{D}^{\alpha}_{a,t}\left({}_{RL}\mathrm{D}^{\beta}_{a,t}f(t)\right)$, ${}_{RL}\mathrm{D}^{\alpha+\beta}_{t,b}f(t)$, and ${}_{RL}\mathrm{D}^{\alpha}_{t,b}\left({}_{RL}\mathrm{D}^{\beta}_{t,b}f(t)\right)$ exist, then*

$$ {}_{RL}\mathrm{D}^{\alpha}_{a,t}\left({}_{RL}\mathrm{D}^{\beta}_{a,t}f(t)\right) = {}_{RL}\mathrm{D}^{\alpha+\beta}_{a,t}f(t) - \sum_{j=1}^{n}\left[{}_{RL}\mathrm{D}^{\beta-j}_{a,t}f(t)\right]_{t=a}\frac{(t-a)^{-\alpha-j}}{\Gamma(1-\alpha-j)}, \tag{1.27}$$

$$ {}_{RL}\mathrm{D}^{\alpha}_{t,b}\left({}_{RL}\mathrm{D}^{\beta}_{t,b}f(t)\right) = {}_{RL}\mathrm{D}^{\alpha+\beta}_{a,t}f(t) - \sum_{j=1}^{n}\left[{}_{RL}\mathrm{D}^{\beta-j}_{t,b}f(t)\right]_{t=b}\frac{(b-t)^{-\alpha-j}}{\Gamma(1-\alpha-j)}. \tag{1.28}$$

Furthermore,

$$\begin{cases} {}_{RL}\mathrm{D}^{\alpha}_{a,t}\left({}_{RL}\mathrm{D}^{\beta}_{a,t}f(t)\right) = {}_{RL}\mathrm{D}^{\beta}_{a,t}\left({}_{RL}\mathrm{D}^{\alpha}_{a,t}f(t)\right) = {}_{RL}\mathrm{D}^{\alpha+\beta}_{a,t}f(t), \\ {}_{RL}\mathrm{D}^{\alpha}_{t,b}\left({}_{RL}\mathrm{D}^{\beta}_{t,b}f(t)\right) = {}_{RL}\mathrm{D}^{\beta}_{t,b}\left({}_{RL}\mathrm{D}^{\alpha}_{t,b}f(t)\right) = {}_{RL}\mathrm{D}^{\alpha+\beta}_{t,b}f(t), \end{cases} \tag{1.29}$$

if $f(t)$ satisfies the following homogeneous conditions

$$\left[{}_{RL}\mathrm{D}^{\beta-j}_{a,t}f(t)\right]_{t=a} = \left[{}_{RL}\mathrm{D}^{\beta-j}_{b,t}f(t)\right]_{t=b}\left[{}_{RL}\mathrm{D}^{\alpha-k}_{a,t}f(t)\right]_{t=a} = \left[{}_{RL}\mathrm{D}^{\alpha-k}_{b,t}f(t)\right]_{t=b} = 0. \tag{1.30}$$

where j=1,..., n, k=1,..., m.

In some situations, (1.30) is substituted for $f^{(j)}(a) = f^{(j)}(b) = 0$, $j = 0, 1, \cdots, r-1$, $r = \max\{m, n\}$, but this is not mathematically reasonable.

If α or β is a positive integer in (1.27) and (1.28) (for example, $\alpha = m$ is a positive integer), then

$$ {}_{RL}\mathrm{D}_{a,t}^{m}\left({}_{RL}\mathrm{D}_{a,t}^{\beta} f(t)\right) = {}_{RL}\mathrm{D}_{a,t}^{m+\beta} f(t). \tag{1.31} $$

Eq. (1.31) can be derived directly from (1.27) by letting $\alpha = m$, where one can see that

$$ \sum_{j=1}^{n} \left[{}_{RL}\mathrm{D}_{a,t}^{\beta-j} f(t)\right]_{t=a} \frac{(t-a)^{-m-j}}{\Gamma(1-m-j)} = 0, $$

due to $\frac{1}{\Gamma(1-m-j)} = 0$ for the nonpositive integer $(1-m-j)$. For the operator ${}_{RL}\mathrm{D}_{a,t}^{\beta}\left({}_{RL}\mathrm{D}_{a,t}^{m}\right)$, the relation

$$ {}_{RL}\mathrm{D}_{a,t}^{\beta}\,{}_{RL}\mathrm{D}_{a,t}^{m} = {}_{RL}\mathrm{D}_{a,t}^{\beta+m} $$

generally does not hold. Actually, we have

$$ \begin{aligned} {}_{RL}\mathrm{D}_{a,t}^{\beta}\left({}_{RL}\mathrm{D}_{a,t}^{m} f(t)\right) &= {}_{RL}\mathrm{D}_{a,t}^{\beta+m} f(t) - \sum_{j=1}^{m} \frac{f^{(m-j)}(a)}{\Gamma(1-\beta-j)}(t-a)^{-\beta-j} \\ &= {}_{RL}\mathrm{D}_{a,t}^{\beta+m} f(t) - \sum_{j=0}^{m-1} \frac{f^{(j)}(a)}{\Gamma(1+j-\beta-m)}(t-a)^{j-\beta-m}. \end{aligned} \tag{1.32} $$

Of course, the above relation can be directly deduced from (1.27) by letting $\beta = m$ and $\alpha = \beta$.

In most real applications, the fractional order between 0 and 2 is of great interest. Now we consider some properties of a special case, which also has much simpler forms. Here, we must suppose that the function $f(t)$ is sufficiently smooth on $[a, b]$ and a certain number of derivatives of $f(t)$ are bounded.

From (1.27) we have

$$ {}_{RL}\mathrm{D}_{a,t}^{\alpha}\left({}_{RL}\mathrm{D}_{a,t}^{\beta} f(t)\right) = {}_{RL}\mathrm{D}_{a,t}^{\alpha+\beta} f(t), \quad 0 < \alpha, \beta < 1, \tag{1.33} $$

where we have used

$$ \left[{}_{RL}\mathrm{D}_{a,t}^{\mu-1} f(t)\right]_{t=a} = 0, \quad 0 < \mu < 1 $$

due to the sufficiently smooth assumption. Actually from (1.26), (1.27) and (1.28), for any $0 < \beta < 1, \alpha \in \mathbb{R}$, we have

$$ {}_{RL}\mathrm{D}_{a,t}^{\alpha}\left({}_{RL}\mathrm{D}_{a,t}^{\beta} f(t)\right) = {}_{RL}\mathrm{D}_{a,t}^{\alpha+\beta} f(t), \quad {}_{RL}\mathrm{D}_{t,b}^{\alpha}\left({}_{RL}\mathrm{D}_{t,b}^{\beta} f(t)\right) = {}_{RL}\mathrm{D}_{t,b}^{\alpha+\beta} f(t). \tag{1.34} $$

In the following, we introduce several properties for the Caputo derivative operator in the real line. And the properties in complex planes can be found in [76]. Using (1.12) and (1.26), we have the following properties for the Caputo derivative operator.

Proposition 1.1.5 *Let $\alpha > 0$, $n-1 < \beta < n$, n is a positive integer, $f \in C^n[a,b]$. Then*

$$\mathrm{D}_{a,t}^{-\alpha}\left({}_C\mathrm{D}_{a,t}^{\beta} f(t)\right) = {}_C\mathrm{D}_{a,t}^{\beta-\alpha} f(t), \quad \alpha \neq \beta, \tag{1.35}$$

$$\mathrm{D}_{t,b}^{-\alpha}\left({}_C\mathrm{D}_{t,b}^{\beta} f(t)\right) = {}_C\mathrm{D}_{t,b}^{\beta-\alpha} f(t), \quad \alpha \neq \beta, \tag{1.36}$$

where ${}_C\mathrm{D}_{a,t}^{\beta-\alpha} = \mathrm{D}_{a,t}^{\beta-\alpha}$ *and* ${}_C\mathrm{D}_{t,b}^{\beta-\alpha} = \mathrm{D}_{t,b}^{\beta-\alpha}$ *if* $\beta < \alpha$. *Especially*

$$\mathrm{D}_{a,t}^{-\beta}\left({}_C\mathrm{D}_{a,t}^{\beta} f(t)\right) = f(t) - \sum_{k=0}^{n-1} \frac{f^{(k)}(a)}{\Gamma(k+1)}(t-a)^k, \tag{1.37}$$

$$\mathrm{D}_{t,b}^{-\beta}\left({}_C\mathrm{D}_{t,b}^{\beta} f(t)\right) = f(t) - \sum_{k=0}^{n-1} \frac{f^{(k)}(b)}{\Gamma(k+1)}(b-t)^k. \tag{1.38}$$

Proof. We first prove (1.35) and (1.37). By the Taylor series expansion, one has

$$f(t) = \sum_{k=0}^{n-1} \frac{f^{(k)}(a)}{\Gamma(k+1)}(t-a)^k + \mathrm{D}_{a,t}^{-n} f^{(n)}(t) = \phi(t) + \mathrm{D}_{a,t}^{-n} f^{(n)}(t).$$

Using (1.12) and (1.26) gives

$$\begin{aligned}
&\mathrm{D}_{a,t}^{-\alpha}\left({}_C\mathrm{D}_{a,t}^{\beta} f(t)\right) = \mathrm{D}_{a,t}^{-\alpha}\left({}_{RL}\mathrm{D}_{a,t}^{\beta}[f(t)-\phi(t)]\right)\\
&= {}_{RL}\mathrm{D}_{a,t}^{\beta-\alpha}[f(t)-\phi(t)] - \sum_{j=1}^{n}\left[{}_{RL}\mathrm{D}_{a,t}^{\beta-j}(f(t)-\phi(t))\right]_{t=a} \frac{(t-a)^{-\alpha-j}}{\Gamma(1-\alpha-j)}\\
&= {}_{RL}\mathrm{D}_{a,t}^{\beta-\alpha}[f(t)-\phi(t)] - \sum_{j=1}^{n}\left[{}_{RL}\mathrm{D}_{a,t}^{\beta-j-n} f^{(n)}(t)\right]_{t=a} \frac{(t-a)^{-\alpha-j}}{\Gamma(1-\alpha-j)}\\
&= {}_{RL}\mathrm{D}_{a,t}^{\beta-\alpha}[f(t)-\phi(t)],
\end{aligned} \tag{1.39}$$

where we use $\left[{}_{RL}\mathrm{D}_{a,t}^{\beta-j-n} f^{(n)}(t)\right]_{t=a} = 0$ when $(\beta-j-n) < 0$ and $f^{(n)}(t)$ is bounded. If $\beta = \alpha$, then ${}_{RL}\mathrm{D}_{a,t}^{\beta-\alpha}[f(t)-\phi(t)] = f(t)-\phi(t)$. If $\beta \neq \alpha$, then ${}_{RL}\mathrm{D}_{a,t}^{\beta-\alpha}[f(t)-\phi(t)] = {}_C\mathrm{D}_{a,t}^{\beta-\alpha} f(t)$. The proofs of (1.35) and (1.37) are completed. The proofs of (1.36) and (1.38) can be similarly given. All this completes the proof. □

For the operators ${}_C\mathrm{D}_{a,t}^{\beta}\mathrm{D}_{a,t}^{-\alpha}$ and ${}_C\mathrm{D}_{t,b}^{\beta}\mathrm{D}_{t,b}^{-\alpha}$, $\alpha,\beta > 0$, we have

Proposition 1.1.6 *Suppose that $\alpha,\beta > 0$, $f(t)$ is sufficiently smooth,* ${}_C\mathrm{D}_{a,t}^{\beta}\mathrm{D}_{a,t}^{-\alpha} f(t)$, ${}_C\mathrm{D}_{a,t}^{\beta-\alpha} f(t)$, ${}_C\mathrm{D}_{t,b}^{\beta}\mathrm{D}_{t,b}^{-\alpha} f(t)$, *and* ${}_C\mathrm{D}_{t,b}^{\beta-\alpha} f(t)$ *exist. Then*

$${}_C\mathrm{D}_{a,t}^{\beta}\left(\mathrm{D}_{a,t}^{-\alpha} f(t)\right) = \begin{cases} \mathrm{D}_{a,t}^{-(\alpha-\beta)} f(t), \quad \beta \leq \alpha \text{ or } \alpha < \beta,\ \alpha \in \mathbb{N}, \\ {}_C\mathrm{D}_{a,t}^{\beta-\alpha} f(t) + \displaystyle\sum_{k=0}^{n-m} \frac{f^{(k)}(a)}{\Gamma(k+1+\alpha-\beta)}(t-a)^{k+\alpha-\beta}, \\ \quad \alpha < \beta,\ m-1 < \alpha < m,\ n-1 < \beta < n,\ m,n \in \mathbb{N}. \end{cases} \tag{1.40}$$

$$
{}_C\mathrm{D}_{t,b}^{\beta}\left(\mathrm{D}_{t,b}^{-\alpha}f(t)\right)=\begin{cases}\mathrm{D}_{t,b}^{-(\alpha-\beta)}f(t), \quad \beta\leq\alpha \text{ or } \alpha<\beta,\ \alpha\in\mathbb{N},\\ {}_C\mathrm{D}_{t,b}^{\beta-\alpha}f(t)+\displaystyle\sum_{k=0}^{n-m}\frac{f^{(k)}(b)}{\Gamma(k+1+\alpha-\beta)}(b-t)^{k+\alpha-\beta},\\ \alpha<\beta,\ m-1<\alpha<m,\ n-1<\beta<n,\ m,n\in\mathbb{N}.\end{cases} \tag{1.41}
$$

Proof. We first prove (1.40). For $\beta\leq\alpha$, we have

$$
\begin{aligned}
{}_C\mathrm{D}_{a,t}^{\beta}\left(\mathrm{D}_{a,t}^{-\alpha}f(t)\right)=&\mathrm{D}_{a,t}^{-(n-\beta)}\,{}_{RL}\mathrm{D}_{a,t}^{n}\mathrm{D}_{a,t}^{-\alpha}f(t)=\mathrm{D}_{a,t}^{-(n-\beta)}\mathrm{D}_{a,t}^{-(\alpha-m)}\mathrm{D}_{a,t}^{-(m-n)}f(t)\\
=&\mathrm{D}_{a,t}^{-(n-\beta)-(\alpha-n)}f(t)=\mathrm{D}_{a,t}^{-(\alpha-\beta)}f(t),
\end{aligned}
$$

where we have used $n\leq m$ and (1.15).

For the positive integer α, i.e., $\alpha=m, m<\beta$, we have

$$
{}_C\mathrm{D}_{a,t}^{\beta}\left(\mathrm{D}_{a,t}^{-m}f(t)\right)=\mathrm{D}_{a,t}^{-(n-\beta)}\,{}_{RL}\mathrm{D}_{a,t}^{n}\mathrm{D}_{a,t}^{-m}f(t)=\mathrm{D}_{a,t}^{-(n-\beta)}\,{}_{RL}\mathrm{D}_{a,t}^{n-m}f(t).
$$

Using $n>m$ and (1.32) yields

$$
\begin{aligned}
&{}_C\mathrm{D}_{a,t}^{\beta}\left(\mathrm{D}_{a,t}^{-m}f(t)\right)=\mathrm{D}_{a,t}^{-(n-\beta)}\,{}_{RL}\mathrm{D}_{a,t}^{n-m}f(t)\\
=&{}_{RL}\mathrm{D}_{a,t}^{\beta-m}f(t)-\sum_{j=0}^{n-m-1}\frac{f^{(j)}(a)}{\Gamma(1+j+m-\beta)}(t-a)^{m-\beta+j}\\
=&{}_{RL}\mathrm{D}_{a,t}^{\beta-m}(f(t)-\phi(t))+{}_{RL}\mathrm{D}_{a,t}^{\beta-m}\phi(t)-\sum_{j=0}^{n-m-1}\frac{f^{(j)}(a)}{\Gamma(1+j+m-\beta)}(t-a)^{m-\beta+j}\\
=&{}_{RL}\mathrm{D}_{a,t}^{\beta-m}(f(t)-\phi(t))={}_C\mathrm{D}_{a,t}^{\beta-m}f(t),
\end{aligned}
$$

where $\phi(t)=\sum_{j=0}^{n-m-1}\frac{f^{(j)}(a)}{\Gamma(j+1)}(t-a)^j$.

For $\alpha<\beta, n-1<\beta<n$, and $m-1<\alpha<m$, one has $n-m<n-\alpha<n-m+1$. By using the Taylor series expansion, one has

$$
f(t)=\sum_{k=0}^{n-m}\frac{f^{(k)}(a)}{\Gamma(k+1)}(t-a)^k+\mathrm{D}_{a,t}^{-(n-m)}f^{(n-m)}(t)=\phi(t)+\mathrm{D}_{a,t}^{-(n-m)}f^{(n-m)}(t).
$$

Hence,

$$
f(t)-\phi(t)=\mathrm{D}_{a,t}^{-(n-m)}f^{(n-m)}(t).
$$

Therefore,

$$
\mathrm{D}_{a,t}^{-\alpha}[f(t)-\phi(t)]=\mathrm{D}_{a,t}^{-(n-m)-\alpha}f^{(n-m)}(t),
$$

which implies that ${}_{RL}\mathrm{D}_{a,t}^{k}\left[\mathrm{D}_{a,t}^{-\alpha}(f(t)-\phi(t))\right]_{t=a}=0$ for $k=0,1,\cdots,n-m$. Combining

(1.12) and (1.25) gives

$$\begin{aligned}
{}_C\mathrm{D}_{a,t}^{\beta}\left(\mathrm{D}_{a,t}^{-\alpha} f(t)\right) &= {}_C\mathrm{D}_{a,t}^{\beta}\mathrm{D}_{a,t}^{-\alpha}[f(t)-\phi(t)] + {}_C\mathrm{D}_{a,t}^{\beta}\mathrm{D}_{a,t}^{-\alpha}\phi(t)\\
&= {}_{RL}\mathrm{D}_{a,t}^{\beta}\mathrm{D}_{a,t}^{-\alpha}[f(t)-\phi(t)] + {}_C\mathrm{D}_{a,t}^{\beta}\mathrm{D}_{a,t}^{-\alpha}\phi(t)\\
&= {}_{RL}\mathrm{D}_{a,t}^{\beta-\alpha}[f(t)-\phi(t)] + {}_C\mathrm{D}_{a,t}^{\beta}\mathrm{D}_{a,t}^{-\alpha}\phi(t)\\
&= {}_C\mathrm{D}_{a,t}^{\beta-\alpha} f(t) + \sum_{k=0}^{n-m} \frac{f^{(k)}(a)}{\Gamma(k+1+\alpha-\beta)}(t-a)^{k+\alpha-\beta}.
\end{aligned} \tag{1.42}$$

The proof of (1.40) is completed. The proof of (1.41) can be similarly given. All this ends the proof. □

For the Riemann–Liouville derivative operators, we have the relation ${}_{RL}\mathrm{D}_{a,t}^{m}\left({}_{RL}\mathrm{D}_{a,t}^{\beta} f(t)\right) = {}_{RL}\mathrm{D}_{a,t}^{m+\beta} f(t)$ for any $\beta > 0$ and any nonnegative integer m. While for the Caputo derivative operators, we have

$${}_C\mathrm{D}_{a,t}^{\beta}\,{}_C\mathrm{D}_{a,t}^{m} f(t) = {}_C\mathrm{D}_{a,t}^{m+\beta} f(t), \quad \beta > 0, m \in \mathbb{N}. \tag{1.43}$$

For ${}_C\mathrm{D}_{a,t}^{m}\,{}_C\mathrm{D}_{a,t}^{\beta}$, $n-1 < \beta < n$, m and n are positive integers, one has

$$\begin{aligned}
{}_C\mathrm{D}_{a,t}^{m}\,{}_C\mathrm{D}_{a,t}^{\beta} f(t) &= {}_{RL}\mathrm{D}_{a,t}^{m}\,{}_{RL}\mathrm{D}_{a,t}^{-(n-\beta)} f^{(n)}(t) = {}_{RL}\mathrm{D}_{a,t}^{m-n-\beta} f^{(n)}(t)\\
&= {}_{RL}\mathrm{D}_{a,t}^{m+\beta} f(t) - \sum_{j=0}^{n-1} \frac{f^{(j)}(a)}{\Gamma(1+j-m-\beta)}(t-a)^{j-m-\beta},
\end{aligned} \tag{1.44}$$

where Eq. (1.32) is used. Denote it by

$$\phi(t) = \sum_{j=0}^{m+n-1} \frac{f^{(j)}(a)}{\Gamma(j+1)}(t-a)^{j}.$$

Then

$$\begin{aligned}
{}_C\mathrm{D}_{a,t}^{m}\,{}_C\mathrm{D}_{a,t}^{\beta} f(t) &= {}_C\mathrm{D}_{a,t}^{m+\beta} f(t) + {}_{RL}\mathrm{D}_{a,t}^{m+\beta}\phi(t) - \sum_{j=0}^{n-1} \frac{f^{(j)}(a)}{\Gamma(1+j-m-\beta)}(t-a)^{j-m-\beta}\\
&= {}_C\mathrm{D}_{a,t}^{m+\beta} f(t) + \sum_{j=n}^{m+n-1} \frac{f^{(j)}(a)}{\Gamma(1+j-m-\beta)}(t-a)^{j-m-\beta}.
\end{aligned} \tag{1.45}$$

By (1.43) and (1.45), the interchange of the Caputo derivative operators in (1.43) is allowed under the following conditions:

$$f^{(j)}(a) = 0, \quad j = n, n+1, \cdots, m+n-1,\ m = 0, 1, 2, \cdots. \tag{1.46}$$

From (1.46) we see that there are no restrictions on the values $f^{(j)}(a) = 0$ ($j = 0, 1, \cdots, n-1$).

For ${}_C\mathrm{D}^{\alpha}_{a,t}{}_C\mathrm{D}^{\beta}_{a,t}$, $m-1<\alpha<m$, $n-1<\beta<n$, m and n are positive integers, one can obtain

$$ {}_C\mathrm{D}^{\alpha}_{a,t}{}_C\mathrm{D}^{\beta}_{a,t}f(t) = {}_C\mathrm{D}^{\alpha+\beta}_{a,t}f(t) + \psi(t). \tag{1.47}$$

In fact, $\psi(t)$ is more complicated than the second term of the right-hand side of (1.27). Here, we present two special cases of ${}_C\mathrm{D}^{\alpha}_{a,t}{}_C\mathrm{D}^{\beta}_{a,t}f(t) = {}_C\mathrm{D}^{\alpha+\beta}_{a,t}f(t)$ with no restrictions on $f(t)$ at $t=a$.

Assume that $f \in C^n[a,t]$, n is a positive integer, $n-1<\beta\leq n$. Hence, $0 \leq n-\beta < 1$. So for $0<n-\beta<1$, one has

$$ {}_C\mathrm{D}^{n-\beta}_{a,t}{}_C\mathrm{D}^{\beta}_{a,t}f(t) = {}_C\mathrm{D}^{n-\beta}_{a,t}\mathrm{D}^{-(n-\beta)}_{a,t}f^{(n)}(t) = f^{(n)}(t), \tag{1.48}$$

where (1.40) has been used.

Let $0<\alpha<1, n-1<\alpha+\beta<n, n-1<\beta<n$, n is a positive integer, $f \in C^n[a,t]$. Then one has

$$ {}_C\mathrm{D}^{\alpha}_{a,t}{}_C\mathrm{D}^{\beta}_{a,t}f(t) = {}_{RL}\mathrm{D}^{\alpha}_{a,t}{}_C\mathrm{D}^{\beta}_{a,t}f(t) = {}_{RL}\mathrm{D}^{\alpha}_{a,t}{}_{RL}\mathrm{D}^{\beta}_{a,t}(f(t)-\phi(t)), \tag{1.49}$$

where $\phi(t) = \sum_{j=0}^{n-1}\frac{f^{(j)}(a)}{\Gamma(j+1)}(t-a)^j$. It is easy to verify that $(f^{(k)} - \phi^{(k)})(a) = 0$ for $k = 0,1,\cdots,n-1$. So

$$\begin{aligned} {}_C\mathrm{D}^{\alpha}_{a,t}{}_C\mathrm{D}^{\beta}_{a,t}f(t) &= {}_{RL}\mathrm{D}^{\alpha}_{a,t}{}_{RL}\mathrm{D}^{\beta}_{a,t}(f(t)-\phi(t)) \\ &= {}_{RL}\mathrm{D}^{\alpha+\beta}_{a,t}(f(t)-\phi(t)) \\ &= {}_C\mathrm{D}^{\alpha+\beta}_{a,t}f(t). \end{aligned} \tag{1.50}$$

1.2 Some Other Properties of Fractional Derivatives

In this section, we introduce some more interesting properties of fractional integration and differentiation. These properties include the linearity, the Leibniz rule, the behaviors near and far from the lower terminal, the Laplace transform, and the Fourier transform.

From the definitions of the fractional integrals (see (1.1) and (1.2)) and derivatives (see (1.3)–(1.9)) in the previous section, it is easy to verify that the fractional integrals and derivatives are *linear operators*, i.e.,

$$ D^{\alpha}(\lambda f(t) + \mu g(t)) = \lambda D^{\alpha}f(t) + \mu D^{\alpha}g(t), \tag{1.51}$$

where D^{α} denotes any fractional integral or derivative hereafter.

1.2.1 Leibniz Rule for Fractional Derivatives

Next, we investigate the *Leibniz rule* for the fractional derivative. Let $f(t)$ and $g(t)$ be two functions with derivatives up to n. Then the Leibniz rule for evaluating the n-th derivative of $g(t)f(t)$ gives:

$$\frac{\mathrm{d}^n}{\mathrm{d}t^n}\big(g(t)f(t)\big) = \sum_{k=0}^{n}\binom{n}{k}g^{(k)}(t)f^{(n-k)}(t). \tag{1.52}$$

Let us replace n with the real-valued parameter α in the right-hand side of (1.52), and denote

$$\Omega_n^{\alpha}(t) = \sum_{k=0}^{n}\binom{\alpha}{k}g^{(k)}(t)f^{(\alpha-k)}(t),$$

where $f^{(\alpha-k)}(t) = {}_{RL}\mathrm{D}_{a,t}^{\alpha-k}f(t)$ or $f^{(\alpha-k)}(t) = {}_{GL}\mathrm{D}_{a,t}^{\alpha-k}f(t)$. We now wonder if there exists a positive integer n such that Ω_n^{α} is just the α-th order derivative of $g(t)f(t)$. This is not the case when α is not an integer. In fact, ${}_{RL}\mathrm{D}_{a,t}^{\alpha}(g(t)f(t))$ has the following form [124]

$${}_{RL}\mathrm{D}_{a,t}^{\alpha}(g(t)f(t)) = \sum_{k=0}^{n}\binom{\alpha}{k}g^{(k)}(t)f^{(\alpha-k)}(t) - R_n^{\alpha}(t) = \Omega_n^{\alpha}(t) - R_n^{\alpha}(t), \tag{1.53}$$

where $n \geq \alpha + 1$ and

$$R_n^{\alpha}(t) = \frac{1}{n!\Gamma(-\alpha)}\int_a^t (t-\tau)^{-\alpha-1}f(\tau)\mathrm{d}\tau\int_{\tau}^t(\tau-\xi)^n g^{(n+1)}(\xi)\mathrm{d}\xi. \tag{1.54}$$

Let $\xi = \tau + \varsigma(t-\tau)$ and $\tau = a + \eta(t-a)$, we obtain the following expression of $R_n^{\alpha}(t)$ as:

$$R_n^{\alpha}(t) = \frac{(-1)^n(t-a)^{n-\alpha+1}}{n!\Gamma(-\alpha)}\int_0^1 F_a(t,\zeta,\eta)\,\mathrm{d}\zeta\,\mathrm{d}\eta, \tag{1.55}$$

where

$$F_a(t,\zeta,\eta) = f(a+\eta(t-a))g^{(n+1)}(a+(t-a)(\zeta+\eta-\zeta\eta)).$$

From (1.55), one obtains

$$\lim_{n\to\infty} R_n^{\alpha}(t) = 0,$$

if $f(t)$ and $g(t)$ together with their derivatives are continuous in $[a,t]$. Under these conditions the *Leibniz rule for fractional differentiation* takes the form:

$${}_{RL}\mathrm{D}_{a,t}^{\alpha}(g(t)f(t)) = \sum_{k=0}^{\infty}\binom{\alpha}{k}g^{(k)}(t)f^{(\alpha-k)}(t). \tag{1.56}$$

Obviously, the above rule (1.56) is especially useful when $g(t)$ is a polynomial.

1.2.2 Fractional Derivative of a Composite Function

The Leibniz rule for the fractional derivative can be used to obtain the fractional derivative for a composite function.

Let $f(t) = 1$ in (1.56), then

$$ {}_{RL}\mathrm{D}_{a,t}^{\alpha} g(t) = \frac{(t-a)^{-\alpha}}{\Gamma(1-\alpha)} g(t) + \sum_{k=1}^{\infty} \binom{\alpha}{k} \frac{(t-a)^{k-\alpha}}{\Gamma(1-\alpha+k)} g^{(k)}(t). \tag{1.57} $$

Suppose that $g(t)$ is a composite function, i.e., $g(t) = F(h(t))$. Then the k-th derivative of $g(t)$ is evaluated with the help of the Faà di Bruno formula [124]

$$ \frac{\mathrm{d}^k}{\mathrm{d}t^k} F(h(t)) = k! \sum_{m=1}^{k} F^{(m)}(h(t)) \sum \prod_{r=1}^{k} \frac{1}{a_r!} \left(\frac{h^{(r)}(t)}{r!} \right)^{a_r}, \tag{1.58} $$

where the sum $\sum$ extends over all combinations of non-negative integer values of $a_1, a_2, \cdots, a_k$ such that

$$ \sum_{r=1}^{k} r a_r = k \quad \text{and} \quad \sum_{r=1}^{k} a_r = m. $$

Inserting $g(t) = F(h(t))$, (1.58) into (1.57) gives the formula for *the fractional derivative of a composite function* as:

$$ \begin{aligned} {}_{RL}\mathrm{D}_{a,t}^{\alpha} F(h(t)) = & \frac{(t-a)^{-\alpha}}{\Gamma(1-\alpha)} F(h(t)) \\ & + \sum_{k=1}^{\infty} \binom{\alpha}{k} \frac{k!(t-a)^{k-\alpha}}{\Gamma(1-\alpha+k)} \sum_{m=1}^{k} F^{(m)}(h(t)) \sum \prod_{r=1}^{k} \frac{1}{a_r!} \left(\frac{h^{(r)}(t)}{r!} \right)^{a_r}, \end{aligned} \tag{1.59} $$

where the sum $\sum$ and coefficients a_r have the meaning explained as above.

1.2.3 Behaviors Near and Far from the Lower Terminal

For the sufficiently smooth function $f(t)$, for example, $f(t)$ is a polynomial, the classical derivative of $f(t)$ exists and is bounded, but this is not the case for the fractional derivative operators.

For the simplicity of the theoretical analysis, we suppose that $f(t)$ has an arbitrary order derivative at the lower terminal ($t = a$). Therefore, $f(t)$ can be represented by the Taylor series

$$ f(t) = \sum_{k=0}^{n} \frac{f^{(k)}(a)}{\Gamma(1+k)} (t-a)^k + \mathrm{D}_{a,t}^{-n} f^{(n)}(t). \tag{1.60} $$

Applying the Riemann–Liouville derivative operator on both sides of (1.60) and using term-by-term differentiation yields

$$ {}_{RL}\mathrm{D}_{a,t}^{\alpha} f(t) = \sum_{k=0}^{n} \frac{f^{(k)}(a)}{\Gamma(1+k-\alpha)} (t-a)^{k-\alpha} + \mathrm{D}_{a,t}^{\alpha-n} f^{(n)}(t), \quad \alpha < n. \tag{1.61} $$

As $t \to a+0$,

$$ {}_{RL}\mathrm{D}_{a,t}^{\alpha} f(t) \sim \frac{f(a)}{\Gamma(1-\alpha)} (t-a)^{-\alpha}. \tag{1.62} $$

Hence

$$\lim_{t\to a+0} {}_{RL}\mathrm{D}^{\alpha}_{a,t}f(t) = \begin{cases} 0, & \alpha < 0, \\ f(a), & \alpha = 0, \\ \infty, & \alpha > 0. \end{cases} \tag{1.63}$$

The simplest example for illustrating (1.63) is to set $f(t) = 1$. This means that the fractional initial value(s) in the sense of Riemann–Liouville must be very carefully given.

Next, we study the behavior of the fractional derivative far from the lower terminal (or, upper terminal), i.e., $t \to \infty$. First consider ${}_{RL}\mathrm{D}^{\alpha}_{a,t}f(t)$, $m-1 \le \alpha < m$, where $f(t)$ is sufficiently smooth. By the definition of the Riemann–Liouville derivative operator we have

$$\begin{aligned} {}_{RL}\mathrm{D}^{\alpha}_{a,t}f(t) =& \frac{1}{\Gamma(m-\alpha)}\frac{\mathrm{d}^m}{\mathrm{d}t^m}\int_a^t (t-s)^{m-\alpha-1}f(s)\mathrm{d}s \\ =& \frac{1}{\Gamma(m-\alpha)}\frac{\mathrm{d}^m}{\mathrm{d}t^m}\int_0^t (t-s)^{m-\alpha-1}f(s)\mathrm{d}s \\ & - \frac{1}{\Gamma(m-\alpha)}\frac{\mathrm{d}^m}{\mathrm{d}t^m}\int_0^a (t-s)^{m-\alpha-1}f(s)\mathrm{d}s \\ =& {}_{RL}\mathrm{D}^{\alpha}_{0,t}f(t) - \frac{1}{\Gamma(m-\alpha)}\frac{\mathrm{d}^m}{\mathrm{d}t^m}\int_0^a (t-s)^{m-\alpha-1}f(s)\mathrm{d}s. \end{aligned} \tag{1.64}$$

For sufficiently large t ($|t| \gg |a|$) and bounded $f(t)$ ($|f(t)| \le C$), one has

$$\begin{aligned} & \left|\frac{1}{\Gamma(m-\alpha)}\frac{\mathrm{d}^m}{\mathrm{d}t^m}\int_0^a (t-s)^{m-\alpha-1}f(s)\mathrm{d}s\right| \\ =& \left|\frac{1}{\Gamma(1-\alpha)}\int_0^a (t-s)^{-\alpha-1}f(s)\mathrm{d}s\right| \\ \le& \frac{C}{|\Gamma(-\alpha)|}\left[(t-a)^{-\alpha} - t^{-\alpha}\right] \to 0 \quad (t\to\infty). \end{aligned} \tag{1.65}$$

Therefore, for large enough t we have

$${}_{RL}\mathrm{D}^{\alpha}_{a,t}f(t) \approx {}_{RL}\mathrm{D}^{\alpha}_{0,t}f(t), \quad (|t| \gg |a|). \tag{1.66}$$

For the sufficiently smooth $f(t)$ and $\alpha > 0$, one has also the following relation [124]

$${}_{RL}\mathrm{D}^{\alpha}_{a,t}f(t) \approx {}_{RL}\mathrm{D}^{\alpha}_{0,t}f(t) + \frac{a\Gamma(\alpha+1)\sin(\alpha\pi)f(0)}{\pi t^{\alpha+1}}, \quad (|t| \gg |a|). \tag{1.67}$$

For the Caputo derivative operator ${}_{C}\mathrm{D}^{\alpha}_{a,t}$, we get

$${}_{C}\mathrm{D}^{\alpha}_{a,t}f(t) \approx {}_{C}\mathrm{D}^{\alpha}_{0,t}f(t), \quad (|t| \gg |a|) \tag{1.68}$$

if $|f^{(m)}(t)|$ is bounded, $m-1 < \alpha < m$.

For the fixed t and sufficiently large $|a|$, i.e., $a \to -\infty$, we can similarly obtain

$$\begin{aligned} {}_{RL}\mathrm{D}_{a,t}^{\alpha} f(t) =& \frac{1}{\Gamma(m-\alpha)} \frac{\mathrm{d}^m}{\mathrm{d}t^m} \int_a^t (t-s)^{m-\alpha-1} f(s)\,\mathrm{d}s \\ =& {}_{RL}\mathrm{D}_{t-a,t}^{\alpha} f(t) - \frac{1}{\Gamma(m-\alpha)} \frac{\mathrm{d}^m}{\mathrm{d}t^m} \int_a^{t-a} (t-s)^{m-\alpha-1} f(s)\,\mathrm{d}s. \end{aligned} \tag{1.69}$$

Similar to (1.65), we can obtain

$$\frac{1}{\Gamma(m-\alpha)} \frac{\mathrm{d}^m}{\mathrm{d}t^m} \int_a^{t-a} (t-s)^{m-\alpha-1} f(s)\,\mathrm{d}s \to 0 \tag{1.70}$$

when $|f^{(m-1)}(t)|$ is bounded and $a \to -\infty$. Hence,

$${}_{RL}\mathrm{D}_{a,t}^{\alpha} f(t) \approx {}_{RL}\mathrm{D}_{t-a,t}^{\alpha} f(t), \quad |a| \gg |t|. \tag{1.71}$$

In a similar manner, we can obtain

$${}_{C}\mathrm{D}_{a,t}^{\alpha} f(t) \approx {}_{C}\mathrm{D}_{t-a,t}^{\alpha} f(t), \quad |a| \gg |t| \tag{1.72}$$

if $|f^{(m)}(t)|$ is bounded, $m-1 < \alpha < m$.

1.2.4 Laplace Transforms of Fractional Derivatives

The *Laplace transform* of a given function $f(t)$ is defined as

$$F(s) = L\{f(t); s\} = \int_0^{\infty} e^{-st} f(t)\,\mathrm{d}t. \tag{1.73}$$

The existence of the integral (1.73) requires that the function $f(t)$ must be of exponential order μ such that for positive constants M and T

$$|f(t)| \le M e^{\mu t} \quad \text{holds for all} \quad t > T.$$

The original function $f(t)$ in (1.73) can be restored from $F(s)$ with the help of the *inverse Laplace transform*

$$f(t) = L^{-1}\{F(s); t\} = \frac{1}{2\pi i} \int_{c-i\infty}^{c+i\infty} e^{st} F(s)\,\mathrm{d}s, \quad c = Re(s) > c_0, \tag{1.74}$$

where c_0 lies in the right half plane of the absolute convergence of the Laplace integral (1.73).

Next, we present two important properties that will be useful in obtaining the Laplace transform of the fractional derivative operators.

The first property states that the Laplace transform of the convolution

$$f(t) * g(t) = \int_0^t f(t-s) g(s)\,\mathrm{d}s = \int_0^t f(s) g(t-s)\,\mathrm{d}s \tag{1.75}$$

is given as

$$L\{f(t) * g(t); s\} = F(s)G(s), \tag{1.76}$$

where $F(s)$ and $G(s)$ are Laplace transforms of $f(t)$ and $g(t)$, respectively, and $f(t)$ and $g(t)$ are equal to zero for $t < 0$.

The second property states that the Laplace transform of $f^{(n)}(t)$ is given by:

$$L\{f^{(n)}(t); s\} = s^n L\{f(t); s\} - \sum_{k=0}^{n-1} s^{n-k-1} f^{(k)}(0) = s^n L\{f(t); s\} - \sum_{k=0}^{n-1} s^k f^{(n-k-1)}(0), \tag{1.77}$$

which can be obtained from the definition of the Laplace transform (1.73) by integrating by parts under the assumption that the corresponding integrals exist (for instance, $f^{(n-1)}$ is bounded).

Next, let us start with the *Laplace transform of the fractional integral.* Let $\alpha > 0$ and $g(t) = t^{\alpha-1}$. Then the fractional integral $\mathrm{D}_{0,t}^{-\alpha} f(t)$ can be rewritten as

$$\mathrm{D}_{0,t}^{-\alpha} f(t) = \frac{1}{\Gamma(\alpha)} \int_0^t (t-s)^{\alpha-1} f(s)\,\mathrm{d}s = \frac{1}{\Gamma(\alpha)} t^{\alpha-1} * f(t). \tag{1.78}$$

It is easy to calculate that

$$G(s) = L\{t^{\alpha-1}; s\} = \Gamma(\alpha) s^{-\alpha}. \tag{1.79}$$

Hence

$$L\{\mathrm{D}_{0,t}^{-\alpha} f(t); s\} = \frac{1}{\Gamma(\alpha)} L\{t^{\alpha-1} * f(t); s\} = s^{-\alpha} L\{f(t); s\} = s^{-\alpha} F(s). \tag{1.80}$$

Now let us turn to the Laplace transform of the Riemann–Liouville derivative operator with order $\alpha, m-1 \leq \alpha < m$. Let

$$g(t) = \mathrm{D}_{0,t}^{-(m-\alpha)} f(t). \tag{1.81}$$

Then

$$_{RL}\mathrm{D}_{0,t}^{\alpha} f(t) = g^{(m)}(t). \tag{1.82}$$

Applying (1.77) gives

$$L\{_{RL}\mathrm{D}_{0,t}^{\alpha} f(t); s\} = L\{g^{(m)}(t); s\} = s^m L\{g(t); s\} - \sum_{k=0}^{m-1} s^k g^{(m-k-1)}(0). \tag{1.83}$$

By (1.80) one has

$$L\{g(t); s\} = L\{\mathrm{D}_{0,t}^{-(m-\alpha)} f(t); s\} = s^{-(m-\alpha)} L\{f(t); s\}. \tag{1.84}$$

Combining (1.81)–(1.84) gives the *Laplace transform of the Riemann–Liouville derivative* as

$$L\{_{RL}\mathrm{D}_{0,t}^{\alpha} f(t); s\} = s^{\alpha} L\{f(t); s\} - \sum_{k=0}^{m-1} s^k \left[{}_{RL}\mathrm{D}_{0,t}^{\alpha-k-1} f(t)\right]_{t=0}, \quad m-1 \leq \alpha < m. \tag{1.85}$$

Next, let us turn attention to the Laplace transform of the Caputo derivative operator. The α-th order Caputo derivative of $f(t)$ can be written as

$$ {}_C\mathrm{D}_{0,t}^{\alpha} f(t) = \mathrm{D}_{0,t}^{-(m-\alpha)} g(t), \quad g(t) = f^{(m)}(t). \tag{1.86}$$

Using (1.80) and (1.77) gives

$$\begin{aligned} L\{{}_C\mathrm{D}_{0,t}^{\alpha} f(t); s\} =& L\{\mathrm{D}_{0,t}^{-(m-\alpha)} g(t); s\} = s^{-(m-\alpha)} L\{g(t); s\} \\ =& s^{-(m-\alpha)} \left[s^m L\{f(t); s\} - \sum_{k=0}^{m-1} s^{m-k-1} f^{(k)}(0) \right] \\ =& s^{\alpha} L\{f(t); s\} - \sum_{k=0}^{m-1} s^{\alpha-k-1} f^{(k)}(0). \end{aligned} \tag{1.87}$$

Therefore, the *Laplace transform of the Caputo derivative operator* reads as

$$L\{{}_C\mathrm{D}_{0,t}^{\alpha} f(t); s\} = s^{\alpha} L\{f(t); s\} - \sum_{k=0}^{m-1} s^{\alpha-k-1} f^{(k)}(0), \quad m-1 < \alpha \le m. \tag{1.88}$$

1.2.5 Fourier Transforms of Fractional Derivatives

The *Fourier transform* of a continuous function $f(t)$ that is absolutely integrable in $(-\infty, \infty)$ is defined by

$$F\{f(t); \omega\} = \int_{-\infty}^{\infty} e^{-i\omega t} f(t) \,\mathrm{d}t. \tag{1.89}$$

The original function $f(t)$ in (1.89) can be restored from $F\{f(t); \omega\}$ with the help of the *inverse Fourier transform*:

$$f(t) = \frac{1}{2\pi} \int_{-\infty}^{\infty} F\{f(t); \omega\} e^{i\omega t} f(t) \,\mathrm{d}\omega. \tag{1.90}$$

Similar to (1.75) and (1.76), the Fourier transform of the convolution

$$f(t) * g(t) = \int_{-\infty}^{\infty} f(t-s) g(s) \,\mathrm{d}s = \int_{-\infty}^{\infty} f(s) g(t-s) \,\mathrm{d}s \tag{1.91}$$

satisfies

$$F\{f(t) * g(t); \omega\} = F\{f(t); \omega\} F\{g(t); \omega\}. \tag{1.92}$$

A useful property of the Fourier transform is the Fourier transform of the derivatives of $f(t)$. If $f^{(k)}(t)$ $(k = 0, 1, 2, \cdots, n-1)$ vanish as $t \to \pm\infty$, then the Fourier transform of $f^{(n)}(t)$ is given as

$$F\{f^{(n)}(t); \omega\} = (i\omega)^n F\{f(t); \omega\}. \tag{1.93}$$

Next, we investigate the *Fourier transform of the fractional integral operator* $\mathrm{D}_{a,t}^{-\alpha}$ with $a = -\infty$ and $0 < \alpha < 1$. Let

$$h_+(t) = \begin{cases} \dfrac{t^{\alpha-1}}{\Gamma(\alpha)}, & t > 0, \\ 0, & t = 0. \end{cases} \tag{1.94}$$

Then

$$\mathrm{D}_{-\infty,t}^{-\alpha} f(t) = h_+(t) * f(t). \tag{1.95}$$

It is easy to calculate that

$$F\{h_+(t);\omega\} = (i\omega)^{-\alpha}. \tag{1.96}$$

Therefore

$$\begin{aligned} F\{\mathrm{D}_{-\infty,t}^{-\alpha} f(t);\omega\} =& F\{h_+(t) * f(t);\omega\} = F\{h_+(t);\omega\}F\{f(t);\omega\} \\ =& (i\omega)^{-\alpha} F\{f(t);\omega\}. \end{aligned} \tag{1.97}$$

For the right fractional integral operator $\mathrm{D}_{t,\infty}^{-\alpha}$, one has

$$\mathrm{D}_{t,\infty}^{-\alpha} f(t) = h_+(-t) * f(t). \tag{1.98}$$

Note that

$$F\{h_+(-t);\omega\} = (-i\omega)^{-\alpha}. \tag{1.99}$$

Hence

$$F\{\mathrm{D}_{t,\infty}^{-\alpha} f(t);\omega\} = F\{h_+(-t);\omega\}F\{f(t);\omega\} = (-i\omega)^{-\alpha} F\{f(t);\omega\}. \tag{1.100}$$

Next, we discuss the *Fourier transform for the fractional derivatives*. Suppose that $m-1 < \alpha < m$, $f(t)$ is sufficiently smooth and $f^{(k)}(-\infty)$ $(k = 0, 1, \cdots, m-1)$ are bounded. Then from (1.10) and (1.11), we see that the left Riemann–Liouville derivative, the left Grünwald–Letnikov derivative, and the left Caputo derivative have the same form:

$$\left.\begin{array}{r} {}_{GL}\mathrm{D}_{-\infty,t}^{\alpha} f(t) \\ {}_{RL}\mathrm{D}_{-\infty,t}^{\alpha} f(t) \\ {}_{C}\mathrm{D}_{-\infty,t}^{\alpha} f(t) \end{array}\right\} = \mathrm{D}_{-\infty,t}^{-(m-\alpha)} f^{(m)}(t), \quad m-1 < \alpha < m. \tag{1.101}$$

One can similarly obtain

$$\left.\begin{array}{r} {}_{GL}\mathrm{D}_{t,\infty}^{\alpha} f(t) \\ {}_{RL}\mathrm{D}_{t,\infty}^{\alpha} f(t) \\ {}_{C}\mathrm{D}_{t,\infty}^{\alpha} f(t) \end{array}\right\} = (-1)^m \mathrm{D}_{t,\infty}^{-(m-\alpha)} f^{(m)}(t), \quad m-1 < \alpha < m. \tag{1.102}$$

Now, let us turn to the evaluation of the Fourier transform of (1.101). From (1.97) and (1.102) one has

$$\begin{aligned} &F\{{}_{RL}\mathrm{D}_{-\infty,t}^{\alpha} f(t);\omega\} = F\{{}_{RL}\mathrm{D}_{-\infty,t}^{-(m-\alpha)} f^{(m)}(t);\omega\} \\ =& (i\omega)^{-(m-\alpha)} F\{f^{(m)}(t);\omega\} = (i\omega)^{-(m-\alpha)}(i\omega)^m F\{f(t);\omega\} \\ =& (i\omega)^{\alpha} F\{f(t);\omega\}. \end{aligned} \tag{1.103}$$

We can similarly obtain

$$F\{{}_{RL}\mathrm{D}_{t,\infty}^{\alpha} f(t);\omega\} = (-i\omega)^{\alpha} F\{f(t);\omega\}. \tag{1.104}$$

1.3 Some Other Fractional Derivatives and Extensions

In this section, we introduce some extensions of the fractional derivatives.

1.3.1 Marchaud Fractional Derivative

Consider the Riemann–Liouville derivative (1.5) with $a = -\infty$ and $0 < \alpha < 1$. For convenience, let us suppose that $f(t)$ is a sufficiently "good" function, for example, $f(t)$ is continuously differentiable with its derivatives, $f'(t)$, vanishing at the infinity as $|t|^{\alpha-1-\epsilon}$, $\epsilon > 0$. Then we obtain

$$\begin{aligned} {}_{RL}\mathrm{D}^{\alpha}_{-\infty,t}f(t) &= \frac{1}{\Gamma(1-\alpha)}\frac{\mathrm{d}}{\mathrm{d}t}\int_{-\infty}^{t}(t-s)^{-\alpha}f(s)\,\mathrm{d}s \\ &= \frac{1}{\Gamma(1-\alpha)}\int_{-\infty}^{t}(t-s)^{-\alpha}f'(s)\,\mathrm{d}s \\ &= \frac{1}{\Gamma(1-\alpha)}\int_{0}^{\infty}\xi^{-\alpha}f'(t-\xi)\,\mathrm{d}\xi \\ &= \frac{\alpha}{\Gamma(1-\alpha)}\int_{0}^{\infty}\frac{f(t)-f(t-\xi)}{\xi^{1+\alpha}}\,\mathrm{d}\xi. \end{aligned} \tag{1.105}$$

One can similarly get

$${}_{RL}\mathrm{D}^{\alpha}_{t,\infty}f(t) = \frac{\alpha}{\Gamma(1-\alpha)}\int_{0}^{\infty}\frac{f(t)-f(t+\xi)}{\xi^{1+\alpha}}\,\mathrm{d}\xi \tag{1.106}$$

for sufficiently "good" function $f(t)$.

From the structures of (1.105) and (1.106), the so called Marchaud fractional derivatives can be derived. The left and right *Marchaud fractional operators* with order α $(0 < \alpha < 1)$ are defined as

$${}_{M}\mathrm{D}^{\alpha}_{+}f(t) = \frac{\alpha}{\Gamma(1-\alpha)}\int_{0}^{\infty}\frac{f(t)-f(t-\xi)}{\xi^{1+\alpha}}\,\mathrm{d}\xi, \tag{1.107}$$

and

$${}_{M}\mathrm{D}^{\alpha}_{-}f(t) = \frac{\alpha}{\Gamma(1-\alpha)}\int_{0}^{\infty}\frac{f(t)-f(t+\xi)}{\xi^{1+\alpha}}\,\mathrm{d}\xi. \tag{1.108}$$

If $f(t)$, with its derivatives $f^{(k)}(t)$ $(k = 1,2,\cdots,m)$, is continuous and vanishes at infinity as $|t|^{\alpha-1-\epsilon}$, $\epsilon > 0$, $m-1 < \alpha < m$, then it follows from (1.5) that

$${}_{RL}\mathrm{D}^{\alpha}_{-\infty,t}f(t) = \frac{\mathrm{d}^{m-1}}{\mathrm{d}t^{m-1}}\left[\mathrm{D}^{\{\alpha\}}_{-\infty,t}f(t)\right] = \frac{\{\alpha\}}{\Gamma(1-\{\alpha\})}\int_{0}^{\infty}\frac{f^{(m-1)}(t)-f^{(m-1)}(t-\xi)}{\xi^{1+\{\alpha\}}}\,\mathrm{d}\xi, \tag{1.109}$$

where $\{\alpha\} = \alpha - m + 1$, and (1.105) is used.

Hence, for any $\alpha > 0$, we can define the left and right Marchaud fractional derivatives as

$$ {}_{M}\mathrm{D}_{+}^{\alpha} f(t) = \frac{\mathrm{d}^{m-1}}{\mathrm{d}t^{m-1}} \left[\mathrm{D}_{+}^{\{\alpha\}} f(t)\right] = \frac{\{\alpha\}}{\Gamma(1-\{\alpha\})} \int_0^{\infty} \frac{f^{(m-1)}(t) - f^{(m-1)}(t-\xi)}{\xi^{1+\{\alpha\}}} \mathrm{d}\xi \tag{1.110} $$

and

$$ {}_{M}\mathrm{D}_{-}^{\alpha} f(t) = \frac{\mathrm{d}^{m-1}}{\mathrm{d}t^{m-1}} \left[\mathrm{D}_{-}^{\{\alpha\}} f(t)\right] = \frac{\{\alpha\}}{\Gamma(1-\{\alpha\})} \int_0^{\infty} \frac{f^{(m-1)}(t) - f^{(m-1)}(t+\xi)}{\xi^{1+\{\alpha\}}} \mathrm{d}\xi, \tag{1.111} $$

respectively, where $\{\alpha\} = \alpha - m + 1$, $m - 1 < \alpha < m$.

1.3.2 The Finite Parts of Integrals

From (1.105), we know that if $f(t)$ is "good," then ${}_{RL}\mathrm{D}_{-\infty,t}^{\alpha} f(t) = \mathrm{D}_{+}^{\alpha} f(t)$. One can also find that ${}_{RL}\mathrm{D}_{-\infty,t}^{\alpha} f(t)$ can be obtained from $\mathrm{D}_{-\infty,t}^{-\alpha} f(t)$ if we replace α with $-\alpha$. We know that

$$ \frac{\alpha}{\Gamma(1-\alpha)} \int_0^{\infty} \frac{f(t-\xi)}{\xi^{1+\alpha}} \mathrm{d}\xi $$

is divergent to infinity. In spite of this, its *finite part in the sense of Hadamard* is introduced below.

Definition 6 *Let a function $\Phi(t)$ be integrable on an interval (ϵ, A) for any $A\ (> \epsilon > 0)$. The function $\Phi(t)$ is said to possess the Hadamard property at the point $t = 0$ if there exist constants a_k, b and $\lambda_k > 0$ such that*

$$ \int_{\epsilon}^{A} \Phi(t)\,\mathrm{d}t = \sum_{k=1}^{N} a_k \epsilon^{-\lambda_k} + b \ln\frac{1}{\epsilon} + J_0(\epsilon), \tag{1.112} $$

where $\lim\limits_{\epsilon\to 0} J_0(\epsilon)$ exists and is finite. Set

$$ \text{p.f.} \int_0^{A} \Phi(t)\,\mathrm{d}t = \lim_{\epsilon\to 0} J_0(\epsilon). \tag{1.113} $$

The limit (1.113) is called a finite part of the divergent integral $\int_0^A \Phi(t)\,\mathrm{d}t$ in the Hadamard sense or simply an integral in the Hadamard sense. The constructive realization of the function $J_0(\epsilon)$ is sometimes called a regularization of the integral $\int_0^A \Phi(t)\,\mathrm{d}t$.

It is not difficult to find that the constants a_k, b and $\lambda_k > 0$ in (1.112) are not dependent on A. Hence, one can easily obtain

$$ \text{p.f.} \int_0^{\infty} \Phi(t)\,\mathrm{d}t = \text{p.f.} \int_0^{A} \Phi(t)\,\mathrm{d}t + \int_A^{\infty} \Phi(t)\,\mathrm{d}t, \tag{1.114} $$

where A is arbitrarily chosen with $A > 0$.

Next we introduce several properties of the finite part of integrals due to Hadamard.

Lemma 1.3.1 ([134]) *Let* $0 < \alpha < 1$ *and* $f(t)$ *be locally Hölderian of order* $\lambda > \alpha$. *Then the function* $\Phi(s) = f(t-s)s^{-1-\alpha}$ *possesses the Hadamard property at the point* $s = 0$ *for each* t *and if* $|f(s)| \le c|s|^{\alpha-\epsilon}$, $\epsilon > 0$, *as* $s \to -\infty$, *then*

$$\text{p.f.} \int_0^\infty \frac{f(t-s)}{s^{1+\alpha}} \mathrm{d}s = \int_0^\infty \frac{f(t-s)-f(t)}{s^{1+\alpha}} \mathrm{d}s.$$

Lemma 1.3.2 ([134]) *Let* $f(t) \in C^m$ *and* $f(t)$ *be locally Hölderian of order* λ, $0 < \lambda < 1$. *Then the function* $\Phi(s) = f(t-s)s^{-1-\alpha}$ *possesses the Hadamard property at the point* $s = 0$ *for each* t *and* $\alpha < m + \lambda$. *If* $|f(s)| \le c|s|^{\alpha-\epsilon}$ *also holds,* $\epsilon > 0$ *for* $s \to -\infty$, *then*

$$\begin{aligned}\frac{1}{\Gamma(-\alpha)}\text{p.f.} \int_0^\infty \frac{f(t-s)}{s^{1+\alpha}} \mathrm{d}s &= \frac{1}{\Gamma(-\alpha)} \int_0^1 \frac{f(t-s) - \sum_{k=0}^m (-1)^k \frac{s^k}{k!} f^{(k)}(t)}{s^{1+\alpha}} \mathrm{d}s \\ &+ \frac{1}{\Gamma(-\alpha)} \int_0^\infty \frac{f(t-s)}{s^{1+\alpha}} \mathrm{d}s + \sum_{k=0}^m \frac{(-1)^k}{k!} \frac{f^{(k)}(t)}{\Gamma(-\alpha)(k-\alpha)},\end{aligned} \tag{1.115}$$

where $\alpha < m + \lambda, \alpha \neq 0, 1, 2, \cdots$.

Theorem 1 ([134]) *Let* $f(t)$ *satisfy the assumption of Lemma 1.3.2 with* $m - 1 < \alpha \le m$. *Then the Liouville fractional derivative* ${}_{RL}\mathrm{D}^{\alpha}_{-\infty,t} f(t)$ *coincides with* (1.115) *for any* $\alpha > 0, \alpha \neq 1, 2, \cdots$.

1.3.3 Directional Integrals and Derivatives in $\mathbb{R}^2$

Definition 7 *Let* $\alpha > 0$, $\theta \in [0, 2\pi)$ *be given. The* α*-th order fractional integral in the direction of* θ *is given by*

$$\mathrm{D}_\theta^{-\alpha} u(x,y) = \frac{1}{\Gamma(\alpha)} \int_0^\infty \xi^{\alpha-1} u(x - \xi\cos\theta, y - \xi\sin\theta)\mathrm{d}\xi. \tag{1.116}$$

Remark 1.3.1 *It is easy to see that for special directions as* $\theta = 0$, $\pi/2$, π *and* $3\pi/2$, *the directional operator is reduced to left and right Riemann–Liouville integral operators, i.e.,*

$$\begin{aligned}\mathrm{D}_0^{-\alpha} u(x,y) &= \mathrm{D}_{-\infty,x}^{-\alpha} u(x,y), \\ \mathrm{D}_\pi^{-\alpha} u(x,y) &= \mathrm{D}_{x,\infty}^{-\alpha} u(x,y), \\ \mathrm{D}_{\pi/2}^{-\alpha} u(x,y) &= \mathrm{D}_{-\infty,y}^{-\alpha} u(x,y), \\ \mathrm{D}_{3\pi/2}^{-\alpha} u(x,y) &= \mathrm{D}_{y,\infty}^{-\alpha} u(x,y).\end{aligned} \tag{1.117}$$

The directional derivatives can be similarly defined as in (1.5).

Definition 8 *Let* n *be a positive integer satisfying* $n - 1 \le \alpha < n$, $\theta \in [0, 2\pi)$. *Then the* α*-th order fractional derivative in the direction* θ *is defined by*

$$\mathrm{D}_\theta^\alpha u(x,y) = \mathrm{D}_\theta^n \mathrm{D}_\theta^{-(n-\alpha)} u(x,y), \tag{1.118}$$

where D_θ^n *is given by*

$$\mathrm{D}_\theta^n u(x,y) = \left(\cos\theta \frac{\partial}{\partial x} + \sin\theta \frac{\partial}{\partial y}\right)^n u(x,y). \tag{1.119}$$

Next, we list some properties of the directional integrals and derivatives, which are similar to those of the fractional integrals and derivatives.

Proposition 1.3.1 ([131]) *The fractional directional integral operator satisfies the following semi-group properties*

$$\mathrm{D}_\theta^{-\alpha}\mathrm{D}_\theta^{-\beta}u(x,y) = \mathrm{D}_\theta^{-\alpha-\beta}u(x,y), \tag{1.120}$$

where $\alpha, \beta > 0$, $\theta \in [0,2\pi)$, $u \in L^2(\mathbb{R}^2)$.

Proposition 1.3.2 ([131]) *For* $\alpha > 0$, $\theta \in [0,2\pi)$, $u \in L^2(\mathbb{R}^2)$, *the following relation holds*

$$\mathrm{D}_\theta^{\alpha}\mathrm{D}_\theta^{-\alpha}u(x,y) = u(x,y). \tag{1.121}$$

Proposition 1.3.3 ([131]) *The fractional directional integral operator* $\mathrm{D}_\theta^{-\alpha}$ *satisfies the following Fourier transform property*

$$F\{\mathrm{D}_\theta^{-\alpha}u(x,y);\omega\} = (i\omega_1\cos\theta + i\omega_2\sin\theta)^{-\alpha}F\{u(x,y);\omega\}, \tag{1.122}$$

where $\omega = (\omega_1, \omega_2)$ *and*

$$F\{u(x,y);\omega\} = \int_{\mathbb{R}^2} e^{-i(\omega_1 x + \omega_2 y)}u(x,y)\,\mathrm{d}x\,\mathrm{d}y.$$

Proposition 1.3.4 ([131]) *For* $u \in C_0^\infty(\Omega), \Omega \in \mathbb{R}^2$ *and* $\alpha > 0$, *we have*

$$F\{\mathrm{D}_\theta^{\alpha}u(x,y);\omega\} = (i\omega_1\cos\theta + i\omega_2\sin\theta)^{\alpha}F\{u(x,y);\omega\}. \tag{1.123}$$

1.3.4 Partial Fractional Derivatives

Similar to the classical partial derivatives, we can also define the *partial fractional derivatives* [134]. For example, let $0 < \alpha_1, \alpha_2 < 1$, the partial fractional derivative ${}_{RL}\mathrm{D}_{x^{\alpha_1}y^{\alpha_2}}^{\alpha_1+\alpha_2}u(x,y)$ is defined by

$$\begin{aligned}
&{}_{RL}\mathrm{D}_{x^{\alpha_1}y^{\alpha_2}}^{\alpha_1+\alpha_2}u(x,y) = {}_{RL}\mathrm{D}_{0,y}^{\alpha_2}\left[{}_{RL}\mathrm{D}_{0,x}^{\alpha_1}u(x,y)\right]\\
=&{}_{RL}\mathrm{D}_{0,y}^{\alpha_2}\left[\frac{1}{\Gamma(1-\alpha_1)}\frac{\partial}{\partial x}\int_0^x (x-s)^{-\alpha_1}u(s,y)\,\mathrm{d}s\right]\\
=&\frac{1}{\Gamma(1-\alpha_1)\Gamma(1-\alpha_2)}\frac{\partial^2}{\partial x\partial y}\int_0^x\int_0^y (x-s)^{-\alpha_1}(y-\tau)^{-\alpha_2}u(s,\tau)\,\mathrm{d}\tau\,\mathrm{d}s.
\end{aligned} \tag{1.124}$$

Obviously, if $u(x,y)$ is "good" enough, then one can easily obtain

$${}_{RL}\mathrm{D}_{x^{\alpha_1}y^{\alpha_2}}^{\alpha_1+\alpha_2}u(x,y) = {}_{RL}\mathrm{D}_{y^{\alpha_2}x^{\alpha_1}}^{\alpha_2+\alpha_1}u(x,y).$$

For any $\alpha_1, \alpha_2 > 0$, we can give the following definition of the partial fractional derivative.

Definition 9 *The partial fractional derivative operator* ${}_{RL}\mathrm{D}_{x^{\alpha_1}y^{\alpha_2}}^{\alpha_1+\alpha_2}$ *with order* $(\alpha_1+\alpha_2)$ *is defined by*

$$\begin{aligned}{}_{RL}\mathrm{D}_{x^{\alpha_1}y^{\alpha_2}}^{\alpha_1+\alpha_2}u(x,y) =&\frac{1}{\Gamma(m-\alpha_1)\Gamma(n-\alpha_2)}\frac{\partial^{m+n}}{\partial x^m \partial y^n}\\ &\times\int_0^x\int_0^y (x-s)^{m-\alpha_1-1}(y-\tau)^{n-\alpha_2-1}u(s,\tau)\mathrm{d}\tau\mathrm{d}s,\end{aligned} \tag{1.125}$$

where $m-1<\alpha_1<m$, $n-1<\alpha_2<n$, m, n *are positive integers.*

Similar to (1.125), the definition of the partial fractional derivative in the Riemann–Liouville sense can be given below.

Definition 10 *The partial fractional derivative operator* ${}_{RL}\mathrm{D}_{x_1^{\alpha_1}x_2^{\alpha_2}\cdots x_\ell^{\alpha_\ell}}^{\alpha_1+\alpha_2+\cdots+\alpha_\ell}$ *with order* $(\alpha_1+\alpha_2+\cdots+\alpha_\ell)$ *is defined by*

$$\begin{aligned}&{}_{RL}\mathrm{D}_{x_1^{\alpha_1}x_2^{\alpha_2}\cdots x_\ell^{\alpha_\ell}}^{\alpha_1+\alpha_2+\cdots+\alpha_\ell}u(x_1,\cdots,x_\ell) = \frac{1}{\prod_{k=1}^{\ell}\Gamma(m_k-\alpha_k)}\frac{\partial^{m_1+m_2+\cdots+m_\ell}}{\partial x_1^{m_1}\partial x_2^{m_2}\cdots\partial x_\ell^{m_\ell}}\\ &\times\int_0^{x_1}\cdots\int_0^{x_\ell}(x_\ell-\xi_\ell)^{m_\ell-\alpha_\ell-1}\cdots(x_1-\xi_1)^{m_1-\alpha_1-1}u(\xi_1,\cdots,\xi_\ell)\mathrm{d}\xi_1\cdots\mathrm{d}\xi_\ell,\end{aligned} \tag{1.126}$$

where $m_k-1<\alpha_k<m_k$ $(k=1,2,\cdots,\ell)$, m_k *are positive integers.*

We can define the partial fractional derivative in the Caputo sense.

Definition 11 *The partial fractional derivative operator* ${}_{C}\mathrm{D}_{x_1^{\alpha_1}x_2^{\alpha_2}\cdots x_\ell^{\alpha_\ell}}^{\alpha_1+\alpha_2+\cdots+\alpha_\ell}$ *with order* $(\alpha_1+\alpha_2+\cdots+\alpha_\ell)$ *is defined by*

$$\begin{aligned}&{}_{C}\mathrm{D}_{x_1^{\alpha_1}x_2^{\alpha_2}\cdots x_\ell^{\alpha_\ell}}^{\alpha_1+\alpha_2+\cdots+\alpha_\ell}u(x,y)\\ =&\frac{1}{\prod_{k=1}^{\ell}\Gamma(m_k-\alpha_k)}\int_0^{x_1}\cdots\int_0^{x_\ell}(x_\ell-\xi_\ell)^{m_\ell-\alpha_\ell-1}\cdots(x_1-\xi_1)^{m_1-\alpha_1-1}\\ &\times\frac{\partial^{m_1+m_2+\cdots+m_\ell}}{\partial\xi_1^{m_1}\partial\xi_2^{m_2}\cdots\partial\xi_\ell^{m_\ell}}u(\xi_1,\cdots,\xi_\ell)\mathrm{d}\xi_1\cdots\mathrm{d}\xi_\ell,\end{aligned} \tag{1.127}$$

where $m_k-1<\alpha_k<m_k$ $(k=1,2,\cdots,\ell)$, m_k *are positive integers.*

Similar to (1.14), one has

$$\begin{aligned}
&\lim_{\alpha_i\to m_i^-} {}_{RL}\mathrm{D}^{\alpha_1+\alpha_2+\cdots+\alpha_\ell}_{x_1^{\alpha_1}x_2^{\alpha_2}\cdots x_\ell^{\alpha_\ell}} u(x_1,\cdots,x_\ell) = {}_{RL}\mathrm{D}^{\alpha_1+\cdots+\alpha_{i-1}+\alpha_{i+1}+\cdots+\alpha_\ell}_{x_1^{\alpha_1}\cdots x_{i-1}^{\alpha_{i-1}}x_{i+1}^{\alpha_{i+1}}\cdots x_\ell^{\alpha_\ell}} \frac{\partial^{m_i}}{\partial x_i^{m_i}} u(x_1,\cdots,x_\ell),\\
&\lim_{\alpha_i\to (m_i-1)^+} {}_{RL}\mathrm{D}^{\alpha_1+\alpha_2+\cdots+\alpha_\ell}_{x_1^{\alpha_1}x_2^{\alpha_2}\cdots x_\ell^{\alpha_\ell}} u(x_1,\cdots,x_\ell) = {}_{RL}\mathrm{D}^{\alpha_1+\cdots+\alpha_{i-2}+\alpha_i+\cdots+\alpha_\ell}_{x_1^{\alpha_1}\cdots x_{i-2}^{\alpha_{i-2}}x_i^{\alpha_i}\cdots x_\ell^{\alpha_\ell}} \frac{\partial^{m_i-1}}{\partial x_i^{m_i-1}} u(x_1,\cdots,x_\ell);\\
&\lim_{\alpha_i\to m_i^-} {}_{C}\mathrm{D}^{\alpha_1+\alpha_2+\cdots+\alpha_\ell}_{x_1^{\alpha_1}x_2^{\alpha_2}\cdots x_\ell^{\alpha_\ell}} u(x_1,\cdots,x_\ell) = {}_{C}\mathrm{D}^{\alpha_1+\cdots+\alpha_{i-1}+\alpha_{i+1}+\cdots+\alpha_\ell}_{x_1^{\alpha_1}\cdots x_{i-1}^{\alpha_{i-1}}x_{i+1}^{\alpha_{i+1}}\cdots x_\ell^{\alpha_\ell}} \frac{\partial^{m_i}}{\partial x_i^{m_i}} u(x_1,\cdots,x_\ell),\\
&\lim_{\alpha_i\to (m_i-1)^+} {}_{C}\mathrm{D}^{\alpha_1+\alpha_2+\cdots+\alpha_\ell}_{x_1^{\alpha_1}x_2^{\alpha_2}\cdots x_\ell^{\alpha_\ell}} u(x_1,\cdots,x_\ell) = {}_{C}\mathrm{D}^{\alpha_1+\cdots+\alpha_{i-2}+\alpha_i+\cdots+\alpha_\ell}_{x_1^{\alpha_1}\cdots x_{i-2}^{\alpha_{i-2}}x_i^{\alpha_i}\cdots x_\ell^{\alpha_\ell}} \frac{\partial^{m_i-1}}{\partial x_i^{m_i-1}} u(x_1,\cdots,x_\ell)\\
&\qquad - {}_{C}\mathrm{D}^{\alpha_1+\cdots+\alpha_{i-2}+\alpha_i+\cdots+\alpha_\ell}_{x_1^{\alpha_1}\cdots x_{i-2}^{\alpha_{i-2}}x_i^{\alpha_i}\cdots x_\ell^{\alpha_\ell}} \frac{\partial^{m_i-1}}{\partial x_i^{m_i-1}} u(x_1,\cdots,x_{i-2},0,x_i,\cdots,x_\ell).
\end{aligned} \tag{1.128}$$

1.4 Physical Meanings

It is known that classical calculus (or "calculus" for brevity) means integration and differentiation. So fractional calculus also means fractional integration and fractional differentiation. Different from the typical derivative, there are more than six kinds of definitions of fractional derivatives. They are not mutually equivalent. Among these definitions, the Riemann–Liouville derivative and the Caputo derivative which are defined on the basis of fractional integral (or, Riemann–Liouville integral) are most frequently used. Stochastic experts, pure mathematicians and physicists would rather use the former, while applied mathematicians and engineers prefer to utilize the latter, mainly due to their respective research backgrounds [46, 107].

In the following, we explain that fractional calculus is not the mathematical generalization of classical calculus.

For the Caputo derivative, if we fix t and let the order $\alpha \in (n-1,n)$ vary, so we have

$$\lim_{\alpha\to(n-1)^+} {}_C\mathrm{D}^\alpha_{0,t}x(t) = x^{(n-1)}(t) - x^{(n-1)}(0), \ \lim_{\alpha\to n^-} {}_C\mathrm{D}^\alpha_{0,t}x(t) = x^{(n)}(t).$$

It follows that the Caputo derivative is not the mathematical extension of a typical derivative. More explanations can be found in [77].

Next consider a simple function below,

$$x(t) = \begin{cases} 1-t, t\in[0,1], \\ t-1, t\in(1,1+t_0). \end{cases}$$

${}_{RL}\mathrm{D}^\alpha_{0,t}x(t)$ $(\alpha\in(0,1))$ exists on the interval $(0,1+t_0]$, while $x'(t)$ exists on the domain $[0,1)\cup(1,1+t_0]$. If ${}_{RL}\mathrm{D}^\alpha_{0,t}x(t)$ $(\alpha\in(0,1))$ is the generalization of $x'(t)$, then the interval $(0,1+t_0]$ should be "bigger" than the domain $[0,1)\cup(1,1+t_0]$. But it is not true. So the Riemann–Liouville derivative is not the mathematical generalization of the typical derivative, either. More illustrations are given in [81, 91].

Overall, fractional calculus, closely related to classical calculus, is not direct generalization of classical calculus in the sense of rigorous mathematics. In the following, we give possible physical and geometrical interpretation.

Recalling the integral $A = \int_a^b f(x)dx$, from the viewpoint of geometry, it means the area of the domain $\{(x,y) \mid a \le x < b, 0 \le y \le f(x)\}$ presuming that $f(x) \ge 0$. From the viewpoint of physics, it implies the displacement from a to b if $f(x)$ indicates the velocity at time x. The geometrical and physical meaning of the derivative is well known to us. For example, $f'(x)$ indicates the slope of the curve $f(x)$ at x. On the other hand, if $s(t)$ is the displacement at time t, then, $s'(t)$ stands for the velocity at time t, $s''(t)$ the acceleration at time t. Now, we give a possible interpretation of the fractional calculus.

The fractional integral with order α

$$\mathrm{D}_{a,t}^{-\alpha} f(t) = {}_{RL}\mathrm{D}_{a,t}^{-\alpha} f(t) = \frac{1}{\Gamma(\alpha)} \int_a^t (t-s)^{\alpha-1} f(s)\,\mathrm{d}s$$

can be rewritten as

$$\mathrm{D}_{a,t}^{-\alpha} f(t) = \int_a^t f(\tau)\mathrm{d}Y_\alpha(\tau),$$

where

$$Y_\alpha(\tau) = \begin{cases} -\frac{(t-\tau)^\alpha}{\Gamma(\alpha+1)}, & \tau \in [a,t], \\ 0, & \tau < a. \end{cases}$$

This is the standard *Stieltjes integral.* $Y_\alpha(\tau)$ is a monotonously increasing function in $(-\infty, t]$. The positive number α is an index characterizing the singularity: the smaller α, the stronger singularity the integral. If $Y_\alpha(\tau) = \tau$, the above integral is reduced to a typical one. So $\mathrm{D}_{a,t}^{-\alpha} f(t)$ indicates the generalized area in the sense of length $Y_\alpha(\tau)$ (geometrical meaning) or the generalized displacement in the sense of $Y_\alpha(\tau)$ if $f(t)$ means the velocity at time t (physical meaning).

The Riemann–Liouville derivative with order $\alpha \in (0,1)$ can be written as

$${}_{RL}\mathrm{D}_{a,t}^{\alpha} f(t) = \frac{\mathrm{d}}{\mathrm{d}t} \int_a^t f(\tau)\mathrm{d}Y_{1-\alpha}(\tau),$$

where

$$Y_{1-\alpha}(\tau) = \begin{cases} -\frac{(t-\tau)^{1-\alpha}}{\Gamma(2-\alpha)}, & \tau \in [a,t], \\ 0, & \tau < a. \end{cases}$$

Obviously, $Y_{1-\alpha}(\tau)$ is a monotonously increasing function in $(-\infty, t]$. So ${}_{RL}\mathrm{D}_{a,t}^{\alpha} f(t)$ indicates the generalized slope in the sense of length $Y_{1-\alpha}(\tau)$ if $f(t)$ means the slope (geometrical meaning) or the generalized velocity in the sense of length $Y_{1-\alpha}(\tau)$ if $f(t)$ means the velocity (physical meaning). If $Y_{1-\alpha}(\tau) = \tau$, it is reduced to the classical case.

Similarly, the interpretation of the Caputo derivative can be also given. In effect, the Caputo derivative with order $\alpha \in (0,1)$ is written as

$$ {}_C\mathrm{D}_{a,t}^{\alpha} f(t) = \int_a^t f'(\tau) \mathrm{d}Y_{1-\alpha}(\tau), $$

where

$$ Y_{1-\alpha}(\tau) = \begin{cases} -\frac{(t-\tau)^{1-\alpha}}{\Gamma(2-\alpha)}, & \tau \in [a,t], \\ 0, & \tau < a. \end{cases} $$

So ${}_C\mathrm{D}_{a,t}^{\alpha} f(t)$ indicates the generalized displacement of the curve $f(t)$ in the sense of length $Y_{1-\alpha}(\tau)$ (physical meaning) if $f(t)$ means the displacement, or represents the generalized curve in the sense of length $Y_{1-\alpha}(\tau)$ if $f(t)$ is a curve (geometrical meaning). If $Y_{1-\alpha}(\tau) = \tau$, the above integral is reduced to the typical one.

1.5 Fractional Initial and Boundary Problems

How to determine the definite conditions for fractional differential systems seems to be a ticklish matter [91]. But after careful analysis, one can grasp it. In the following, we first study the Caputo case.

Consider the following Caputo-type differential equation

$$ {}_C\mathrm{D}_{0,t}^{\alpha} x(t) = f(x,t), \ n-1 < \alpha < n \in Z^+. $$

Noticing that ${}_C D_{0,t}^{\alpha} x(t) = \mathrm{D}_{0,t}^{-(n-\alpha)} D^n x(t) = \mathrm{D}_{0,t}^{-(n-\alpha)} x^{(n)}(t)$, acting ${}_{RL}\mathrm{D}_{0,t}^{n-\alpha}$ in both sides of the above equation yields

$$ x^{(n)}(t) = {}_{RL}\mathrm{D}_{0,t}^{n-\alpha} f(x,t) = \frac{1}{\Gamma(\alpha-n+1)} \frac{\mathrm{d}}{\mathrm{d}t} \int_0^t (t-\tau)^{\alpha-n} f(x(\tau),\tau) \mathrm{d}\tau =: F(x,t). $$

It immediately follows that the initial value conditions of the Caputo-type differential equation are given as $x^{(k)}(0) = x_0^{(k)}, k = 0,1,\ldots,n-1$. So the initial value condition(s), boundary value condition(s), the initial and boundary value conditions of the Caputo-type (ordinary or partial) differential equation are the same as those of the classical (ordinary or partial) differential equation. Therefore we do not discuss the definite conditions of the Caputo-type differential equation any more.

In the following, we consider the Riemann–Liouville type differential equation,

$$ {}_{RL}\mathrm{D}_{0,t}^{\alpha} x(t) = f(x,t), \ n-1 < \alpha < n \in Z^+. $$

Noting ${}_{RL}\mathrm{D}_{0,t}^{\alpha} x(t) = \frac{\mathrm{d}^n}{\mathrm{d}t^n} {}_{RL}\mathrm{D}_{0,t}^{-(n-\alpha)} x(t)$, the above equation can be changed into

$$ \frac{\mathrm{d}^n}{\mathrm{d}t^n} \left({}_{RL}\mathrm{D}_{0,t}^{-(n-\alpha)} x(t) \right) = f(x,t), $$

therefore the initial value conditions of the Riemann–Liouville type differential equation should be given as

$$\frac{\mathrm{d}^k}{\mathrm{d}t^k}\left({}_{RL}\mathrm{D}_{0,t}^{-(n-\alpha)}x(t)\right)_{t=0} = {}_{RL}\mathrm{D}_{0,t}^{k+\alpha-n}x(t)|_{t=0} = x_0^{(k)},\ k = 0,1,\ldots,n-1.$$

When $k = 0$, the corresponding initial value condition is the integral initial value condition. Such definite conditions for this Riemann–Liouville type differential equation makes the Cauchy problem well-posed. Here "being well-posed" means the existence, uniqueness and stability of its solution. Please note that the above initial value conditions can not be replaced by the classical initial value conditions $x^{(k)}(0) = x_0^{(k)}, k = 0,1,\ldots,n-1$; otherwise, the problem will be ill-posed. In effect, if $x(t)$ is a solution to the above Riemann–Liouville-type differential equation, then $x(0+) = \lim_{x\to 0^+} x(t)$ is often unbounded. Under ${}_{RL}\mathrm{D}_{0,t}^{k+\alpha-n}x(t)|_{t=0} = 0$, $k = 0,1,\ldots,n-1$, the initial value conditions of the Riemann–Liouville type differential equation is often transformed as $x^{(k)}(0) = 0, k = 0,1,\ldots,n-1$. However, this is not mathematically true. If some extra conditions are imposed, then we can do it like this. Ref. [39] seems to give a suitable choice.

Based on the above analysis, we can give two point boundary value conditions for the Riemann–Liouville type differential equation as follows,

$${}_{RL}\mathrm{D}_{a,x}^{\alpha}y(x) = f(x,y),\ x \in (a,b),\ 1 < \alpha < 2,$$

its boundary value conditions can be given as:

i) ${}_{RL}\mathrm{D}_{a,x}^{\alpha-2}y(x)|_{x=a} = c_1$, ${}_{RL}\mathrm{D}_{x,b}^{\alpha-2}y(x)|_{x=b} = c_2$,

or,

ii) ${}_{RL}\mathrm{D}_{a,x}^{\alpha-1}y(x)|_{x=a} = c_3$, ${}_{RL}\mathrm{D}_{x,b}^{\alpha-1}y(x)|_{x=b} = c_4$,

or,

iii) ${}_{RL}\mathrm{D}_{a,x}^{\alpha-2}y(x)|_{x=a} = c_1$, ${}_{RL}\mathrm{D}_{x,b}^{\alpha-1}y(x)|_{x=b} = c_4$,

or,

iv) ${}_{RL}\mathrm{D}_{a,x}^{\alpha-1}y(x)|_{x=a} = c_3$, ${}_{RL}\mathrm{D}_{x,b}^{\alpha-2}y(x)|_{x=b} = c_2$.

Next, we only list the *initial-boundary value problem* of the Riemann–Liouville type partial differential equation,

$${}_{RL}\mathrm{D}_{0,t}^{\alpha}u(x,t) = {}_{RL}\mathrm{D}_{a,x}^{\beta}u(x,t),\ t > 0,\ x \in (a,b),\ \alpha \in (0,1), \beta \in (1,2),$$

where ${}_{RL}\mathrm{D}_{0,t}^{\alpha}u(x,t)$ means the α-th order partial derivative of $u(x,t)$ with respect to t and ${}_{RL}\mathrm{D}_{a,x}^{\beta}u(x,t)$ means the β-th order partial derivative of $u(x,t)$ with respect to x. The initial value condition of this fractional partial differential equation is given as

$${}_{RL}\mathrm{D}_{0,t}^{\alpha-1}u(x,t)|_{t=0} = \phi(x).$$

Its boundary value conditions can be of Dirichlet type (the first class), or of Neumann type (the second class), or of Rubin type (the third class), which are presented below:

i) *Dirichlet type*

$${}_{RL}\mathrm{D}_{a,x}^{\beta-2}u(x,t)|_{x=a} = \xi_1(t),\ {}_{RL}\mathrm{D}_{x,b}^{\beta-2}u(x,t)|_{x=b} = \xi_2(t).$$

ii) *Neumann type*

$$ {}_{RL}\mathrm{D}_{a,x}^{\beta-1}u(x,t)|_{x=a}=\xi_1(t),\ {}_{RL}\mathrm{D}_{x,b}^{\beta-1}u(x,t)|_{x=b}=\xi_2(t). $$

iii) *Rubin type*

$$ \left({}_{RL}\mathrm{D}_{a,x}^{\beta-1}u(x,t)-\sigma_1\,{}_{RL}\mathrm{D}_{a,x}^{\beta-2}u(x,t)\right)_{x=a}=\xi_1(t),\ \sigma_1>0, $$

$$ \left({}_{RL}\mathrm{D}_{x,b}^{\beta-1}u(x,t)+\sigma_2\,{}_{RL}\mathrm{D}_{x,b}^{\beta-2}u(x,t)\right)_{x=b}=\xi_2(t),\ \sigma_2>0. $$

If $\alpha\in(1,2)$ and β remains unchanged, then two initial value conditions are needed which read as below,

$$ {}_{RL}\mathrm{D}_{0,t}^{\alpha-2}u(x,t)|_{t=0}=\phi(x),\ {}_{RL}\mathrm{D}_{0,t}^{\alpha-1}u(x,t)|_{t=0}=\psi(x). $$

The boundary value conditions are unchanged.

For the fractional partial differential equation, the definite conditions need to be properly proposed, otherwise the corresponding initial-boundary value problems will be ill-posed.

Chapter 2

Numerical Methods for Fractional Integral and Derivatives

In the previous chapter, the important properties of the most frequently used fractional integral and fractional derivatives are introduced. In this chapter, we mainly construct the efficient algorithms for Riemann–Liouville integrals, Riemann–Liouville derivatives, Caputo derivatives, and Riesz derivatives, etc.

2.1 Approximations to Fractional Integrals

The fractional integral operator plays an important role in fractional calculus, which is useful for converting the fractional differential equations into integral equations with a weakly singular kernel. So it is necessary to study the numerical methods for approximating fractional integrals. This section introduces numerical approaches used to approximate the fractional integrals based on the polynomial interpolation.

Suppose that $f(t) \in C(I), I = [0,T]$. Let Δt be the step size with $\Delta t = T/n_T, n_T \in \mathbb{N}$, and denote by $t_k = k\Delta t$. Next, we investigate how to numerically calculate the following integral

$$\mathrm{D}_{0,t}^{-\alpha} f(t) = \frac{1}{\Gamma(\alpha)} \int_0^t (t-s)^{\alpha-1} f(s)\,\mathrm{d}s, \quad \alpha > 0. \tag{2.1}$$

One way to numerically calculate (2.1) is to approximate $f(t)$ by a certain function $\tilde{f}(t)$ in order that $\mathrm{D}_{0,t}^{-\alpha}\tilde{f}(t)$ can easily be calculated exactly. We naturally think of the polynomial approximation of $f(t)$ on the interval $[0,T]$. Theoretically speaking, $\mathrm{D}_{0,t}^{-\alpha}\tilde{f}(t)$ can be calculated exactly if $\tilde{f}(t)$ is a polynomial. For $t = t_n$, $n \in \mathbb{N}$, we rewrite $\left[\mathrm{D}_{0,t}^{-\alpha} f(t)\right]_{t=t_n}$ as the following form

$$\begin{aligned} \left[\mathrm{D}_{0,t}^{-\alpha} f(t)\right]_{t=t_n} &= \frac{1}{\Gamma(\alpha)} \int_0^{t_n} (t_n - s)^{\alpha-1} f(s)\,\mathrm{d}s \\ &= \frac{1}{\Gamma(\alpha)} \sum_{k=0}^{n-1} \int_{t_k}^{t_{k+1}} (t_n - s)^{\alpha-1} f(s)\,\mathrm{d}s. \end{aligned} \tag{2.2}$$

Next, we introduce the numerical methods based on the polynomial interpolation to calculate (2.2).

2.1.1 Numerical Methods Based on Polynomial Interpolation

This subsection extends the numerical methods for the classical integrals to the fractional integrals.

- **Fractional Rectangular Formula**

On each subinterval $[t_k, t_{k+1}], k = 0, 1, \cdots, n-1$, the function $f(t)$ is approximated by a constant, i.e.,

$$f(t)|_{[t_k,t_{k+1})} \approx \tilde{f}(t)|_{[t_k,t_{k+1})} = f(t_k), \tag{2.3}$$

one obtains

$$\begin{aligned}\left[\mathrm{D}_{0,t}^{-\alpha} f(t)\right]_{t=t_n} &= \frac{1}{\Gamma(\alpha)} \sum_{k=0}^{n-1} \int_{t_k}^{t_{k+1}} (t_n - s)^{\alpha-1} f(s)\,\mathrm{d}s \\ &\approx \frac{1}{\Gamma(\alpha)} \sum_{k=0}^{n-1} \int_{t_k}^{t_{k+1}} (t_n - s)^{\alpha-1} f(t_k)\,\mathrm{d}s \\ &= \sum_{k=0}^{n-1} b_{n-k-1} f(t_k),\end{aligned} \tag{2.4}$$

where

$$b_k = \frac{\Delta t^\alpha}{\Gamma(\alpha+1)} \left[(k+1)^\alpha - k^\alpha\right]. \tag{2.5}$$

Hence,

$$\left[\mathrm{D}_{0,t}^{-\alpha} f(t)\right]_{t=t_n} \approx \sum_{k=0}^{n-1} b_{n-k-1} f(t_k). \tag{2.6}$$

Similar to the classical left rectangular formula, we call (2.6) the ***left fractional rectangular formula***.

Similarly, if

$$f(t)|_{(t_k,t_{k+1}]} \approx \tilde{f}(t)|_{(t_k,t_{k+1}]} = f(t_{k+1}), \tag{2.7}$$

then we get the following ***right fractional rectangular formula***

$$\left[\mathrm{D}_{0,t}^{-\alpha} f(t)\right]_{t=t_n} \approx \sum_{k=0}^{n-1} b_{n-k-1} f(t_{k+1}). \tag{2.8}$$

The formulae (2.7) and (2.8) can be seen as the special cases of the following ***weighted fractional rectangular formula***

$$\left[\mathrm{D}_{0,t}^{-\alpha} f(t)\right]_{t=t_n} \approx \sum_{k=0}^{n-1} b_{n-k-1} \left[\theta f(t_k) + (1-\theta) f(t_{k+1})\right], \quad 0 \le \theta \le 1. \tag{2.9}$$

Of course, one can also obtain the following formula

$$\left[\mathrm{D}_{0,t}^{-\alpha} f(t)\right]_{t=t_n} \approx \sum_{k=0}^{n-1} b_{n-k-1} f(t_k + (1-\theta)\Delta t), \quad 0 \le \theta \le 1. \tag{2.10}$$

Remark 2.1.1 *If $\theta = 1$ (or 0) in (2.9) (or (2.10)), the **left fractional rectangular formula** (2.6) (or **right fractional rectangular formula** (2.8)) will be recovered. If $\alpha = 1$ (or 1/2) in (2.9) (or (2.10)), then the formula (2.9) (or (2.10)) is reduced to the **composite trapezoidal formula** (or **midpoint formula**) for the classical integral [127].*

- **Fractional Trapezoidal Formula**

On each subinterval $[t_k, t_{k+1}]$, $f(t)$ is approximated by the following piecewise polynomial with degree of order one

$$f(t)|_{[t_k,t_{k+1}]} \approx \tilde{f}(t)|_{[t_k,t_{k+1}]} = \frac{t_{k+1}-t}{t_{k+1}-t_k} f(t_k) + \frac{t-t_k}{t_{k+1}-t_k} f(t_{k+1}), \tag{2.11}$$

one obtains the ***fractional trapezoidal formula*** as follows

$$\begin{aligned}\left[\mathrm{D}_{0,t}^{-\alpha} f(t)\right]_{t=t_n} &\approx \left[\mathrm{D}_{0,t}^{-\alpha} \tilde{f}(t)\right]_{t=t_n} \\ &= \frac{1}{\Gamma(\alpha)} \sum_{k=0}^{n-1} \int_{t_k}^{t_{k+1}} (t_n - t)^{\alpha-1} \left(\frac{t_{k+1}-t}{t_{k+1}-t_k} f(t_k) + \frac{t-t_k}{t_{k+1}-t_k} f(t_{k+1}) \right) \mathrm{d}t \\ &= \sum_{k=0}^{n} a_{k,n} f(t_k),\end{aligned} \tag{2.12}$$

where

$$a_{k,n} = \frac{1}{\Gamma(\alpha)} \begin{cases} \int_0^{t_1} (t_n - t)^{\alpha-1} \frac{t_1-t}{t_1-t_0} \,\mathrm{d}t, & k = 0, \\ \int_{t_k}^{t_{k+1}} (t_n - t)^{\alpha-1} \frac{t_{k+1}-t}{t_{k+1}-t_k} \,\mathrm{d}t + \int_{t_{k-1}}^{t_k} (t_n - t)^{\alpha-1} \frac{t-t_{k-1}}{t_k-t_{k-1}} \,\mathrm{d}t, & 1 \le k \le n-1, \\ \int_{t_{n-1}}^{t_n} (t_n - t)^{\alpha-1} \frac{t-t_{k-1}}{t_n-t_{n-1}} \,\mathrm{d}t, & k = n. \end{cases} \tag{2.13}$$

By simple calculation, one has

$$a_{k,n} = \frac{\Delta t^\alpha}{\Gamma(\alpha+2)} \begin{cases} (n-1)^{\alpha+1} - (n-1-\alpha)n^\alpha, & k = 0, \\ (n-k+1)^{\alpha+1} + (n-1-k)^{\alpha+1} - 2(n-k)^{\alpha+1}, & 1 \le k \le n-1, \\ 1, & k = n. \end{cases} \tag{2.14}$$

- **Fractional Simpson's Formula**

On each subinterval $[t_k, t_{k+1}]$, denote by $t_{k+\frac{1}{2}} = \frac{t_k+t_{k+1}}{2}$. Interpolating $f(t)$ at the grid points $\{t_k, t_{k+1/2}, t_{k+1}\}$ on the subinterval $[t_k, t_{k+1}]$, i.e., $f(t)$ is approximated by a piecewise quadratic polynomial on the whole interval $[0, t_n]$, which is given by

$$f(t)|_{[t_k,t_{k+1}]} \approx \tilde{f}(t)|_{[t_k,t_{k+1}]} = \sum_{i \in S} f(t_i) l_{k,i}(t), \tag{2.15}$$

where $\{l_{k,i}(t)\}$ are Lagrangian base functions defined on the grid points $\{t_j, j\in S\}$, $S = \{0, \frac{1}{2}, 1\}$, which are given by

$$l_{k,i}(t) = \prod_{j\in S, j\neq i} \frac{t-t_j}{t_i-t_j}, \quad i\in S.$$

Replacing $f(t)$ in (2.2) by $\tilde{f}(t)$ defined by (2.15), one obtains the ***fractional Simpson's formula*** as follows

$$\left[\mathrm{D}_{0,t}^{-\alpha} f(t)\right]_{t=t_n} \approx \left[\mathrm{D}_{0,t}^{-\alpha} \tilde{f}(t)\right]_{t=t_n} = \sum_{k=0}^{n} c_{k,n} f(t_k) + \sum_{k=0}^{n-1} \hat{c}_{k,n} f(t_{k+\frac{1}{2}}), \tag{2.16}$$

where

$$c_{k,n} = \frac{\Delta t^\alpha}{\Gamma(\alpha+3)} \begin{cases} 4\left[(n+1)^{2+\alpha} - n^{2+\alpha}\right] - (\alpha+2)\left[3(n+1)^{1+\alpha} + n^{1+\alpha}\right] \\ \quad + (\alpha+2)(\alpha+1)(n+1)^\alpha, & k=0, \\ -(\alpha+2)\left[(n+1-k)^{1+\alpha} + 6(n-k)^{1+\alpha} + (n-k-1)^{1+\alpha}\right] \\ \quad + 4\left[(n+1-k)^{2+\alpha} - (n-1-k)^{2+\alpha}\right], & 1\le k\le n-1, \\ 2-\alpha, & k=n, \end{cases} \tag{2.17}$$

and

$$\begin{aligned} \hat{c}_{k,n} = & \frac{4\Delta t^\alpha}{\Gamma(\alpha+3)} \Big\{ (\alpha+2)\left[(n+1-k)^{1+\alpha} + (n-k)^{1+\alpha}\right] \\ & -2\left[(n+1-k)^{2+\alpha} - (n-k)^{2+\alpha}\right] \Big\}, \qquad 0\le k\le n-1. \end{aligned} \tag{2.18}$$

- **Fractional Newton–Cotes Formula**

Theoretically speaking, higher order methods such as fractional Newton–Cotes formulas can be derived if $f(t)$ is approximated by polynomials with higher degrees. On each subinterval $[t_k, t_{k+1}]$, $f(t)$ can be interpolated by a polynomial $p_{k,r}(t)$ of degree r on the grid points $\{t_k = t_0^{(k)}, t_1^{(k)}, \cdots, t_{r-1}^{(k)}, t_r^{(k)} = t_{k+1}\}$. Letting

$$l_{k,i}(t) = \prod_{j=0, j\neq i}^{r} \frac{t-t_j^{(k)}}{t_i^{(k)}-t_j^{(k)}},$$

we get

$$p_{k,r}(t) = \sum_{i=0}^{r} f(t_i^{(k)}) l_{k,i}(t).$$

Setting $\tilde{f}(t)|_{[x_k,x_{k+1}]} = p_{k,r}(t)$, we can calculate $\left[\mathrm{D}_{0,t}^{-\alpha} \tilde{f}(t)\right]_{t=t_n}$ analytically, which yields the ***fractional Newton–Cotes formula***

$$\left[\mathrm{D}_{0,t}^{-\alpha} f(t)\right]_{t=t_n} \approx \left[\mathrm{D}_{0,t}^{-\alpha} \tilde{f}(t)\right]_{t=t_n} = \sum_{k=0}^{n-1}\sum_{i=0}^{r} A_{i,n}^{(k)} f(t_i^{(k)}), \tag{2.19}$$

where

$$A_{i,n}^{(k)} = \frac{1}{\Gamma(\alpha)} \int_{t_k}^{t_{k+1}} (t_n - t)^{\alpha-1} l_{k,i}(t)\mathrm{d}t$$

is computable for big k. Here, we do not give the explicit expression of $A_{i,n}^{(k)}$.

If $f \in C^{r+1}([0,T])$, $r \in \mathbb{N}$, then the error estimate of $\tilde{f}(t)$ on each subinterval $[t_k, t_{k+1}]$ can be expressed by

$$\left[f(t) - \tilde{f}(t)\right]_{[t_k,t_{k+1}]} = \frac{f^{(r+1)}(\xi_k)}{(r+1)!} \prod_{j=0}^{r} \left(t - t_j^{(k)}\right), \quad \xi_k \in [t_k, t_{k+1}].$$

Therefore, we can get that the error estimate of (2.19) is $O(\Delta t^{r+1})$, which can be simply derived by the following calculation

$$\begin{aligned}
&\left|\left[\mathrm{D}_{0,t}^{-\alpha} f(t)\right]_{t=t_n} - \left[\mathrm{D}_{0,t}^{-\alpha} \tilde{f}(t)\right]_{t=t_n}\right| \\
&\le \frac{1}{\Gamma(\alpha)} \sum_{k=0}^{n-1} \int_{t_k}^{t_{k+1}} (t_n - t)^{\alpha-1} |f(t) - p_r^{(k)}(t)|\,\mathrm{d}t \\
&\le \max_{s\in[0,t_n]} \left\{|f^{(r+1)}(s)|\right\} \frac{1}{(r+1)!}\frac{1}{\Gamma(\alpha)} \sum_{k=0}^{n-1} \int_{t_k}^{t_{k+1}} (t_n - t)^{\alpha-1} \prod_{j=0}^{r} \left|t - t_j^{(k)}\right| \mathrm{d}t \qquad (2.20) \\
&\le \max_{s\in[0,t_n]} \left\{|f^{(r+1)}(s)|\right\} \frac{(t_{k+1} - t_k)^{r+1}}{(r+1)!\Gamma(\alpha)} \sum_{k=0}^{n-1} \int_{t_k}^{t_{k+1}} (t_n - t)^{\alpha-1}\,\mathrm{d}t \\
&= \max_{s\in[0,t_n]} \left\{|f^{(r+1)}(s)|\right\} \frac{t_n^{\alpha}}{(r+1)!\Gamma(\alpha+1)} \Delta t^{r+1}.
\end{aligned}$$

Remark 2.1.2 *It is known that the error estimate for the classical composite Newton–Cotes formula is of order $O(\Delta t^{r+2})$ for the odd number r [127]. This is not applicable for the fractional composite Newton–Cotes formula* (2.19), *which is due to the nonsymmetry of the weakly singular kernel $(t_n - t)^{\alpha-1}$ that leads to the nonsymmetry of the remainder term $(t_n - t)^{\alpha-1} \prod_{j=0}^{r} \left(t - t_j^{(k)}\right)$ in the integrand.*

Remark 2.1.3 *The **fractional rectangular formulae*** (2.6) *and* (2.7), *the **fractional trapezoidal formula*** (2.12) *and the **fractional Simpson's formula*** (2.16) *are special cases of the **fractional Newton–Cotes formula*** (2.19). *Hence, the convergence orders of these methods are $O(\Delta t)$,$O(\Delta t^2)$ and $O(\Delta t^3)$ for the noninteger number $\alpha > 0$. If $\alpha = 1$, these formulae reduce to the corresponding classical formulas of the classical integral except that the formulas* (2.9) *and* (2.10) *with $\theta = 1/2$ are reduced to the classical trapezoidal formula and the midpoint formula, respectively.*

- **Cubic Spline Interpolation**

On each subinterval $[t_k, t_{k+1}]$, $k = 0, 1, \cdots, n-1$, $\tilde{f}(t)$ is the *cubic spline interpolation* given by the following expression [73]

$$
\begin{aligned}
\tilde{f}(t)|_{[t_k,t_{k+1}]} =& \left(1 - 2\frac{t-t_k}{t_k - t_{k+1}}\right)\left(\frac{t-t_{k+1}}{t_k - t_{k+1}}\right)^2 f(t_k) + \left(1 - 2\frac{t-t_{k+1}}{t_{k+1} - t_k}\right)\left(\frac{t-t_k}{t_{k+1} - t_k}\right)^2 f(t_{k+1}) \\
&+ (t-t_k)\left(\frac{t-t_{k+1}}{t_k - t_{k+1}}\right)^2 f'(t_k) + (t-t_{k+1})\left(\frac{t-t_k}{t_{k+1} - t_k}\right)^2 f'(t_{k+1}).
\end{aligned} \tag{2.21}
$$

Then, $\left[\mathrm{D}_{0,t}^{-\alpha} f(t)\right]_{t=t_n}$ can be approximated by

$$
\left[\mathrm{D}_{0,t}^{-\alpha} f(t)\right]_{t=t_n} \approx \left[\mathrm{D}_{0,t}^{-\alpha} \tilde{f}(t)\right]_{t=t_n} = \frac{\Delta t^{\alpha}}{\Gamma(\alpha+4)} \left\{ \sum_{j=0}^{n} e_{j,n} f(t_j) + \Delta t \sum_{j=0}^{n} \hat{e}_{j,n} f'(t_j) \right\}, \tag{2.22}
$$

where

$$
\begin{aligned}
e_{0,n} =& -6(n-1)^{2+\alpha}(1+2n+\alpha) + n^{\alpha}\left(12n^3 - 6(3+\alpha)n^2 + (1+\alpha)(2+\alpha)(3+\alpha)\right), \\
e_{j,n} =& 6\left(4(n-j)^{3+\alpha} + (n-j-1)^{2+\alpha}(2j-2n-1-\alpha)\right) \\
&+ (1+n-j)^{2+\alpha}(2j-2n+1+\alpha), \quad j = 1, 2, \cdots, n-1, \\
e_{n,n} =& 6(1+\alpha), \\
\hat{e}_{0,n} =& -2(n-1)^{2+\alpha}(3n+\alpha) + n^{1+\alpha}\left(6n^2 - 4(3+\alpha)n + (2+\alpha)(3+\alpha)\right), \\
\hat{e}_{j,n} =& 2(n-j-1)^{\alpha+2}(3j-3n-\alpha) - 2(n-j+1)^{\alpha+2}(3j-3n+\alpha), \\
\hat{e}_{n,n} =& -2\alpha.
\end{aligned} \tag{2.23}
$$

The error estimate of the formula (2.22) is of order $O(\Delta t^4)$, which is determined by the error of the cubic spine interpolation, see [73].

2.1.2 High-Order Methods Based on Gauss Interpolation

The procedure that leads to the fractional Newton–Cotes formula can be used to generate the following more generalized formula of the form

$$
\left[\mathrm{D}_{0,t}^{-\alpha} f(t)\right]_{t=t_n} \approx \left[\mathrm{D}_{0,t}^{-\alpha} p_N(t)\right]_{t=t_n} = \sum_{k=0}^{N} w_{j,k} f(t_k),
$$

where $p_N(t)$ is an approximate polynomial of $f(t)$ with degree of order N. For example, $p_N(t)$ is an interpolation of $f(t)$ on the collocation points $\{t_k\}_{k=0}^{N}, t_k \in [0, T]$ or an orthogonal projector. When N is big enough, $\{w_{j,k}\}$ are not easy to compute, though they can be calculated exactly.

Next, we introduce an algorithm to compute $\{w_{j,k}\}$ effectively. As is known, any polynomial $p_N(t)$, $t \in [0, T]$ can be written in the following form

$$
p_N(t) = \sum_{j=0}^{N} c_j P_j^{a,b}(2t/L - 1) = \sum_{j=0}^{N} c_j P_j^{a,b}(x) = \hat{p}_N(x), \tag{2.24}
$$

where $t = \frac{L(x+1)}{2} \in [0,L], x \in [-1,1]$, $\{P_j^{a,b}(x)\}$ are *Jacobi orthogonal polynomials* defined on $[-1,1]$ with respect to the weight functions $\omega^{(a,b)}(x) = (1-x)^a(1+x)^b\,(a,b > 0)$. The coefficients $\{c_j\}$ can be easily calculated due to the orthogonal property of Jacobi polynomials (see Theorem 3.9 in [135]). Here we do not present the explicit expression of c_j, which will be illustrated later on when $p_N(t)$ is an interpolation polynomial of $f(t)$.

Next, we consider the αth-order fractional integral of $\hat{p}_N(x)$, i.e.,

$$\mathrm{D}_{-1,x}^{-\alpha}\hat{p}_N(x), \quad x \in [-1,1].$$

In order to derive the fractional integral of $\hat{p}_N(x)$, we need to introduce the Jacobi polynomials. The three-term recurrence relation of *Jacobi polynomials* $\{P_j^{a,b}(x)\}$ is given by [135]

$$\begin{aligned} &P_0^{a,b}(x) = 1, \qquad P_1^{a,b}(x) = \frac{1}{2}(a+b+2)x + \frac{1}{2}(a-b), \\ &P_{j+1}^{a,b}(x) = (A_j^{a,b}x - B_j^{a,b})P_j^{a,b}(x) - C_j^{a,b}P_{j-1}^{a,b}(x), \quad j \geq 1, \end{aligned} \tag{2.25}$$

where

$$\begin{aligned} A_j^{a,b} &= \frac{(2j+a+b+1)(2j+a+b+2)}{2(j+1)(j+a+b+1)}, \\ B_j^{a,b} &= \frac{(b^2-a^2)(2j+a+b+1)}{2(j+1)(j+a+b+1)(2j+a+b)}, \\ C_j^{a,b} &= \frac{(j+a)(j+b)(2j+a+b+2)}{(j+1)(j+a+b+1)(2j+a+b)}. \end{aligned} \tag{2.26}$$

The Jacobi polynomials are orthogonal with the weight function $\omega^{a,b}(x) = (1-x)^a(1+x)^b$, i.e.,

$$\int_{-1}^{1} P_m^{a,b}(x)P_n^{a,b}(x)\omega^{a,b}(x)\,\mathrm{d}x = \begin{cases} 0, & m \neq n, \\ \gamma_n^{a,b}, & m = n, \end{cases} \tag{2.27}$$

where

$$\gamma_n^{a,b} = \frac{2^{a+b+1}\Gamma(n+a+1)\Gamma(n+b+1)}{(2n+a+b+1)n!\Gamma(n+a+b+1)}. \tag{2.28}$$

Some other properties of the Jacobi polynomials are shown as follows

$$P_j^{a,b}(1) = \binom{j+a}{j} = \frac{\Gamma(j+a+1)}{j!\Gamma(a+1)}, \quad P_j^{a,b}(-1) = (-1)^j\frac{\Gamma(j+b+1)}{j!\Gamma(b+1)}. \tag{2.29}$$

$$\frac{\mathrm{d}^m}{\mathrm{d}x^m}P_j^{a,b}(x) = d_{j,m}^{a,b}P_{j-m}^{a+m,b+m}(x), \quad j \geq m, m \in \mathbb{N}, \tag{2.30}$$

where

$$d_{j,m}^{a,b} = \frac{\Gamma(j+m+a+b+1)}{2^m\Gamma(j+a+b+2)}. \tag{2.31}$$

$$P_j^{a,b}(x)=\widehat{A}_j^{a,b}\frac{\mathrm{d}}{\mathrm{d}x}P_{j-1}^{a,b}(x)+\widehat{B}_j^{a,b}\frac{\mathrm{d}}{\mathrm{d}x}P_j^{a,b}(x)+\widehat{C}_j^{a,b}\frac{\mathrm{d}}{\mathrm{d}x}P_{j+1}^{a,b}(x),\quad j\geq 1. \tag{2.32}$$

Here

$$\begin{aligned}\widehat{A}_j^{a,b}&=\frac{-2(j+a)(j+b)}{(j+a+b)(2j+a+b)(2j+a+b+1)},\\ \widehat{B}_j^{a,b}&=\frac{2(a-b)}{(2j+a+b)(2j+a+b+2)},\\ \widehat{C}_j^{a,b}&=\frac{2(j+a+b+1)}{(2j+a+b+1)(2j+a+b+2)}.\end{aligned} \tag{2.33}$$

If $j=1$, $\widehat{A}_1^{a,b}$ in (2.32) is set to be zero.

The key to calculating $\mathrm{D}_{-1,x}^{-\alpha}\hat{p}_N(x)$ rests on computing $\mathrm{D}_{-1,x}^{-\alpha}P_j^{a,b}(x)$ effectively when $\hat{p}_N(x)$ has the expression $\hat{p}_N(x)=\sum_{j=0}^{N}c_jP_j^{a,b}(x)$. Let

$$\widehat{P}_j^{a,b,\alpha}(x)=\mathrm{D}_{-1,x}^{-\alpha}P_j^{a,b}(x)=\frac{1}{\Gamma(\alpha)}\int_{-1}^{x}(x-s)^{\alpha-1}P_j^{a,b}(s)\mathrm{d}s,\quad x\in[-1,1].$$

For fixed x, $\widehat{P}_j^{a,b,\alpha}(x)$ $(j=0,1,\cdots,N)$ can be evaluated with $O(N)$ operations, we will give the detailed deduction below.

For simplicity, we denote

$$F_j^{a,b}(x)=P_j^{a,b}(x)=\widehat{A}_j^{a,b}P_{j-1}^{a,b}(x)+\widehat{B}_j^{a,b}P_j^{a,b}(x)+\widehat{C}_j^{a,b}P_{j+1}^{a,b}(x). \tag{2.34}$$

From (2.25), one has

$$\begin{aligned}\widehat{P}_{j+1}^{a,b,\alpha}(x)=&\frac{1}{\Gamma(\alpha)}\int_{-1}^{x}(x-s)^{\alpha-1}P_{j+1}^{a,b}(s)\mathrm{d}s\\ =&\frac{1}{\Gamma(\alpha)}\int_{-1}^{x}(x-s)^{\alpha-1}\left[(A_j^{a,b}s-B_j^{a,b})P_j^{a,b}(s)-C_j^{a,b}P_{j-1}^{a,b}(s)\right]\mathrm{d}s\\ =&(A_j^{a,b}x-B_j^{a,b})\widehat{P}_j^{a,b,\alpha}(x)-C_j^{a,b}\widehat{P}_{j-1}^{a,b,\alpha}(x)+\frac{A_j^{a,b}}{\Gamma(\alpha)}\int_{-1}^{x}(x-s)^{\alpha}P_j^{a,b}(s)\mathrm{d}s.\end{aligned} \tag{2.35}$$

From (2.32) and (2.34), one has

$$\begin{aligned}&\int_{-1}^{x}(x-s)^{\alpha}P_j^{a,b}(s)\mathrm{d}s\\ =&\int_{-1}^{x}(x-s)^{\alpha}\frac{\mathrm{d}}{\mathrm{d}s}P_j^{a,b}(s)\\ =&(x-s)^{\alpha}P_j^{a,b}(s)\Big|_{-1}^{x}+\frac{1}{\alpha}\int_{-1}^{x}(x-s)^{\alpha-1}F_j^{a,b}(s)\mathrm{d}s\\ =&(x+1)^{\alpha}P_j^{a,b}(-1)+\frac{1}{\alpha}\left[\widehat{A}_j^{a,b}\widehat{P}_{j-1}^{a,b}(x)+\widehat{B}_j^{a,b}\widehat{P}_j^{a,b}(x)+\widehat{C}_j^{a,b}\widehat{P}_{j+1}^{a,b}(x)\right].\end{aligned} \tag{2.36}$$

Note that $\widehat{P}_0^{a,b,\alpha}(x)$ and $\widehat{P}_1^{a,b,\alpha}(x)$ can be obtained very easily. Therefore, we can derive the recurrence formula to calculate $\widehat{P}_j^{a,b,\alpha}(x)$ from (2.35) and (2.36) as follows

$$\begin{cases} \widehat{P}_0^{a,b,\alpha}(x) = \dfrac{(x+1)^\alpha}{\Gamma(\alpha+1)}, \\ \widehat{P}_1^{a,b,\alpha}(x) = \dfrac{a+b+2}{2}\left(\dfrac{x(x+1)^\alpha}{\Gamma(\alpha+1)} - \dfrac{\alpha(x+1)^{\alpha+1}}{\Gamma(\alpha+2)}\right) + \dfrac{a-b}{2}\widehat{P}_0^{a,b,\alpha}(x), \\ \widehat{P}_{j+1}^{a,b,\alpha}(x) = \dfrac{A_j^{a,b}x - B_j^{a,b} - \alpha A_j^{a,b}\widehat{B}_j^{a,b}}{1+\alpha A_j^{a,b}\widehat{C}_j^{a,b}}\widehat{P}_j^{a,b,\alpha}(x) - \dfrac{C_j^{a,b} + \alpha A_j^{a,b}\widehat{A}_j^{a,b}}{1+\alpha A_j^{a,b}\widehat{C}_j^{a,b}}\widehat{P}_{j-1}^{a,b,\alpha}(x) \\ \qquad + \dfrac{\alpha A_j^{a,b}\left(\widehat{A}_j^{a,b}P_{j-1}^{a,b}(-1) + \widehat{B}_j^{a,b}P_j^{a,b}(-1) + \widehat{C}_j^{a,b}P_{j+1}^{a,b}(-1)\right)}{\Gamma(\alpha+1)\left(1+\alpha A_j^{a,b}\widehat{C}_j^{a,b}\right)}(x+1)^\alpha, j\geq 1. \end{cases} \tag{2.37}$$

Hence, $\mathrm{D}_{-1,x}^{-\alpha}\hat{p}_N(x) = \sum_{j=0}^N c_j\widehat{P}_j^{a,b,\alpha}(x)$ can be evaluated effectively with $O(N)$ operations for a fixed number x.

Denoting by $t = \frac{L(x+1)}{2} \in [0,L], x \in [-1,1]$, we can easily derive

$$\begin{aligned} \mathrm{D}_{0,t}^{-\alpha}p_N(t) =& \frac{1}{\Gamma(\alpha)}\int_0^t (t-s)^{\alpha-1}p_N(s)\,\mathrm{d}s \\ =& \left(\frac{L}{2}\right)^\alpha \frac{1}{\Gamma(\alpha)}\int_{-1}^x (x-s)^{\alpha-1}\,\hat{p}_N(s)\,\mathrm{d}s \\ =& \left(\frac{L}{2}\right)^\alpha D_{-1,x}^{\alpha}\hat{p}_N(x), \quad x = 2t/L-1. \end{aligned} \tag{2.38}$$

Hence, we have the desired formula to calculate $D_{0,t}^{\alpha}p_N(t)$ as follows.

$$\mathrm{D}_{0,t}^{-\alpha}p_N(t) = \left(\frac{L}{2}\right)^\alpha \sum_{j=0}^N c_j\widehat{P}_j^{a,b,\alpha}(2t/L-1). \tag{2.39}$$

If $p_N(t)$ is the *Jacobi–Gauss–Lobatto interpolation* of $f(t)$, then we have the explicit expression of c_j defined in (2.24) (see also (2.39)) as follows

$$c_j = \frac{1}{\delta_j^{a,b}}\sum_{k=0}^N f(t_k)P_j^{a,b}(x_k)\omega_k = \frac{1}{\delta_j^{a,b}}\sum_{k=0}^N f(L(x_k+1)/2)P_j^{a,b}(x_k)\omega_k, \tag{2.40}$$

where ω_k is the weight with respect to the Jacobi–Gauss–Lobatto point x_k, and

$$\delta_j^{a,b} = \begin{cases} \gamma_j^{a,b}, & j = 0,1,\cdots,N-1, \\ \left(2+\dfrac{a+b+1}{N}\right)\gamma_N^{a,b}. \end{cases}$$

$\gamma_j^{a,b}$ in the above equation is defined by (2.28). The *Jacobi–Gauss–Lobatto points* $\{x_k\}$ are defined as the roots of the following polynomial

$$(1-x^2)\frac{\mathrm{d}}{\mathrm{d}x}P_N^{a,b}(x).$$

Readers can refer to [135] for more properties of the Jacobi polynomials.

We briefly denote the matrix $\widehat{D}_L^{(-\alpha,a,b)} \in \mathbb{R}^{(N+1)\times(N+1)}$ with

$$\left(\widehat{D}_L^{(-\alpha,a,b)}\right)_{i,j} = \widehat{P}_j^{a,b,\alpha}(x_i), \quad i,j = 0,1,\cdots,N. \tag{2.41}$$

Then from $\mathrm{D}^{\alpha}_{-1,x} p_N(x) = (\widehat{P}_0^{a,b,\alpha}(x), \widehat{P}_1^{a,b,\alpha}(x), .., \widehat{P}_N^{a,b,\alpha}(x))(c_0, c_1, \cdots, c_N)^T$, one can obtain

$$\begin{pmatrix} \left[\mathrm{D}_{0,t}^{-\alpha} p_N(t)\right]_{t=t_0} \\ \left[\mathrm{D}_{0,t}^{-\alpha} p_N(t)\right]_{t=t_1} \\ \vdots \\ \left[\mathrm{D}_{0,t}^{-\alpha} p_N(t)\right]_{t=t_N} \end{pmatrix} = \left(\frac{L}{2}\right)^{\alpha} \begin{pmatrix} \left[\mathrm{D}_{-1,x}^{-\alpha} \hat{p}_N(x)\right]_{x=x_0} \\ \left[\mathrm{D}_{-1,x}^{-\alpha} \hat{p}_N(x)\right]_{x=x_1} \\ \vdots \\ \left[\mathrm{D}_{-1,x}^{-\alpha} \hat{p}_N(x)\right]_{x=x_N} \end{pmatrix} = \left(\frac{L}{2}\right)^{\alpha} \left(\widehat{D}_L^{(-\alpha,a,b)}\right) \begin{pmatrix} c_0 \\ c_1 \\ \vdots \\ c_N \end{pmatrix}.$$

If the Jacobi–Gauss–Lobatto point x_i in (2.41) is replaced by other collocation points, i.e., *Jacobi–Gauss point* or *Jacobi–Gauss–Radau point*, one can also obtain the corresponding matrices as $\widehat{D}_L^{(-\alpha,a,b)}$.

The above formula (2.39) has the rapid convergence rate if $f(t) \in C^r([0,T])$. The error bound for (2.39) is given by

$$\left|\mathrm{D}_{0,t}^{-\alpha} f(t) - \mathrm{D}_{0,t}^{-\alpha} p_N(t)\right| \leq C N^{3/4-r},$$

when $a = b = 0$ and $p_N(t)$ is the interpolation of $f(t)$ on the Legendre–Gauss–Lobatto points $\{t_k\}, k = 0,1,\cdots,N$. Of course, if $p_N(t)$ is the interpolation of $f(t)$ on any other Gauss points, then the spectral accuracy can be still achieved under the condition that $f(t)$ is suitably smooth. See [9, 135] for more error estimates of the Jacobi–Gauss interpolations and orthogonal projections.

2.1.3 Fractional Linear Multistep Methods

The *fractional linear multistep methods* (FLMMs) based on the convolution quadrature were studied by Lubich [104], who got pth order ($p = 1,2,\cdots,6$) approximation of $\mathrm{D}_{0,t}^{-\alpha} f(t)$. The FLMMs have a very close relationship to the classical linear multistep methods (LMMs). If $f(t)$ is suitably smooth, then the pth-order ($p = 1,2,\cdots,6$) FLMMs for $\left[\mathrm{D}_{0,t}^{-\alpha} f(t)\right]_{t=t_n}$ are given by

$$\left[\mathrm{D}_{0,t}^{-\alpha} f(t)\right]_{t=t_n} = \Delta t^{\alpha} \sum_{j=0}^{n} \omega_{n-j}^{(\alpha)} f(t_j) + \Delta t^{\alpha} \sum_{j=0}^{s} \omega_{n,j}^{(\alpha)} f(t_j) + O(\Delta t^p), \tag{2.42}$$

where $\{\omega_j^{(\alpha)}\}$ are called *convolution weights* defined by the coefficients of Taylor expansions of the following *generating functions*

$$w^{(\alpha)}(z) = \left[\sum_{j=1}^{p} \frac{1}{j}(1-z)^j\right]^{-\alpha}, \quad p = 1, 2, \cdots, 6, \tag{2.43}$$

$$w^{(\alpha)}(z) = (1-z)^{-\alpha}\left[\gamma_0 + \gamma_1(1-z) + \gamma_2(1-z)^2 + \cdots + \gamma_{p-1}(1-z)^{p-1}\right], \tag{2.44}$$

$$w^{(\alpha)}(z) = \left(\frac{1}{2}\frac{1+z}{1-z}\right)^{\alpha}, \quad p = 2, \tag{2.45}$$

in which $\{\gamma_k\}$ in (2.44) satisfy the following relation

$$\left(\frac{\ln z}{z-1}\right)^{-\alpha} = \sum_{k=0}^{\infty} \gamma_k (1-z)^k, \quad \gamma_0 = 1, \gamma_1 = -\frac{\alpha}{2}.$$

The starting weights $\{w_{n,k}^{(\beta)}\}$ are chosen such that the asymptotic behavior of the function $f(t)$ near the origin ($t = 0$) is taken into account [28]. One way to determine $\{w_{n,k}^{(\beta)}\}$ for the suitably smooth function $u(t)$ is given [28, 104] by

$$\sum_{k=1}^{p-1} \omega_{n,k}^{(\alpha)} k^q = \frac{\Gamma(q+1)}{\Gamma(q+\alpha+1)} n^{q+\alpha} - \sum_{k=1}^{n} \omega_{n-k}^{(\alpha)} k^q, \quad q = 0, 1, \cdots, p-1. \tag{2.46}$$

The above choices of $\{w_{n,k}^{(\alpha)}\}$ imply that (2.42) is exact for $f(t) = t^{\mu}, \mu = 0, 1, \cdots, p-1$.

If $f(t)$ is not suitably smooth with expression $f(t) = \sum_{k=0}^{s} f_k t^{\sigma(k)} + t^{\mu}\phi(t)$, where $\phi(t)$ is smooth and $\mu \geq p-1 \geq \sigma(k)$, then one can still construct the *pth-order FLMMs* as in (2.42). In such a case, we can obtain the *starting weights* $\{w_{n,k}^{(\beta)}\}$ in the following way

$$\left[\mathrm{D}_{0,t}^{-\alpha} t^{\sigma(j)}\right]_{t=t_n} = \Delta t^{\alpha} \sum_{k=0}^{s} \omega_{n,k}^{(\alpha)} (k\Delta t)^{\sigma(j)} + \Delta t^{\alpha} \sum_{k=0}^{n} \omega_{n-j}^{(\alpha)} (k\Delta t)^{\sigma(j)} \tag{2.47}$$

through inserting $f(t) = t^{\sigma(j)}$ into (2.42) and letting (2.42) be exact. We rewrite (2.47) as the following equivalent form

$$\sum_{k=0}^{s} \omega_{n,k}^{(\alpha)} k^{\sigma(j)} = \frac{\Gamma(\sigma(j)+1)}{\Gamma(\sigma(j)+1+\alpha)} n^{\sigma(j)+\alpha} - \sum_{k=0}^{n} \omega_{n-k}^{(\alpha)} k^{\sigma(j)}, \quad j = 0, 1, \cdots, s. \tag{2.48}$$

The derivation of the FLMMs for the fractional integral is more complicated than that of the classical LMMs; readers can refer to [104] for detailed information.

Remark 2.1.4 *Let $w_p(z) = \sum_{j=1}^{p} \frac{1}{j}(1-z)^j$. Then $w_p(z)$ is just the generating function of the $(p+1)$-point backward difference method. For any $\alpha \in \mathbb{R}$, the coefficients $\omega_j^{(\alpha)}$ of the Taylor expansions of the generating function* (2.43) *can be easily and effectively calculated by the fast Fourier transform method. Of course, there exists a recurrence formula to calculate $\omega_j^{(\alpha)}$, which is given below [28]*

$$\omega_j^{(\alpha)} = \frac{1}{j u_0} \sum_{i=0}^{j-1} (\alpha(j-i) - i)\omega_i^{(\alpha)} u_{j-i},$$

where u_j satisfies $w_p(z) = \sum_{j=0}^{p} u_j z^j$.

Remark 2.1.5 *If $\alpha < 0$, α is not an integer, $D_{0,t}^{-\alpha} f(t)$ is just the finite-part integral [124], which is equivalent to the $(-\alpha)$th-order Riemann–Liouville derivative of $f(t)$. In this case, the formula (2.42) is the corresponding pth-order approximation of the $(-\alpha)$th Riemann–Liouville derivative of the given function $f(t)$.*

The coefficients γ_n in (2.44) can be calculated by the following formula (see Theorem 2 in [54])

$$\gamma_0 = 1, \quad \gamma_n = \sum_{j=1}^{n} \left(\frac{(1-\alpha)j - n}{n(j+1)} \right) \gamma_{n-j}, \quad n = 1, 2, \cdots. \tag{2.49}$$

The first six coefficients are given by

$$\begin{aligned}
\gamma_0 &= 1, \\
\gamma_1 &= -\frac{\alpha}{2}, \\
\gamma_2 &= \frac{1}{8}\alpha^2 - \frac{5}{24}\alpha, \\
\gamma_3 &= -\frac{1}{48}\alpha^3 + \frac{5}{48}\alpha^2 - \frac{1}{8}\alpha, \\
\gamma_4 &= \frac{1}{384}\alpha^4 - \frac{5}{192}\alpha^3 + \frac{97}{1152}\alpha^2 - \frac{251}{2880}\alpha, \\
\gamma_5 &= -\frac{1}{3840}\alpha^5 + \frac{5}{1152}\alpha^4 - \frac{61}{2304}\alpha^3 + \frac{401}{5760}\alpha^2 - \frac{19}{288}\alpha.
\end{aligned} \tag{2.50}$$

2.2 Approximations to Riemann–Liouville Derivatives

For a class of functions, both the Grünwald–Letnikov derivative and the Riemann–Liouville derivative are equivalent, especially for applications. Therefore, the Riemann–Liouville definition is suitable for the problem formulation, where the Grünwald–Letnikov definition is utilized to obtain the numerical solution [106, 124]. This section mainly focuses on the approximation of the Riemann–Liouville derivative.

We mainly consider the numerical methods for the Riemann–Liouville derivative with fractional order $0 < \alpha < 1$ and $1 < \alpha < 2$, which has special importance in real applications, such as the modeling of the anomalous diffusion [92, 113, 114]. In this section, we investigate the numerical discretization of the Riemann–Liouville operator. Here, we only introduce the numerical methods for the left Riemann–Liouville derivatives; the methods for the right Riemann–Liouville derivatives can be similarly obtained.

2.2.1 Grünwald–Letnikov Type Approximation

- **Grünwald–Letnikov Approximation**

If $f(t)$ is suitably smooth, the Grünwald–Letnikov derivative is equivalent to the Riemann–Liouville derivative. Therefore, using Eq. (1.3) to approximate the Riemann–Liouville derivative is natural. Denoting by $\omega_j^{(\alpha)} = (-1)^j\binom{\alpha}{j}$, one gets

$$\left[{}_{RL}\mathrm{D}_{0,t}^{\alpha}f(t)\right]_{t=t_n} \approx \frac{1}{\Delta t^{\alpha}}\sum_{j=0}^{n}\omega_j^{(\alpha)}f(t_{n-j}). \tag{2.51}$$

The above formula (2.51) is convergent of order 1 for any $\alpha > 0$ [124]. We call (2.51) the *standard Grünwald–Letnikov formula*, which may contribute to unstable numerical schemes in solving FDEs [111] for $1 < \alpha < 2$. The shifted Grünwald–Letnikov formula is useful for constructing the stable numerical schemes. The *right shifted Grünwald–Letnikov formula* (p shifts, $p \in \mathbb{N}$) to approximate the left Riemann–Liouville derivative is defined by

$$\left[{}_{RL}\mathrm{D}_{0,t}^{\alpha}f(t)\right]_{t=t_n} \approx \frac{1}{\Delta t^{\alpha}}\sum_{j=0}^{n+p}\omega_j^{(\alpha)}f(t_{n-j+p}). \tag{2.52}$$

The above shifted Grünwald–Letnikov formula gives the first-order accuracy; the best performance comes from minimizing $|p-\alpha/2|$ [111, 118]. If $1 < \alpha \leq 2$, the optimal choice is $p = 1$. The case of $\alpha = 2$ reduces to the second order central difference method for the second order classical derivative.

Theorem 2 *If $1 < \alpha < 2$, $f(t) = t^{\mu}$, μ is a nonnegative integer, $t = t_n = n\Delta t$, then the following relations hold [140]*

$$\begin{aligned}
\frac{1}{\Delta t^{\alpha}}\sum_{j=0}^{n}\omega_j^{(\alpha)}f(t_{n-j}) &= \left[{}_{RL}\mathrm{D}_{0,t}^{\alpha}f(t)\right]_{t=t_n} + (1-\alpha)\frac{(-\alpha)t^{-1-\alpha}}{2\Gamma(1-\alpha)}\Delta t + O(\Delta t^2), \quad \mu = 0,\\
\frac{1}{\Delta t^{\alpha}}\sum_{j=0}^{n}\omega_j^{(\alpha)}f(t_{n-j}) &= \left[{}_{RL}\mathrm{D}_{0,t}^{\alpha}f(t)\right]_{t=t_n} + (-\alpha)\frac{\Gamma(\mu+1)t^{\mu-1-\alpha}}{2\Gamma(\mu-\alpha)}\Delta t + O(\Delta t^2), \quad \mu > 0,
\end{aligned} \tag{2.53}$$

and

$$\begin{aligned}
\frac{1}{\Delta t^{\alpha}}\sum_{j=0}^{n+1}\omega_j^{(\alpha)}f(t_{n-j+1}) &= \left[{}_{RL}\mathrm{D}_{0,t}^{\alpha}f(t)\right]_{t=t_n} + (3-\alpha)\frac{(-\alpha)t^{-1-\alpha}}{2\Gamma(1-\alpha)}\Delta t + O(\Delta t^2), \quad \mu = 0,\\
\frac{1}{\Delta t^{\alpha}}\sum_{j=0}^{n+1}\omega_j^{(\alpha)}f(t_{n-j+1}) &= \left[{}_{RL}\mathrm{D}_{0,t}^{\alpha}f(t)\right]_{t=t_n} + (2-\alpha)\frac{\Gamma(\mu+1)t^{\mu-1-\alpha}}{2\Gamma(\mu-\alpha)}\Delta t + O(\Delta t^2), \quad \mu > 0.
\end{aligned} \tag{2.54}$$

The leading terms of the truncation errors for the standard Grünwald–Letnikov formula (2.51) and the shifted (one shift) Grünwald–Letnikov formula (2.52) are slightly different.

From Theorem 2, we find that the Grünwald–Letnikov formula (2.52) does not have first-order accuracy for the smooth function $f(t)$ if $f(0) \neq 0$. The remedy is to use the following technique

$$\begin{aligned}\left[{}_{RL}\mathrm{D}_{0,t}^{\alpha} f(t)\right]_{t=t_n} &= \left[{}_{RL}\mathrm{D}_{0,t}^{\alpha} (f(t)-f(0))\right]_{t=t_n} + \frac{f(0)t_n^{-\alpha}}{\Gamma(1-\alpha)} \\ &\approx \frac{1}{\Delta t^{\alpha}} \sum_{j=0}^{n+p} \omega_j^{(\alpha)} \left[f(t_{n-j+p}) - f(0)\right] + \frac{f(0)t_n^{-\alpha}}{\Gamma(1-\alpha)}.\end{aligned} \tag{2.55}$$

The advantage of such a remedy is that the above formula (2.55) is exact when $f(t)$ is a constant.

For $f(t) = t^{\mu}, \mu > 0$, we can obtain from (2.53) and (2.54) that

$$\begin{aligned}(2-\alpha)\frac{1}{\Delta t^{\alpha}} \sum_{j=0}^{n} \omega_j^{(\alpha)} f(t_{n-j}) =& (2-\alpha)\left[{}_{RL}\mathrm{D}_{0,t}^{\alpha} f(t)\right]_{t=t_n} \\ &+ (2-\alpha)(-\alpha)\frac{\Gamma(\mu+1)t^{\mu-1-\alpha}}{2\Gamma(\mu-\alpha)}\Delta t + O(\Delta t^2)\end{aligned} \tag{2.56}$$

and

$$\begin{aligned}\alpha\frac{1}{\Delta t^{\alpha}} \sum_{j=0}^{n+1} \omega_j^{(\alpha)} f(t_{n-j+1}) =& \alpha\left[{}_{RL}\mathrm{D}_{0,t}^{\alpha} f(t)\right]_{t=t_n} \\ &+ \alpha(2-\alpha)\frac{\Gamma(\mu+1)t^{\mu-1-\alpha}}{2\Gamma(\mu-\alpha)}\Delta t + O(\Delta t^2).\end{aligned} \tag{2.57}$$

Eliminating the term $\alpha(2-\alpha)\frac{\Gamma(\mu+1)t^{\mu-1-\alpha}}{2\Gamma(\mu-\alpha)}\Delta t$ from (2.56) and (2.57) yields

$$\left[{}_{RL}\mathrm{D}_{0,t}^{\alpha} f(t)\right]_{t=t_n} = \frac{1}{\Delta t^{\alpha}}\left[\frac{(2-\alpha)}{2}\sum_{j=0}^{n} \omega_j^{(\alpha)} f(t_{n-j}) + \frac{\alpha}{2}\sum_{j=0}^{n+1} \omega_j^{(\alpha)} f(t_{n-j+1})\right] + O(\Delta t^2). \tag{2.58}$$

Hence, for a suitably smooth function $f(t)$ with $f(0) = 0$, a second-order method (2.58) is obtained to approximate ${}_{RL}\mathrm{D}_{0,t}^{\alpha} f(t)$.

Similar to (2.55), we can obtain the following second-order method

$$\begin{aligned}\left[{}_{RL}\mathrm{D}_{0,t}^{\alpha} f(t)\right]_{t=t_n} =& \frac{1}{\Delta t^{\alpha}}\left[\frac{(2-\alpha)}{2}\sum_{j=0}^{n} \omega_j^{(\alpha)} (f(t_{n-j}) - f(0)) + \frac{\alpha}{2}\sum_{j=0}^{n+1} \omega_j^{(\alpha)} (f(t_{n-j+1}) - f(0))\right] \\ &+ \frac{f(t_0)t_n^{-\alpha}}{\Gamma(1-\alpha)} + O(\Delta t^2)\end{aligned} \tag{2.59}$$

for any suitably smooth function $f(t)$.

Tadjeran et al. [145] proved that the shifted Grünwald–Letnikov formula (2.52) has *first-order accuracy* for suitably smooth function $f(t)$ satisfying $f(0) = 0$ by the

Fourier transform method. Tian et al. [146] proposed a class of *second-order methods* as in (2.58) to discretize the Riemann–Liouville derivative of $f(t)$ satisfying $f(0)=0$, which are given as follows

$$\left[{}_{RL}\mathrm{D}_{0,t}^{\alpha}f(t)\right]_{t=t_n} = \frac{1}{\Delta t^{\alpha}}\left[\frac{(\alpha-2q)}{2(p-q)}\sum_{j=0}^{n}\omega_j^{(\alpha)}f(t_{n-j+p}) + \frac{2p-\alpha}{2(p-q)}\sum_{j=0}^{n+1}\omega_j^{(\alpha)}f(t_{n-j+q})\right] + O(\Delta t^2), \tag{2.60}$$

where p, q are integers.

- **Fractional Linear Multistep Methods**

In Subsection 2.1.3, we know that the FLMMs for the fractional integral are introduced. The formula (2.42) is also suitable for the discretization of the Riemann–Liouville derivative. The FLMMs for the αth-order Riemann–Liouville derivative are presented below

$$\left[{}_{RL}\mathrm{D}_{0,t}^{\alpha}f(t)\right]_{t=t_n} = \Delta t^{-\alpha}\sum_{j=0}^{n}\omega_{n-j}^{(-\alpha)}f(t_j) + \Delta t^{-\alpha}\sum_{j=0}^{s}\omega_{n,j}^{(-\alpha)}f(t_j) + O(\Delta t^p) \tag{2.61}$$

where $\omega_j^{(-\alpha)}$ are called convolution weights defined by the coefficients of Taylor expansions of the generating functions $w^{(-\alpha)}(z)$, which can be derived from (2.43)–(2.45) with α being replaced by $(-\alpha)$. The starting weights $\omega_{n,j}^{(-\alpha)}$ are chosen such that (2.61) is exact for some $f(t)=t^{\mu}$, which is determined the same way as $\omega_{n,j}^{(\alpha)}$, defined in (2.42).

If $f(t)$ is suitably smooth and $f^{(k)}(0)=0, k=0,1,\cdots,p-1$, then one can remove $\Delta t^{-\alpha}\sum_{j=0}^{s}\omega_{n,j}^{(-\alpha)}f(t_j)$ in (2.61) to obtain the corresponding discretization with the same pth-order accuracy.

2.2.2 L1, L2 and L2C Methods

This subsection concerns the numerical methods for the Riemann–Liouville derivative with fractional order $0<\alpha<1$ and $1<\alpha<2$. The classical L1 method [71, 96, 118, 144] is suitable for the case of $0<\alpha<1$. The L2 and L2C methods are suitable for the case of $1<\alpha<2$. Next, we simply introduce the construction of these methods.

- **L1 method**

Here we introduce the detailed derivation and theoretical analysis of the *L1 method*, since the L1 method is often used by some researchers for the discretization of the time fractional differential equations, which can lead to *unconditionally stable algorithms* [56, 65, 66, 71, 96, 130, 144, 166, 176].

By (1.12), we get

$$ {}_{RL}\mathrm{D}_{0,t}^{\alpha} f(t) = {}_{C}\mathrm{D}_{0,t}^{\alpha} f(t) + \frac{f(0)}{\Gamma(1-\alpha)} t^{-\alpha}. $$

Letting $t = t_n$ and $0 < \alpha < 1$, one gets

$$ \begin{aligned} \left[{}_{C}\mathrm{D}_{0,t}^{\alpha} f(t) \right]_{t=t_n} &= \frac{1}{\Gamma(1-\alpha)} \int_{t_0}^{t_n} (t_n - s)^{-\alpha} f'(s) \mathrm{d}s \\ &= \frac{1}{\Gamma(1-\alpha)} \sum_{k=0}^{n-1} \int_{t_k}^{t_{k+1}} (t_n - s)^{-\alpha} f'(s) \mathrm{d}s \\ &\approx \frac{1}{\Gamma(1-\alpha)} \sum_{k=0}^{n-1} \int_{t_k}^{t_{k+1}} (t_n - s)^{-\alpha} \frac{f(t_{k+1}) - f(t_k)}{\Delta t} \mathrm{d}s \\ &= \sum_{k=0}^{n-1} b_{n-k-1} \left(f(t_{k+1}) - f(t_k) \right), \end{aligned} \tag{2.62} $$

where

$$ t_0 = 0, \quad b_k = \frac{\Delta t^{-\alpha}}{\Gamma(2-\alpha)} \left[(k+1)^{1-\alpha} - k^{1-\alpha} \right]. $$

Therefore,

$$ \left[{}_{RL}\mathrm{D}_{0,t}^{\alpha} f(t) \right]_{t=t_n} \approx \frac{f(0) t_n^{-\alpha}}{\Gamma(1-\alpha)} + \sum_{k=0}^{n-1} b_{n-k-1} \left[f(t_{k+1}) - f(t_k) \right]. \tag{2.63} $$

The above L1 method (2.63) has the following error estimate [71, 96, 144]

$$ \left| \frac{f(0) t_n^{-\alpha}}{\Gamma(1-\alpha)} + \sum_{k=0}^{n-1} b_{n-k-1} \left[f(t_{k+1}) - f(t_k) \right] - \left[{}_{RL}\mathrm{D}_{0,t}^{\alpha} f(t) \right]_{t=t_n} \right| \leq C \Delta t^{2-\alpha}, $$

where C is a positive constant only dependent on α and f.

The derivative of the classical L1 method can be extended to the more general case on the *nonuniform grids* [172].

Let $\{s_j\}$ be the any division of $[0,T]$ with $0 = s_0 \leq s_1 \leq \cdots \leq s_{N-1} \leq s_N = T$ and $\tau_j = s_{j+1} - s_j$. Then one has

$$ \begin{aligned} \int_{s_0}^{s_n} (s_n - s)^{-\alpha} f'(s) \mathrm{d}s &= \sum_{k=0}^{n-1} \int_{s_k}^{s_{k+1}} (s_n - s)^{-\alpha} f'(s) \mathrm{d}s \\ &= \sum_{k=0}^{n-1} \int_{s_k}^{s_{k+1}} (s_n - s)^{-\alpha} \frac{f(s_{k+1}) - f(s_k)}{\tau_k} \mathrm{d}s + \hat{R}^n \\ &= \sum_{k=0}^{n-1} a_{k+1}^{n} \left(f(s_{k+1}) - f(s_k) \right) + \hat{R}^n, \end{aligned} \tag{2.64} $$

where

$$a_{k+1}^n = \frac{1}{\tau_k}\int_{s_k}^{s_{k+1}}(s_n - s)^{-\alpha}\,\mathrm{d}s = \frac{1}{(1-\alpha)\tau_k}\left[(s_n - s_k)^{1-\alpha} - (s_n - s_{k+1})^{1-\alpha}\right].$$

Hence, one derives

$$\begin{aligned}\left[{}_C\mathrm{D}_{0,t}^{\alpha}f(t)\right]_{t=s_n} &= \sum_{k=0}^{n-1} b_{k+1}^n\left(f(s_{k+1}) - f(s_k)\right) + R^n,\\ \left[{}_{RL}\mathrm{D}_{0,t}^{\alpha}f(t)\right]_{t=s_n} &= \sum_{k=0}^{n-1} b_{k+1}^n\left(f(s_{k+1}) - f(s_k)\right) + \frac{f(0)s_n^{-\alpha}}{\Gamma(1-\alpha)} + R^n,\end{aligned} \tag{2.65}$$

in which $R^n = \hat{R}^n/\Gamma(1-\alpha)$, and

$$b_{k+1}^n = \frac{a_{k+1}^n}{\Gamma(1-\alpha)} = \frac{1}{\Gamma(2-\alpha)\tau_k}\left[(s_n - s_k)^{1-\alpha} - (s_n - s_{k+1})^{1-\alpha}\right].$$

It can be proved that if $\tau_{\max}/\tau_{\min} \le C_0$, $\tau_{\max} = \max_{0\le j\le N-1}\{\tau_j\}$ and $\tau_{\min} = \min_{0\le j\le N-1}\{\tau_j\}$, then $|R^n| \le C(\tau_{\max})^{2-\alpha}$; see [172].

Theorem 3 *Let* $0 < \alpha < 1$ *and* $f(t) \in C^2[0,T]$. *Then it holds*

$$|\hat{R}^n| = \left|\int_{t_0}^{t_n}(s_n - s)^{-\alpha}f'(s)\,\mathrm{d}s - \sum_{k=0}^{n-1}a_{k+1}^n\left(f(s_{k+1}) - f(s_k)\right)\right| \le C(\tau_{\max})^{2-\alpha}\max_{0\le t\le T}|f''(t)|, \tag{2.66}$$

where C *is only dependent on* α *and* $\tau_{\max}/\tau_{\min}$.

Obviously, when $\{s_j = t_j\}$ are *uniform grids*, then the method (2.65) is reduced to the classical L1 method, see (2.62) and (2.63). Next, we introduce a special case with $s_0 = t_0, s_j = t_{j+1/2} = (t_j + t_{j+1})/2, j = 0,1,2,\cdots$. In such a case, (2.65) is reduced to

$$\begin{aligned}\left[{}_C\mathrm{D}_{0,t}^{\alpha}f(t)\right]_{t=t_{n+1/2}} =& b_0 f(t_{n+1/2}) - \sum_{j=1}^{n}(b_{n-j} - b_{n-j+1})f(t_{j-1/2})\\ &- (b_n - B_n)f(t_{1/2}) - B_n f(t_0) + O(\Delta t^{2-\alpha}),\end{aligned} \tag{2.67}$$

where

$$b_n = \frac{\Delta t^{-\alpha}}{\Gamma(2-\alpha)}\left[(n+1)^{1-\alpha} - n^{1-\alpha}\right],\quad B_n = \frac{2\Delta t^{-\alpha}}{\Gamma(2-\alpha)}\left[(n+1/2)^{1-\alpha} - n^{1-\alpha}\right]. \tag{2.68}$$

Replacing $f(t_{j-1/2})$ with $(f(t_j) + f(t_{j-1}))/2$ in (2.67) yields

$$\begin{aligned}\left[{}_C\mathrm{D}_{0,t}^{\alpha}f(t)\right]_{t=t_{n+1/2}} =& \frac{b_0}{2}(f(t_{n+1}) + f(t_n)) - \frac{1}{2}\sum_{j=1}^{n}(b_{n-j} - b_{n-j+1})(f(t_{j-1}) + f(t_j))\\ &- \frac{1}{2}(b_n - B_n)(f(t_0) + f(t_1)) - B_n f(t_0) + O(\Delta t^{2-\alpha}).\end{aligned} \tag{2.69}$$

By (2.69) and ${}_{C}\mathrm{D}_{0,t}^{\alpha}f(t) = {}_{RL}\mathrm{D}_{0,t}^{\alpha}(f(t)-f(0))$, we obtain

$$\begin{aligned}\left[{}_{RL}\mathrm{D}_{0,t}^{\alpha}f(t)\right]_{t=t_{n+1/2}} =& \frac{b_0}{2}(f(t_{n+1})+f(t_n)) - \frac{1}{2}\sum_{j=1}^{n}(b_{n-j}-b_{n-j+1})(f(t_{j-1})+f(t_j)) \\ &- \frac{1}{2}(b_n - B_n)(f(t_0)+f(t_1)) - A_n f(t_0) + O(\Delta t^{2-\alpha}),\end{aligned} \tag{2.70}$$

where $A_n = B_n - \frac{(1-\alpha)(n+1/2)^{-\alpha}}{\Gamma(2-\alpha)\Delta t^{\alpha}}$, and b_n and B_n are defined in (2.68). For simplicity, we call the method using (2.69) and (2.70) the *modified L1 method*.

We will find that the discretization (2.70) is useful to obtain the Crank–Nicolson method for the time-fractional subdiffusion equation [166], which can be seen as a natural extension of the classical Crank–Nicolson method.

- **L2 and L2C Methods**

The L2 method and its variant *L2C method* [105, 118] are used to discretize the Riemann–Liouville derivative of order α ($1 < \alpha < 2$), which can be obtained in a way similar to that of the L1 method. For the Caputo derivative with order $1 < \alpha < 2$, one has

$$\begin{aligned}\left[{}_{C}\mathrm{D}_{0,t}^{\alpha}f(t)\right]_{t=t_n} &= \frac{1}{\Gamma(2-\alpha)}\int_{t_0}^{t_n}(t_n-s)^{1-\alpha}f''(s)\,\mathrm{d}s \\ &= \frac{1}{\Gamma(2-\alpha)}\sum_{k=0}^{n-1}\int_{t_k}^{t_{k+1}}(t_n-s)^{1-\alpha}f''(s)\,\mathrm{d}s \\ &= \frac{1}{\Gamma(2-\alpha)}\sum_{k=0}^{n-1}\int_{t_k}^{t_{k+1}}s^{1-\alpha}f''(t_n-s)\,\mathrm{d}s.\end{aligned} \tag{2.71}$$

On each subinterval $[t_k, t_{k+1}]$, one gets

$$\int_{t_k}^{t_{k+1}}s^{1-\alpha}f''(t_n-s)\,\mathrm{d}s \approx \frac{f(t_n-t_{k+1})-2f(t_n-t_k)+f(t_n-t_{k-1})}{\Delta t^2}\int_{t_k}^{t_{k+1}}s^{1-\alpha}\,\mathrm{d}s.$$

Hence, one has

$$\begin{aligned}\left[{}_{C}\mathrm{D}_{0,t}^{\alpha}f(t)\right]_{t=t_n} &\approx \frac{1}{\Gamma(2-\alpha)}\frac{f(t_{n-k-1})-2f(t_{n-k})+f(t_{n-k+1})}{\Delta t^2}\int_{t_k}^{t_{k+1}}s^{1-\alpha}\,\mathrm{d}s \\ &= \sum_{k=-1}^{n}W_k f(t_{n-k}),\end{aligned} \tag{2.72}$$

which leads to the following *L2 method* for Riemann–Liouville derivative

$$\left[{}_{RL}\mathrm{D}_{0,t}^{\alpha}f(t)\right]_{t=t_n} \approx \frac{f(0)t_n^{-\alpha}}{\Gamma(1-\alpha)} + \frac{f'(0)t_n^{1-\alpha}}{\Gamma(2-\alpha)} + \sum_{k=-1}^{n}W_k f(t_{n-k}), \tag{2.73}$$

where

$$W_k = \frac{\Delta t^{-\alpha}}{\Gamma(3-\alpha)} \begin{cases} 1, & k=-1, \\ 2^{2-\alpha}-3, & k=0, \\ (k+2)^{2-\alpha}-3(k+1)^{2-\alpha}+3k^{2-\alpha}-(k-1)^{2-\alpha}, & 1 \le k \le n-2, \\ -2n^{2-\alpha}+3(n-1)^{2-\alpha}-(n-2)^{2-\alpha}, & k=n-1, \\ n^{2-\alpha}-(n-1)^{2-\alpha}, & k=n. \end{cases} \quad (2.74)$$

On the other hand, we have ${}_C\mathrm{D}_{0,t}^{\alpha} f(t) = {}_C\mathrm{D}_{0,t}^{\alpha-1} f'(t)$. Hence, the L1 method can be used to discretize the $(\alpha-1)$-order Caputo derivative of $f'(t)$. We use (2.67) to discretize ${}_CD_{0,t}^{\alpha} f'(t)$, which yields

$$\begin{aligned} \left[{}_C\mathrm{D}_{0,t}^{\alpha} f(t)\right]_{t=t_{n+1/2}} =& b_0 f'(t_{n+1/2}) - \sum_{j=1}^{n} (b_{n-j}-b_{n-j+1}) f'(t_{j-1/2}) \\ & -(b_n - B_n) f'(t_{1/2}) - B_n f'(t_0) + O(\Delta t^{3-\alpha}), \end{aligned} \quad (2.75)$$

where b_n and B_n are defined by

$$b_n = \frac{\Delta t^{1-\alpha}}{\Gamma(3-\alpha)}\left[(n+1)^{2-\alpha} - n^{2-\alpha}\right], \quad B_n = \frac{2\Delta t^{1-\alpha}}{\Gamma(3-\alpha)}\left[(n+1/2)^{2-\alpha} - n^{2-\alpha}\right]. \quad (2.76)$$

Obviously, $f'(t_{j-1/2})$ satisfies $f'(t_{j-1/2}) = \frac{f(t_j)-f(t_{j-1})}{\Delta t} + O(\Delta t^2) = \delta_t f(t_{j-1/2}) + O(\Delta t^2)$. Hence, we can derive the following discretization

$$\begin{aligned} \left[{}_C\mathrm{D}_{0,t}^{\alpha} f(t)\right]_{t=t_{n+1/2}} =& b_0 \delta_t f(t_{n+1/2}) - \sum_{j=1}^{n} (b_{n-j}-b_{n-j+1}) \delta_t f(t_{j-1/2}) \\ & -(b_n - B_n) \delta_t f(t_{1/2}) - B_n f'(t_0) + O(\Delta t^{3-\alpha}). \end{aligned} \quad (2.77)$$

The L2C method can be derived by letting

$$\begin{aligned} & \int_{t_k}^{t_{k+1}} s^{1-\alpha} f''(t_n - s)\,\mathrm{d}s \\ \approx & \frac{f(t_n - t_{k+2}) - f(t_n - t_{k+1}) + f(t_n - t_{k-1}) - f(t_n - t_k)}{2\Delta t^2} \int_{t_k}^{t_{k+1}} s^{1-\alpha}\,\mathrm{d}s. \end{aligned}$$

So the *L2C method* for Riemann–Liouville derivative is given by

$$\left[{}_{RL}\mathrm{D}_{0,t}^{\alpha} f(t)\right]_{t=t_n} \approx \frac{f(0) t_n^{-\alpha}}{\Gamma(1-\alpha)} + \frac{f'(0) t_n^{1-\alpha}}{\Gamma(2-\alpha)} + \sum_{k=-1}^{n+1} \hat{W}_k f(t_{n-k}), \quad (2.78)$$

where

$$\hat{W}_k = \frac{\Delta t^{-\alpha}}{2\Gamma(3-\alpha)} \begin{cases} 1, & k=-1, \\ 2^{2-\alpha}-2, & k=0, \\ 3^{2-\alpha}-2^{2-\alpha}, & k=1, \\ (k+2)^{2-\alpha}-2(k+1)^{2-\alpha}+2(k-1)^{2-\alpha}-(k-2)^{2-\alpha}, & 2\le k\le n-2, \\ -n^{2-\alpha}-(n-3)^{2-\alpha}+2(n-2)^{2-\alpha}, & k=n-1, \\ -n^{2-\alpha}+2(n-1)^{2-\alpha}-(n-2)^{2-\alpha}, & k=n, \\ n^{2-\alpha}-(n-1)^{2-\alpha}, & k=n+1. \end{cases} \tag{2.79}$$

The accuracy of the L2 and L2C methods depends on α. If $\alpha = 1$, the L2 and L2C methods reduce to the backward difference method and the central difference method for the first order derivative, respectively. If $\alpha = 2$, the L2 method reduces to the central difference method for the second order derivative, and the L2C method reduces to

$$\frac{\mathrm{d}^2 f(t_k)}{\mathrm{d}t^2} \approx \frac{f(t_{k+2})+f(t_k)-f(t_{k-1})-f(t_{k+1})}{2\Delta t^2}$$

with accuracy of order 1. In fact, the L2 method converges with order $O(\Delta t^{3-\alpha})$. Experiments show that the L2 method is more accurate than the L2C method for $1<\alpha<1.5$, while the reverse happens for $1.5<\alpha<2$. Near $\alpha = 1.5$, the two methods have almost similar results [105].

Remark 2.2.1 *The numerical methods based on the polynomial interpolation for the fractional integral in the previous section can be directly extended to the Riemann–Liouville derivative. By* (1.12)*, we get*

$$_{RL}\mathrm{D}_{0,t}^{\alpha} f(t) = {}_{C}\mathrm{D}_{0,t}^{\alpha} f(t) + \sum_{k=0}^{m-1} \frac{f^{(k)}(0)t^{k-\alpha}}{\Gamma(k+1-\alpha)}, \quad m-1<\alpha<m. \tag{2.80}$$

Hence, we only need to develop numerical methods for ${}_{C}\mathrm{D}_{0,t}^{\alpha} f(t)$*. From the definition of the Caputo derivative, we find that the* α*th-order* $(m-1<\alpha<m)$ *Caputo derivative of a given function* $f(t)$ *can be seen as the* $(m-\alpha)$*th-order fractional integral of the function* $f^{(m)}(t)$*. Therefore, the numerical methods developed in Section 2.1 can be directly extended to simulate the numerical solutions of the Caputo derivative, which leads to the numerical methods for the Riemann–Liouville derivative. Here, we do not list these methods, which will be discussed in the following section.*

2.3 Approximations to Caputo Derivatives

Since the Riemann–Liouville derivative and the Caputo derivative have the relation (1.12), almost all the numerical methods for the Riemann–Liouville derivative can be theoretically extended to the Caputo derivative if $f(t)$ satisfies suitable smooth conditions. We first list some algorithms that are often used in the simulation of the Caputo derivative in FDEs.

2.3.1 L1, L2 and L2C Methods

- *The L1 method* for the Caputo derivative is given by:

$$\left[{}_C\mathrm{D}_{0,t}^{\alpha} f(t)\right]_{t=t_n} = \sum_{k=0}^{n-1} b_{n-k-1}\left(f(t_{k+1}) - f(t_k)\right) + O(\Delta t^{2-\alpha}), \quad 0 < \alpha < 1, \tag{2.81}$$

where $b_k = \frac{\Delta t^{-\alpha}}{\Gamma(2-\alpha)}\left[(k+1)^{1-\alpha} - k^{1-\alpha}\right]$.

- *The modified L1 method* for the Caputo derivative is given by:

$$\begin{aligned}\left[{}_C\mathrm{D}_{0,t}^{\alpha} f(t)\right]_{t=t_{n+1/2}} =& \frac{b_0}{2}\left(f(t_{n+1}) + f(t_n)\right) - \frac{1}{2}\sum_{j=1}^{n}(b_{n-j} - b_{n-j+1})(f(t_{j-1}) + f(t_j))\\ & - \frac{1}{2}(b_n - B_n)(f(t_0) + f(t_1)) - B_n f(t_0) + O(\Delta t^{2-\alpha}),\end{aligned} \tag{2.82}$$

where $b_n = \frac{\Delta t^{-\alpha}}{\Gamma(2-\alpha)}\left[(n+1)^{1-\alpha} - n^{1-\alpha}\right]$ and $B_n = \frac{2\Delta t^{-\alpha}}{\Gamma(2-\alpha)}\left[(n+1/2)^{1-\alpha} - n^{1-\alpha}\right]$.

- *The L2 method* for the Caputo derivative is given by:

$$\left[{}_C\mathrm{D}_{0,t}^{\alpha} f(t)\right]_{t=t_n} = \sum_{k=-1}^{n} W_k f(t_{n-k}) + O(\Delta t^{3-\alpha}), \quad 1 < \alpha < 2, \tag{2.83}$$

where $\{W_k\}$ are defined by (2.74).

- *The L2C method* for the Caputo derivative is given by:

$$\left[{}_C\mathrm{D}_{0,t}^{\alpha} f(t)\right]_{t=t_n} = \sum_{k=-1}^{n+1} \hat{W}_k f(t_{n-k}) + O(\Delta t^{3-\alpha}), \quad 1 < \alpha < 2, \tag{2.84}$$

where $\{\hat{W}_k\}$ are defined by (2.79).

2.3.2 Approximations Based on Polynomial Interpolation

From the definition of the Caputo derivative, we can find that the αth-order $(m-1 < \alpha < m)$ Caputo derivative of a given function $f(t)$ can be seen as the $(m-\alpha)$th-order fractional integral of the function $f^{(m)}(t)$. Therefore, we can extend the numerical methods developed in Section 2.1 to simulate the numerical solutions of the Caputo derivative. Here, we give the generalized formulae and some of their special cases and modifications.

- **Fractional Rectangular Formula**

From (2.9), one gets the following formula:

$$\left[{}_C\mathrm{D}_{0,t}^{\alpha}f(t)\right]_{t=t_n} \approx \sum_{k=0}^{n-1} w_{n-k-1}\left[\theta f^{(m)}(t_k)+(1-\theta)f^{(m)}(t_{k+1})\right], \quad 0\leq\theta\leq 1, \tag{2.85}$$

where

$$w_k = \frac{\Delta t^{m-\alpha}}{\Gamma(m+1-\alpha)}\left[(k+1)^{m-\alpha}-k^{m-\alpha}\right].$$

If the mth-order derivative of $f(t)$ is known, the formula (2.85) provides easy implementation of the method. In many cases, $f^{(m)}(t)$ is not given, so it is necessary to combine (2.85) and the numerical methods of the classical derivative to give the more convenient formulae. In order to illustrate the numerical method clearly, we denote by

$$\delta_t f(t_k) = \frac{f(t_{k+1})-f(t_k)}{\Delta t},$$
$$\delta_t^2 f(t_k) = \frac{f(t_{k+1})-2f(t_k)+f(t_{k-1})}{\Delta t^2}.$$

Next, we give the two cases with the same accuracy as (2.85).

Case I: If $0<\alpha<1$, then we use $f'(t_k)\approx\delta_t f(t_k)$ to get the following formula

$$\left[{}_C\mathrm{D}_{0,t}^{\alpha}f(t)\right]_{t=t_n} \approx \sum_{k=0}^{n-1} w_{n-k-1}\left[\theta\delta_t f(t_k)+(1-\theta)\delta_t f(t_{k+1})\right], \quad 0\leq\theta\leq 1, \tag{2.86}$$

where

$$w_k = \frac{\Delta t^{1-\alpha}}{\Gamma(2-\alpha)}\left[(k+1)^{1-\alpha}-k^{1-\alpha}\right].$$

Case II: If $1<\alpha<2$, then we use $f''(t_k)\approx\delta_t^2 f(t_k)$ to get the following formula

$$\left[{}_C\mathrm{D}_{0,t}^{\alpha}f(t)\right]_{t=t_n} \approx \sum_{k=0}^{n-1} w_{n-k-1}\left[\theta\delta_t^2 f(t_k)+(1-\theta)\delta_t^2 f(t_{k+1})\right], \quad 0\leq\theta\leq 1, \tag{2.87}$$

where

$$w_k = \frac{\Delta t^{1-\alpha}}{\Gamma(3-\alpha)}\left[(k+1)^{2-\alpha}-k^{2-\alpha}\right].$$

It is easy to see that formulas (2.86) and (2.87) are convergent with order $O(\Delta t)$.

- **Fractional Trapezoidal Formula**

The fractional trapezoidal formula for ${}_C\mathrm{D}_{0,t}^{\alpha}f(t)$ is given by

$$\left[{}_C\mathrm{D}_{0,t}^{\alpha}f(t)\right]_{t=t_n} \approx \sum_{k=0}^{n} a_{k,n}f^{(m)}(t_k), \tag{2.88}$$

where

$$a_{k,n} = \frac{\Delta t^{m-\alpha}}{\Gamma(m+2-\alpha)} \begin{cases} (n-1)^{m-\alpha+1} - (n-1-m+\alpha)n^{m-\alpha}, & k = 0, \\ (n-k+1)^{m-\alpha+1} + (n-1-k)^{m-\alpha+1} \\ \quad -2(n-k)^{m-\alpha+1}, & 1 \le k \le n-1, \\ 1, & k = n. \end{cases} \tag{2.89}$$

Similar to (2.85), we list the two special modifications of (2.88).

Case I: If $0 < \alpha < 1$, one can get the following modified formula of (2.88)

$$\left[{}_C\mathrm{D}_{0,t}^{\alpha} f(t)\right]_{t=t_n} \approx \sum_{k=0}^{n} a_{k,n} \delta_{\hat{t}} f(t_k), \tag{2.90}$$

where $a_{k,n}$ is defined by (2.89), and

$$\delta_{\hat{t}} f(t_k) = \frac{f(t_{k+1}) - f(t_{k-1})}{2\Delta t}.$$

For the suitably smooth function $f(t)$, the formula (2.88) has convergence of order $O(\Delta t^2)$. Since $\delta_{\hat{t}} f(t_k) - f'(t_k) = O(\Delta t^2)$, therefore, the formula (2.90) still keeps convergent of order $O(\Delta t^2)$. In (2.90), $f(t_{-1}) = f(-\Delta t)$ is used. In order to avoid using $f(-\tau)$, one can use $f'(t_0) = \frac{-3f(t_0)+2f(t_1)-f(t_2)}{2\Delta t} + O(\Delta t^2)$ to get the following formula

$$\left[{}_C\mathrm{D}_{0,t}^{\alpha} f(t)\right]_{t=t_n} \approx \sum_{k=1}^{n} a_{k,n} \delta_{\hat{t}} f(t_k) + a_{0,n} \frac{-3f(t_0) + 2f(t_1) - f(t_2)}{2\Delta t}. \tag{2.91}$$

Case II: If $1 < \alpha < 2$, we can get the following second order formula

$$\left[{}_C\mathrm{D}_{0,t}^{\alpha} f(t)\right]_{t=t_n} \approx \sum_{k=0}^{n} a_{k,n} \delta_t^2 f(t_k), \tag{2.92}$$

where $a_{k,n}$ is defined by (2.89), and $f(t_{-1}) = f(-\Delta t)$. Similar to (2.91), one can get

$$\left[{}_C\mathrm{D}_{0,t}^{\alpha} f(t)\right]_{t=t_n} \approx \sum_{k=1}^{n} a_{k,n} \delta_t^2 f(t_k) + a_{0,n} \frac{f(t_1) - 2f(t_0) + f(t_{-1})}{\Delta t^2}. \tag{2.93}$$

- **Fractional Newton–Cotes Formula**

Similar to (2.19), the fractional Newton–Cotes formula for the Caputo derivative are given by

$$\left[{}_C\mathrm{D}_{0,t}^{\alpha} f(t)\right]_{t=t_n} \approx I(\tilde{f}, t_n, \alpha) = \sum_{k=0}^{n-1} \sum_{i=0}^{r} A_{i,n}^{(k,m)} f^{(m)}(t_i^{(k)}), \tag{2.94}$$

where

$$A_{i,n}^{(k,m)} = \frac{1}{\Gamma(\alpha)} \int_{t_k}^{t_{k+1}} (t_n - t)^{m-\alpha-1} l_{k,i}(t).$$

2.3.3 High-Order Methods

Suppose that $p_N(t) = \hat{p}(x)$, $t = L(x+1)/2$ has the following representation

$$p_N(t) = \sum_{j=0}^{N} c_j P_j^{a,b}(2t/L-1) = \sum_{j=0}^{N} c_j P_j^{a,b}(x) = \hat{p}(x), \quad t\in[0,T]. \tag{2.95}$$

Using the property (2.30), we can easily get

$$\begin{aligned}
{}_C\mathrm{D}_{-1,x}^{\alpha} P_j^{a,b}(x) =& \frac{1}{\Gamma(m-\alpha)} \int_{-1}^{x} (x-s)^{m-\alpha-1} \frac{\mathrm{d}^m}{\mathrm{d}s^m} \left[P_j^{a,b}(s)\right] \mathrm{d}s \\
=& \frac{1}{\Gamma(m-\alpha)} \int_{-1}^{x} (x-s)^{m-\alpha-1}\, d_{j,m}^{a,b} P_{j-m}^{a+m,b+m}(s)\,\mathrm{d}s \\
=& d_{j,m}^{a,b} \widehat{P}_{j-m}^{a+m,b+m,m-\alpha}(x),
\end{aligned} \tag{2.96}$$

where $d_{j,m}^{a,b}$ and $\widehat{P}_j^{a+m,b+m,m-\alpha}(x)$ are defined by (2.31) and (2.37), respectively, with $\widehat{P}_j^{a+m,b+m,m-\alpha}(x) = 0$ for $j = 0,1,\cdots,m-1$. On the other hand, we have

$$\begin{aligned}
{}_C\mathrm{D}_{0,L}^{\alpha} p_N(t) =& \frac{1}{\Gamma(m-\alpha)} \int_{0}^{t} (t-s)^{m-\alpha-1} \frac{\mathrm{d}^m}{\mathrm{d}s^m} (p_N(s))\,\mathrm{d}s \\
=& \left(\frac{L}{2}\right)^{-\alpha} \frac{1}{\Gamma(m-\alpha)} \int_{-1}^{\frac{2t}{L}-1} \left(\frac{2t}{L}-1-s\right)^{m-\alpha-1} p_N(s)\mathrm{d}s \\
=& \left(\frac{L}{2}\right)^{-\alpha} {}_C\mathrm{D}_{-1,\frac{2t}{L}-1}^{\alpha} p_N(2t/L-1) = \left(\frac{L}{2}\right)^{-\alpha} {}_C\mathrm{D}_{-1,x}^{\alpha} \hat{p}_N(x).
\end{aligned} \tag{2.97}$$

Therefore, for any $\alpha > 0$, from (2.95)–(2.97) one has [89]

$$ {}_C\mathrm{D}_{0,L}^{\alpha} p_N(t) = \left(\frac{L}{2}\right)^{-\alpha} {}_C\mathrm{D}_{-1,x}^{\alpha} \hat{p}_N(x) = \left(\frac{L}{2}\right)^{-\alpha} \sum_{j=0}^{N} c_j d_{j,m}^{a,b} \widehat{P}_{j-m}^{a+m,b+m,m-\alpha}(x). \tag{2.98}$$

Let t_j $(j = 0,1,\cdots,N)$ be collocation points on $[0,T]$. Then $x_j = 2t_j/L-1$ are collocation points on $[-1,1]$. We can obtain ${}_CD_{0,L}^{\alpha} p_N(t)$ at $t = t_j$ as follows

$$\begin{pmatrix} \left[{}_C\mathrm{D}_{0,t}^{\alpha} p_N(t)\right]_{t=t_0} \\ \left[{}_C\mathrm{D}_{0,t}^{\alpha} p_N(t)\right]_{t=t_1} \\ \vdots \\ \left[{}_C\mathrm{D}_{0,t}^{\alpha} p_N(t)\right]_{t=t_N} \end{pmatrix} = \left(\frac{L}{2}\right)^{-\alpha} \begin{pmatrix} \left[{}_C\mathrm{D}_{-1,x}^{\alpha} \hat{p}_N(x)\right]_{x=x_0} \\ \left[{}_C\mathrm{D}_{-1,x}^{\alpha} \hat{p}_N(x)\right]_{x=x_1} \\ \vdots \\ \left[{}_C\mathrm{D}_{-1,x}^{\alpha} \hat{p}_N(x)\right]_{x=x_N} \end{pmatrix} = \left(\frac{L}{2}\right)^{-\alpha} \left(\widehat{D}_{L,C}^{(\alpha,a,b)}\right) \begin{pmatrix} c_0 \\ c_1 \\ \vdots \\ c_N \end{pmatrix},$$

where the matrix $\widehat{D}_{L,C}^{(\alpha,a,b)}$ is given by

$$\left(\widehat{D}_{L,C}^{(\alpha,a,b)}\right)_{i,j} = d_{j,m}^{a,b} \widehat{P}_{j-m}^{a+m,b+m,m-\alpha}(x_i).$$

If $p_N(t)$ is the Legendre–Gauss–Lobatto interpolation of $f(t)$, $f\in H^r([0,L])$, then the following error estimate holds

$$\left|{}_C\mathrm{D}_{0,t}^{\alpha} f(t) - {}_C\mathrm{D}_{0,t}^{\alpha} p_N(t)\right| \le C N^{3/4+2m-r} \|f\|_{H^r}, \quad r \ge 2m.$$

Remark 2.3.1 *Generally speaking, $p_N(t)$ is not necessarily the interpolation of $f(t)$. $p_N(t)$ can be any approximation of $f(t)$ that is expressed in the form of* (2.95). *For example, $p_N(t)$ can be the orthogonal projection of $f(t)$ [135], the formula* (2.98) *is still valid and efficient with high accuracy if $f(t)$ is suitably smooth.*

Next, we introduce another *operational matrix* to approximate the Caputo derivative, which is based on the explicit expression of the Jacobi polynomials. The Jacobi polynomial $P_j^{a,b}(x), x \in [0,1]$ has the following explicit expression

$$P_j^{a,b}(x) = \frac{\Gamma(j+b+1)}{\Gamma(j+a+b+1)} \sum_{k=0}^{j} (-1)^{j-k} \frac{\Gamma(j+k+a+b+1)}{\Gamma(k+b+1)k!(j-k)!2^k} (1+x)^k. \tag{2.99}$$

It is easy to get

$$_C\mathrm{D}_{-1,x}^{\alpha} P_j^{a,b}(x) = \frac{\Gamma(j+b+1)}{\Gamma(j+a+b+1)} \sum_{k=0}^{j} \frac{(-1)^{j-k}\Gamma(j+k+a+b+1)}{\Gamma(k+b+1)(j-k)!\Gamma(k+1-\alpha)2^k} (1+x)^{k-\alpha}, k \geq \alpha. \tag{2.100}$$

Hence, for any $p_N(t)$, $t \in [0,L]$ of the form $p_N(t) = \hat{p}_N(x) = \sum_{j=0}^{N} c_j P_j^{a,b}(x)\,(x = 2t/L-1)$ and $m-1 < \alpha < m$, $m \in Z^+$, we have

$$\begin{aligned} _C\mathrm{D}_{0,L}^{\alpha} p_N(t) &= \left(\frac{L}{2}\right)^{-\alpha} {_C\mathrm{D}_{-1,x}^{\alpha}} \hat{p}_N(x) = \sum_{j=0}^{N} c_j \frac{\Gamma(j+b+1)}{\Gamma(j+a+b+1)} \\ &\quad \times \sum_{k=m}^{j} \frac{(-1)^{j-k}\Gamma(j+k+a+b+1)}{\Gamma(k+b+1)(j-k)!\Gamma(k+1-\alpha)2^k} (1+x)^{k-\alpha} \\ &= \sum_{j=0}^{N} c_j \frac{\Gamma(j+b+1)}{\Gamma(j+a+b+1)} \sum_{k=m}^{j} \frac{(-1)^{j-k}\Gamma(j+k+a+b+1)}{\Gamma(k+b+1)(j-k)!\Gamma(k+1-\alpha)L^k} t^{k-\alpha}. \end{aligned} \tag{2.101}$$

It is clear that this technique gives an exact expression of ${_C\mathrm{D}_{0,L}^{\alpha}} p_N(t)$, but it seems a little tedious.

Let $\Phi^{a,b}(x) = (P_0^{a,b}(x), P_1^{a,b}(x), \cdots, P_N^{a,b}(x))^T, \mathbf{c} = (c_0, c_1, \cdots, c_N)^T$. Then $\hat{p}_N(x) = \mathbf{c}^T \Phi^{a,b}(x)$. It is known that $\frac{\mathrm{d}}{\mathrm{d}x}\hat{p}_N(x)$ can be simply expressed in the following form

$$\frac{\mathrm{d}}{\mathrm{d}x}\hat{p}_N(x) = \mathbf{c}^T D^{(1)} \Phi^{a,b}(x),$$

where $D^{(1)}$ can be easily derived from (2.32). For example, if $a = b = 0$, then

$$D^{(1)} = (d_{ij}) = \begin{cases} 2(2j+1), & j = i-k, k = 1,3,\cdots,m,\ k \text{ is odd} \\ & \qquad \text{or } k = 1,3,\cdots,m-1,\ k \text{ is even}, \\ 0, \text{ otherwise}. \end{cases} \tag{2.102}$$

Does there exist a matrix $D^{(\alpha)}$ such that ${_C\mathrm{D}_{-1,x}^{\alpha}}\hat{p}_N(x) = \mathbf{c}^T D^{(\alpha)} \Phi^{a,b}(x)$? Obviously this is not true for a noninteger number α, since $\mathbf{c}^T D^{(\alpha)}\Phi^{a,b}(x)$ is a polynomial while ${}_C\mathrm{D}^{\alpha}_{-1,x}\hat{p}_N(x)$ is not a polynomial.

Some researchers construct matrix $D^{(\alpha)}$ such that ${}_C\mathrm{D}^{\alpha}_{-1,x}\hat{p}_N(x) \approx \mathbf{c}^T D^{(\alpha)}\Phi^{a,b}(x)$, for example, see [43, 44, 45, 132]. Their methods are derived from further expanding $(1+x)^{k-\alpha}$ in (2.101) in the series of Jacobi polynomials, i.e.,

$$(1+x)^{k-\alpha} \approx \sum_{l=0}^{N} b_{k,l} P_l^{a,b}(x). \tag{2.103}$$

Hence, one has

$$\begin{aligned}
{}_C\mathrm{D}^{\alpha}_{-1,x}\hat{p}_N(x) \approx & \sum_{j=0}^{N} c_j \frac{\Gamma(j+b+1)}{\Gamma(j+a+b+1)} \sum_{k=m}^{j} \frac{(-1)^{j-k}\Gamma(j+k+a+b+1)}{\Gamma(k+b+1)(j-k)!\Gamma(k+1-\alpha)2^k} \\
& \times \sum_{l=0}^{N} b_{k,l} P_l^{a,b}(x) \\
=& \mathbf{c}^T D^{(\alpha)}\Phi^{a,b}(x).
\end{aligned} \tag{2.104}$$

This approach seems somewhat complicated. For $a = b = 0$, $b_{k,l}$ in (2.103) is given by

$$b_{k,l} = (2l+1)\sum_{r}^{l} \frac{(-1)^{l+r}(l+r)!}{(l-r)(r!)^2(k+r-\alpha+1)}.$$

And the matrix $D^{(\alpha)}$ is given by

$$D^{(\alpha)} = \begin{pmatrix}
0 & 0 & \cdots & 0 \\
\vdots & \vdots & \cdots & \vdots \\
0 & 0 & \cdots & 0 \\
\sum\limits_{k=m}^{m} \theta_{m,0,k} & \sum\limits_{k=m}^{m} \theta_{m,1,k} & \cdots & \sum\limits_{k=m}^{m} \theta_{m,N,k} \\
\vdots & \vdots & \cdots & \vdots \\
\sum\limits_{k=m}^{i} \theta_{i,0,k} & \sum\limits_{k=m}^{i} \theta_{i,1,k} & \cdots & \sum\limits_{k=m}^{i} \theta_{i,N,k} \\
\vdots & \vdots & \cdots & \vdots \\
\sum\limits_{k=m}^{N} \theta_{N,0,k} & \sum\limits_{k=m}^{N} \theta_{N,1,k} & \cdots & \sum\limits_{k=m}^{N} \theta_{N,N,k}
\end{pmatrix}, \tag{2.105}$$

where $\theta_{i,j,k}$ is given by

$$\theta_{i,j,k} = (2j+1)\sum_{l=0}^{j} \frac{(-1)^{i+j+k+l}(i+k)!(l+j)!}{(i-k)!k!\Gamma(k-\alpha+1)(j-l)!(l!)^2(k+l-\alpha+1)}.$$

The operational matrix as $D^{(\alpha)}$ defined by (2.105) based on Chebyshev polynomials is established in [43, 44]. The operational matrix based on generalized Jacobi polynomials is developed in [45]. The operational matrices based on the Legendre wavelets for the fractional integration and Caputo derivative are presented in [128]. The operational matrix based on the B-spline functions is constructed in [70]. For other related works, see [1, 2, 12, 15, 48, 51, 61, 79, 84, 129, 133, 142, 155, 162] and the references cited therein.

2.4 Approximation to Riesz Derivatives

In this section, we derive some high-order algorithms for the *Riesz derivative* with order α $(1 < \alpha < 2)$ defined as follows [68, 157]

$$ {}_{RZ}\mathrm{D}_x^{\alpha} u(x) = \frac{\partial^{\alpha} u(x)}{\partial |x|^{\alpha}} = -\Psi_{\alpha}\left({}_{RL}\mathrm{D}_{a,x}^{\alpha} + {}_{RL}\mathrm{D}_{x,b}^{\alpha} \right) u(x), \tag{2.106} $$

where $\Psi_{\alpha} = \frac{1}{2}\sec\left(\frac{\pi\alpha}{2}\right)$, ${}_{RL}\mathrm{D}_{a,x}^{\alpha}$ and ${}_{RL}\mathrm{D}_{x,b}^{\alpha}$ are the left and right Riemann–Liouville derivatives. We take the mesh points $x_m = a + mh$, $m = 0, 1, \ldots, M$, where $h = (b-a)/M$, i.e., h is the uniform spatial stepsize. The numerical schemes come from a series of papers by Ding and Li, et al. [37, 38, 39]. It should be noted that the high-order algorithms for Riemann–Liouville derivatives are first proposed by Lubich [104], while the high order algorithms for Riesz derivatives are constructed by Ding and Li [37, 38, 39].

2.4.1 High-Order Algorithms (I)

For every α $(1 < \alpha < 2)$, we assume that the left, right Riemann–Liouville derivatives exist and coincide with the left, right Grünwald–Letnikov derivatives under suitable conditions, respectively, where the definitions of the left, right Grünwald–Letnikov derivative with order α are given below [124]

$$ {}_{GL}\mathrm{D}_{a,x}^{\alpha} u(x_m) = \frac{1}{h^{\alpha}} \sum_{k=0}^{m} \varpi_k^{(\alpha)} u(x_{m-k}) + O(h), $$

and

$$ {}_{GL}\mathrm{D}_{x,b}^{\alpha} u(x_m) = \frac{1}{h^{\alpha}} \sum_{k=0}^{M-m} \varpi_k^{(\alpha)} u(x_{m+k}) + O(h), $$

in which $\varpi_k^{(\alpha)} = (-1)^k \binom{\alpha}{k} = \frac{(-1)^k \Gamma(1+\alpha)}{\Gamma(1+k)\Gamma(1+\alpha-k)}$.

So, the Riesz derivative with order $\alpha \in (1,2)$ can be discretized in the following ways.

- **By the standard Grünwald–Letnikov formula**

Based on the above assumption and the equation (1.9), we can obtain the first order approximation formula

$$\frac{\partial^\alpha u(x_m)}{\partial|x|^\alpha} = -\frac{\Psi_\alpha}{h^\alpha}\left(\sum_{k=0}^{m}\varpi_k^{(\alpha)}u(x_{m-k}) + \sum_{k=0}^{M-m}\varpi_k^{(\alpha)}u(x_{m+k})\right) + O(h).$$

- **By the shifted Grünwald–Letnikov formula**

In [111], Meerschaert and Tadjeran show that above standard Grünwald–Letnikov formula is often unstable for time dependent problems. Hence, they propose the following *shifted Grünwald–Letnikov formulas* for the left and right Riemann–Liouville derivatives in order to overcome the instability,

$${}_{RL}\mathrm{D}_{a,x}^{\alpha}u(x_m) = \sum_{k=0}^{m+1}\varpi_k^{(\alpha)}u(x_{m-k+1}) + O(h)$$

and

$${}_{RL}\mathrm{D}_{x,b}^{\alpha}u(x_m) = \sum_{k=0}^{M-m+1}\varpi_k^{(\alpha)}u(x_{m+k-1}) + O(h).$$

Therefore, the modified first order approximation scheme is constructed as follows,

$$\frac{\partial^\alpha u(x_m)}{\partial|x|^\alpha} = -\frac{\Psi_\alpha}{h^\alpha}\left(\sum_{k=0}^{m+1}\varpi_k^{(\alpha)}u(x_{m-k+1}) + \sum_{k=0}^{M-m+1}\varpi_k^{(\alpha)}u(x_{m+k-1})\right) + O(h).$$

- **By the L2 approximation method**

Note that the left, right Riemann–Liouville derivatives can be rewritten as ($1 < \alpha < 2$),

$${}_{RL}\mathrm{D}_{a,x}^{\alpha}u(x) = \sum_{k=0}^{1}\frac{x^{k-\alpha}}{\Gamma(k+1-\alpha)}\frac{\partial^k u(a)}{\partial x^k} + \frac{1}{\Gamma(2-\alpha)}\int_a^x \frac{\partial^2 u(\xi)}{\partial \xi^2}(x-\xi)^{1-\alpha}d\xi$$

and

$${}_{RL}\mathrm{D}_{x,b}^{\alpha}u(x) = \sum_{k=0}^{1}\frac{(b-x)^{k-\alpha}}{\Gamma(k+1-\alpha)}\frac{\partial^k u(b)}{\partial x^k} + \frac{1}{\Gamma(2-\alpha)}\int_x^b \frac{\partial^2 u(\xi)}{\partial \xi^2}(\xi-x)^{1-\alpha}d\xi.$$

Hence, we can obtain a first order scheme for the left and right Riemann–Liouville derivatives [157],

$$\begin{aligned}{}_{RL}\mathrm{D}_{a,x}^{\alpha}u(x_m) &= \frac{1}{\Gamma(3-\alpha)h^\alpha}\left\{\frac{(1-\alpha)(2-\alpha)u(x_0)}{m^\alpha} + \frac{(2-\alpha)\left[u(x_1)-u(x_0)\right]}{m^{\alpha-1}}\right.\\ &\quad\left.+\sum_{k=0}^{m-1}d_k^{(\alpha)}\left[u(x_{m-k+1})-2u(x_{m-k})+u(x_{m-k-1})\right]\right\}\\ &\quad+O(h),\end{aligned}$$

$$\begin{aligned}
{}_{RL}\mathrm{D}_{x,b}^{\alpha}u(x_m) \quad = \quad & \frac{1}{\Gamma(3-\alpha)h^{\alpha}}\left\{\frac{(1-\alpha)(2-\alpha)u(x_M)}{(M-m)^{\alpha}}+\frac{(2-\alpha)\left[u(x_M)-u(x_{M-1})\right]}{m^{\alpha-1}}\right.\\
&\left.+\sum_{k=0}^{M-m-1} d_k^{(\alpha)}\left[u(x_{m+k-1})-2u(x_{m+k})+u(x_{m+k+1})\right]\right\}\\
&+O(h),
\end{aligned}$$

where $d_k^{(\alpha)}=(k+1)^{2-\alpha}-k^{2-\alpha}, k=0,1,\ldots,m-1$, or $k=0,1,\ldots,M-m-1$,

Therefore, applying the above two formulas and (2.106) gives

$$\begin{aligned}
\frac{\partial^{\alpha}u(x_m)}{\partial|x|^{\alpha}} \quad = \quad & -\frac{\Psi_{\alpha}}{\Gamma(3-\alpha)h^{\alpha}}\left\{\frac{(1-\alpha)(2-\alpha)u(x_0)}{m^{\alpha}}+\frac{(2-\alpha)\left[u(x_1)-u(x_0)\right]}{m^{\alpha-1}}\right.\\
&+\sum_{k=0}^{m-1} d_k^{(\alpha)}\left[u(x_{m-k+1})-2u(x_{m-k})+u(x_{m-k-1})\right]\\
&+\frac{(1-\alpha)(2-\alpha)u(x_M)}{(M-m)^{\alpha}}+\frac{(2-\alpha)\left[u(x_M)-u(x_{M-1})\right]}{m^{\alpha-1}}\\
&\left.+\sum_{k=0}^{M-m-1} d_k^{(\alpha)}\left[u(x_{m+k-1})-2u(x_{m+k})+u(x_{m+k+1})\right]\right\}\\
&+O(h),
\end{aligned}$$

in which $d_k^{(\alpha)}$ is defined as above.

- **By the spline interpolation method**

In [139], Sousa proposed a second-order scheme by *linear spline interpolation method* for the left and right Riemann–Liouville derivatives,

$$ {}_{RL}\mathrm{D}_{a,x}^{\alpha}u(x_m)=\frac{1}{\Gamma(4-\alpha)h^{\alpha}}\sum_{k=0}^{m+1}\bar{z}_{m,k}^{(\alpha)}u(x_k)+O(h^2), \tag{2.107}$$

where

$$\bar{z}_{m,k}^{(\alpha)}=\begin{cases}\bar{c}_{m-1,k}-2\bar{c}_{m,k}+\bar{c}_{m+1,k}, & k\le m-1,\\ -2\bar{c}_{m,k}+\bar{c}_{m+1,k}, & k=m,\\ \bar{c}_{m+1,k}, & k=m+1,\\ 0, & k>m+1,\end{cases}$$

in which

$$\bar{c}_{j,k}=\begin{cases}(j-1)^{3-\alpha}-j^{2-\alpha}(j-3+\alpha), & k=0,\\ (j-k+1)^{3-\alpha}-2(j-k)^{3-\alpha}+(j-k-1)^{3-\alpha}, & 1\le k\le j-1,\\ 1, & k=j,\end{cases}$$

and

$$ {}_{RL}\mathrm{D}^{\alpha}_{x,b}u(x_m) = \frac{1}{\Gamma(4-\alpha)h^{\alpha}} \sum_{k=m-1}^{M} \tilde{z}^{(\alpha)}_{m,k}u(x_k) + O(h^2), \tag{2.108} $$

where

$$ \tilde{z}^{(\alpha)}_{m,k} = \begin{cases} 0, \quad k < m-1, \\ \tilde{c}_{m-1,m-1}, \quad k = m-1, \\ -2\tilde{c}_{m,m} + \tilde{c}_{m-1,m}, \quad k = m, \\ \tilde{c}_{m-1,k} - 2\tilde{c}_{m,k} + \tilde{c}_{m+1,k}, \quad m+1 \le k \le M, \end{cases} $$

in which

$$ \tilde{c}_{j,k} = \begin{cases} 1, \quad k = j, \\ (k-j+1)^{3-\alpha} - 2(k-j)^{3-\alpha} + (k-j-1)^{3-\alpha}, \quad j+1 \le k \le M-1, \\ (3-\alpha-M+j)(M-j)^{2-\alpha} + (M-j-1)^{3-\alpha}, \quad k = M, \end{cases} $$

with $j = m-1, m, m+1$.

Combining (2.107), (2.108) and (2.106) gives

$$ \frac{\partial^{\alpha}u(x_m)}{\partial|x|^{\alpha}} = \frac{-\Psi_{\alpha}}{\Gamma(4-\alpha)h^{\alpha}} \sum_{k=0}^{M} z^{(\alpha)}_{m,k}u(x_k) + O(h^2), $$

where

$$ z^{(\alpha)}_{m,k} = \begin{cases} \bar{z}^{(\alpha)}_{m,k}, \quad k < m-1, \\ \bar{z}^{(\alpha)}_{m,m-1} + \tilde{z}^{(\alpha)}_{m,m-1}, \quad k = m-1, \\ \bar{z}^{(\alpha)}_{m,m} + \tilde{z}^{(\alpha)}_{m,m}, \quad k = m, \\ \bar{z}^{(\alpha)}_{m,m+1} + \tilde{z}^{(\alpha)}_{m,m+1}, \quad k = m+1, \\ \tilde{z}^{(\alpha)}_{m,k}, \quad k > m+1. \end{cases} $$

- **By the fractional central difference method**

In [119], Ortigueira introduced a symmetrical *fractional central difference operator* as follows

$$ \Delta^{\alpha}_{h}u(x) = \sum_{k=-\infty}^{\infty} \frac{(-1)^k\Gamma(\alpha+1)}{\Gamma\left(\frac{\alpha}{2}-k+1\right)\Gamma\left(\frac{\alpha}{2}+k+1\right)}u(x-kh). $$

Later on, Çelik and Duman [14] proved that the above symmetrical *fractional central difference operator* for the Riesz fractional derivative has the following estimate

$$ \frac{\partial^{\alpha}u(x_m)}{\partial|x|^{\alpha}} = -\frac{1}{h^{\alpha}}\Delta^{\alpha}_{h}u(x_m) + O(h^2). $$

- **By the weighted and shifted Grünwald–Lentikov formulas**

In [146], Tian and Deng proposed the *second-order and third-order numerical schemes for the left and right Riemann–Liouville derivatives*:

$$ {}_{RL}\mathrm{D}_{a,x}^{\alpha}u(x_m) = \frac{\nu_1}{h^\alpha}\sum_{k=0}^{m+\ell_1}\varpi_k^{(\alpha)}u(x_{m-k+\ell_1}) + \frac{\nu_2}{h^\alpha}\sum_{k=0}^{m+\ell_2}\varpi_k^{(\alpha)}u(x_{m-k+\ell_2}) + O(h^2), $$

and

$$ {}_{RL}\mathrm{D}_{x,b}^{\alpha}u(x_m) = \frac{\nu_1}{h^\alpha}\sum_{k=0}^{M-m+\ell_1}\varpi_k^{(\alpha)}u(x_{m+k-\ell_1}) + \frac{\nu_2}{h^\alpha}\sum_{k=0}^{M-m+\ell_2}\varpi_k^{(\alpha)}u(x_{m+k-\ell_2}) + O(h^2), $$

where ℓ_1 and ℓ_2 are two arbitrary integers and $\ell_1 - \ell_2 \neq 0$, $\nu_1 = \dfrac{\alpha - 2\ell_2}{2(\ell_1 - \ell_2)}$, $\nu_2 = \dfrac{2\ell_1 - \alpha}{2(\ell_1 - \ell_2)}$.

And

$$ \begin{aligned} {}_{RL}\mathrm{D}_{a,x}^{\alpha}u(x_m) &= \frac{\kappa_1}{h^\alpha}\sum_{k=0}^{m+\ell_1}\varpi_k^{(\alpha)}u(x_{m-k+\ell_1}) + \frac{\kappa_2}{h^\alpha}\sum_{k=0}^{m+\ell_2}\varpi_k^{(\alpha)}u(x_{m-k+\ell_2}) \\ &\quad + \frac{\kappa_3}{h^\alpha}\sum_{k=0}^{m+\ell_3}\varpi_k^{(\alpha)}u(x_{m-k+\ell_3}) + O(h^3), \end{aligned} $$

and

$$ \begin{aligned} {}_{RL}\mathrm{D}_{x,b}^{\alpha}u(x_m) &= \frac{\kappa_1}{h^\alpha}\sum_{k=0}^{M-m+\ell_1}\varpi_k^{(\alpha)}u(x_{m+k-\ell_1}) + \frac{\kappa_2}{h^\alpha}\sum_{k=0}^{M-m+\ell_2}\varpi_k^{(\alpha)}u(x_{m+k-\ell_2}) \\ &\quad + \frac{\kappa_3}{h^\alpha}\sum_{k=0}^{M-m+\ell_3}\varpi_k^{(\alpha)}u(x_{m+k-\ell_3}) + O(h^3), \end{aligned} $$

in which ℓ_1, ℓ_2 and ℓ_3 are three arbitrary integers and $(\ell_1 - \ell_2)(\ell_2 - \ell_3)(\ell_1 - \ell_3) \neq 0$, $\kappa_1 = \dfrac{12\ell_2\ell_3 - (6\ell_2 + 6\ell_3 + 1)\alpha + 3\alpha^2}{12(\ell_2\ell_3 - \ell_1\ell_2 - \ell_1\ell_3 + \ell_1^2)}$, $\kappa_2 = \dfrac{12\ell_1\ell_3 - (6\ell_1 + 6\ell_3 + 1)\alpha + 3\alpha^2}{12(\ell_1\ell_3 - \ell_1\ell_2 - \ell_2\ell_3 + \ell_2^2)}$, $\kappa_3 = \dfrac{12\ell_1\ell_2 - (6\ell_1 + 6\ell_2 + 1)\alpha + 3\alpha^2}{12(\ell_1\ell_2 - \ell_1\ell_3 - \ell_2\ell_3 + \ell_3^2)}$.

Naturally, we can obtain the following second-order and third-order numerical formulas for the Riesz fractional derivative

$$\begin{aligned}\frac{\partial^\alpha u(x_m)}{\partial|x|^\alpha} &= -\frac{\Psi_\alpha}{h^\alpha}\left(\nu_1\sum_{k=0}^{m+\ell_1}\varpi_k^{(\alpha)}u(x_{m-k+\ell_1})+\nu_2\sum_{k=0}^{m+\ell_2}\varpi_k^{(\alpha)}u(x_{m-k+\ell_2})\right.\\ &\quad\left.+\nu_1\sum_{k=0}^{M-m+\ell_1}\varpi_k^{(\alpha)}u(x_{m+k-\ell_1})+\nu_2\sum_{k=0}^{M-m+\ell_2}\varpi_k^{(\alpha)}u(x_{m+k-\ell_2})\right)\\ &\quad+O(h^2),\end{aligned}$$

and,

$$\begin{aligned}\frac{\partial^\alpha u(x_m)}{\partial|x|^\alpha} &= -\frac{\Psi_\alpha}{h^\alpha}\left(\kappa_1\sum_{k=0}^{m+\ell_1}\varpi_k^{(\alpha)}u(x_{m-k+\ell_1})+\kappa_2\sum_{k=0}^{m+\ell_2}\varpi_k^{(\alpha)}u(x_{m-k+\ell_2})\right.\\ &\quad+\kappa_3\sum_{k=0}^{m+\ell_3}\varpi_k^{(\alpha)}u(x_{m-k+\ell_3})+\kappa_1\sum_{k=0}^{M-m+\ell_1}\varpi_k^{(\alpha)}u(x_{m+k-\ell_1})\\ &\quad\left.+\kappa_2\sum_{k=0}^{M-m+\ell_2}\varpi_k^{(\alpha)}u(x_{m+k-\ell_2})+\kappa_3\sum_{k=0}^{M-m+\ell_3}\varpi_k^{(\alpha)}u(x_{m+k-\ell_3})\right)\\ &\quad+O(h^3),\end{aligned}$$

respectively.

Here, we construct another *second-order scheme and two kinds of fourth-order numerical schemes for the Riesz derivative*. In order to construct the new computational schemes, we introduce the following theorem.

Lemma 2.4.1 ([49]) *Let $\alpha > 0$, $u(x) \in C_0^\infty(\mathbb{R})$, the Fourier transforms of the left and right Riemann–Liouville derivative are,*

$$\mathcal{F}\left({}_{RL}\mathrm{D}_{-\infty,x}^{\alpha}u(x)\right) = (i\omega)^\alpha\,\hat{u}(\omega),$$

and

$$\mathcal{F}\left({}_{RL}\mathrm{D}_{x,\infty}^{\alpha}u(x)\right) = (-i\omega)^\alpha\,\hat{u}(\omega),$$

where $\hat{u}(\omega)$ denotes the Fourier transform of the function $u(x)$, i.e.,

$$\hat{u}(\omega) = \int_{\mathbb{R}} \exp(-i\omega x)\,u(x)dx.$$

In [148], Tuan and Gorenflo introduce the following *left fractional central difference operator*:

$$ {}_C\Delta_{-h}^{\alpha}u(x) = \sum_{k=0}^{\infty}\varpi_k^{(\alpha)}u\left(x-\left(k-\frac{\alpha}{2}\right)h\right). \qquad (2.109)$$

Similarly, we define the following *right fractional central difference operator*:

$$ {}_C\Delta^{\alpha}_{+h}u(x) = \sum_{k=0}^{\infty} \varpi_k^{(\alpha)} u\left(x + \left(k - \frac{\alpha}{2}\right)h\right). \tag{2.110} $$

Analogous to the integer-order finite difference formula, we define the following fractional average operator,

$$ \mu^{\alpha}_{\pm h}u(x - sh) = \frac{u\left(x \pm \left(s - \frac{\alpha}{2}\right)h\right) + u\left(x \pm \left(s + \frac{\alpha}{2}\right)h\right)}{2}. \tag{2.111} $$

Then we can get the following *fractional left and right average central difference operators* based on (2.109), (2.110) and (2.111), respectively.

$$ \begin{aligned} {}_{AC}\Delta^{\alpha}_{-h}u(x) &= \mu^{\alpha}_{-h}\left({}_C\Delta^{\alpha}_{-h}u(x)\right) \\ &= \sum_{j=0}^{\infty}(-1)^j\binom{\alpha}{j}\mu^{\alpha}_{-h}\left(u\left(x - \left(j - \frac{\alpha}{2}\right)h\right)\right) \\ &= \frac{1}{2}\sum_{j=0}^{\infty}(-1)^j\binom{\alpha}{j}(u(x - jh) + u(x - (j - \alpha)h)) \end{aligned} \tag{2.112} $$

and

$$ \begin{aligned} {}_{AC}\Delta^{\alpha}_{+h}u(x) &= \mu^{\alpha}_{+h}\left({}_C\Delta^{\alpha}_{+h}u(x)\right) \\ &= \sum_{j=0}^{\infty}(-1)^j\binom{\alpha}{j}\mu^{\alpha}_{+h}\left(u\left(x + \left(j - \frac{\alpha}{2}\right)h\right)\right) \\ &= \frac{1}{2}\sum_{j=0}^{\infty}(-1)^j\binom{\alpha}{j}(u(x + jh) + u(x + (j - \alpha)h)). \end{aligned} \tag{2.113} $$

Here, we always assume that $\mu^{\alpha}_{\pm h}$ can commute with the infinite summation.

For the fractional left and right average central difference operators defined in (2.112) and (2.113),we have the following result.

Theorem 4 *Let $u(x)$ and the Fourier transforms of ${}_{RL}D^{\alpha+2}_{-\infty,x}u(x)$ and ${}_{RL}D^{\alpha+2}_{x,+\infty}u(x)$ both be in $L_1(\mathbb{R})$, then*

$$ {}_{RL}\mathrm{D}^{\alpha}_{-\infty,x}u(x) = \frac{{}_{AC}\Delta^{\alpha}_{-h}u(x)}{h^{\alpha}} + O(h^2) \tag{2.114} $$

and

$$ {}_{RL}\mathrm{D}^{\alpha}_{x,+\infty}u(x) = \frac{{}_{AC}\Delta^{\alpha}_{+h}u(x)}{h^{\alpha}} + O(h^2) $$

uniformly hold for $x \in \mathbb{R}$.

Proof. Here, we only prove (2.114). As $u(x,t)$ with respect to x belongs to $L_1(\mathbb{R})$, then the Fourier transform of the fractional average central difference operator (2.112) exists and has the following form

$$\begin{aligned}
&\mathcal{F}\left\{\frac{{}_{AC}\Delta^{\alpha}_{-h}u(x)}{h^{\alpha}};\omega\right\}\\
&=\frac{1}{2h^{\alpha}}\sum_{j=0}^{\infty}(-1)^j\binom{\alpha}{j}(\exp(-i\omega jh)+\exp(-i\omega(j-\alpha)h))\,\hat{u}(\omega)\\
&=\frac{1}{h^{\alpha}}\left(\sum_{j=0}^{\infty}(-1)^j\binom{\alpha}{j}\exp(-ij\omega h)\right)\left(\frac{1+\exp(i\omega\alpha h)}{2}\right)\hat{u}(\omega)\\
&=(i\omega)^{\alpha}\left(\frac{1-\exp(-i\omega h)}{i\omega h}\right)^{\alpha}\left(\frac{1+\exp(i\omega\alpha h)}{2}\right)\hat{u}(\omega).
\end{aligned} \tag{2.115}$$

Note that the function $\left(\frac{1-\exp(-i\omega h)}{i\omega h}\right)^{\alpha}\left(\frac{1+\exp(i\omega\alpha h)}{2}\right)$ has the following Taylor expansion:

$$\left(\frac{1-\exp(-i\omega h)}{i\omega h}\right)^{\alpha}\left(\frac{1+\exp(i\omega\alpha h)}{2}\right)=1+\frac{\alpha(3\alpha+1)}{24}(i\omega h)^2+O(|i\omega h|)^4. \tag{2.116}$$

If we denote

$$\hat{\phi}(\omega,h)=\mathcal{F}\left\{\frac{{}_{AC}\Delta^{\alpha}_{-h}u(x)}{h^{\alpha}};\omega\right\}-\mathcal{F}\left({}_{RL}D^{\alpha}_{-\infty,x}u(x)\right),$$

then from (2.115), (2.116) and Lemma 2.4.1, we have

$$\left|\hat{\phi}(\omega,h)\right|\le C_1h^2\left|(i\omega)^{\alpha+2}\,\hat{u}(\omega)\right|.$$

In light of the condition $\mathcal{F}\left({}_{RL}D^{\alpha+2}_{-\infty,x}u(x)\right)\in L_1(\mathbb{R})$, i.e.,

$$\int_{\mathbb{R}}\left|\mathcal{F}\left({}_{RL}D^{\alpha+2}_{-\infty,x}u(x)\right)\right|d\omega<C_2,$$

we obtain

$$\begin{aligned}
\left|\frac{{}_{AC}\Delta^{\alpha}_{-h}u(x)}{h^{\alpha}} - {}_{RL}D^{\alpha}_{-\infty,x}u(x)\right| &= |\phi(\omega,h)| \\
&= \frac{1}{2\pi}\left|\int_{\mathbb{R}} \exp(i\omega h)\hat{\phi}(\omega,h)d\omega\right| \\
&\leq \frac{1}{2\pi}\int_{\mathbb{R}} \left|\hat{\phi}(\omega,h)\right| d\omega \\
&\leq \frac{C_1}{2\pi}\left(\int_{\mathbb{R}} \left|(i\omega)^{\alpha+2}\,\hat{u}(\omega)\right| d\omega\right) h^2 \\
&= \frac{C_1}{2\pi}\left(\int_{\mathbb{R}} \left|\mathcal{F}\left({}_{RL}D^{\alpha+2}_{-\infty,x}u(x)\right)\right| d\omega\right) h^2 \\
&\leq Ch^2 \\
&= O(h^2).
\end{aligned}$$

where $C = \frac{C_1C_2}{2\pi}$. This finishes the proof. □

Next, we construct two classes of *fourth-order difference schemes for the left and right Riemann–Liouville derivatives* based on (2.112) and (2.113) through the following theorem:

Theorem 5 *Let* $u(x)$ *and the Fourier transforms of* ${}_{RL}\mathrm{D}^{\alpha+4}_{-\infty,x}u(x)$ *and* ${}_{RL}\mathrm{D}^{\alpha+4}_{x,+\infty}u(x)$ *both be in* $L_1(\mathbb{R})$*, then*

$${}_{RL}\mathrm{D}^{\alpha}_{-\infty,x}u(x) = \frac{1}{h^{\alpha}}\left(1 + \frac{\alpha(3\alpha+1)}{24}\delta_x^2\right)^{-1} {}_{AC}\Delta^{\alpha}_{-h}u(x) + O(h^4)$$

and

$${}_{RL}\mathrm{D}^{\alpha}_{x,+\infty}u(x) = \frac{1}{h^{\alpha}}\left(1 + \frac{\alpha(3\alpha+1)}{24}\delta_x^2\right)^{-1} {}_{AC}\Delta^{\alpha}_{+h}u(x) + O(h^4)$$

uniformly hold for $x \in \mathbb{R}$*, where* δ_x^2 *denotes second-order central difference operator and is defined by* $\delta_x^2 u(x_j) = u(x_{j+1}) - 2u(x_j) + u(x_{j-1})$.

Proof. The proof is almost the same as that of Theorem 4, so is omitted here. □

Combining (2.106), Theorems 4 and 5, we can get the following difference schemes for the Riesz derivative

$$\begin{aligned}
\frac{\partial^{\alpha}u(x)}{\partial|x|^{\alpha}} = -\frac{\Psi_{\alpha}}{2h^{\alpha}}&\left[\sum_{j=0}^{\infty}(-1)^j\binom{\alpha}{j}(u(x-jh)+u(x-(j-\alpha)h))\right. \\
&\left.+\sum_{j=0}^{\infty}(-1)^j\binom{\alpha}{j}(u(x+jh)+u(x+(j-\alpha)h))\right] + O(h^2)
\end{aligned} \tag{2.117}$$

and

$$\begin{aligned}\frac{\partial^\alpha u(x)}{\partial|x|^\alpha} = -\frac{\Psi_\alpha}{2h^\alpha}\Bigg[&\sum_{j=0}^{\infty}(-1)^j\binom{\alpha}{j}\left(1+\frac{\alpha(3\alpha+1)}{24}\delta_x^2\right)^{-1}(u(x-jh)+u(x-(j-\alpha)h))\\ &+\sum_{j=0}^{\infty}(-1)^j\binom{\alpha}{j}\left(1+\frac{\alpha(3\alpha+1)}{24}\delta_x^2\right)^{-1}(u(x+jh)+u(x+(j-\alpha)h))\Bigg]+O(h^4).\end{aligned} \tag{2.118}$$

Moreover, let

$$\tilde{u}(x) = \begin{cases} u(x), & x \in [a,b], \\ 0, & x \notin [a,b], \end{cases} \tag{2.119}$$

then formulas (2.117) and (2.118) change into

$$\begin{aligned}\frac{\partial^\alpha u(x)}{\partial|x|^\alpha} = -\frac{\Psi_\alpha}{2h^\alpha}\Bigg[&\sum_{j=0}^{\left[\frac{x-a}{h}\right]}(-1)^j\binom{\alpha}{j}(u(x-jh)+u(x-(j-\alpha)h))\\ &+\sum_{j=0}^{\left[\frac{b-x}{h}\right]}(-1)^j\binom{\alpha}{j}(u(x+jh)+u(x+(j-\alpha)h))\Bigg]+O(h^2),\end{aligned} \tag{2.120}$$

and

$$\begin{aligned}\frac{\partial^\alpha u(x)}{\partial|x|^\alpha} = -\frac{\Psi_\alpha}{2h^\alpha}\Bigg[&\sum_{j=0}^{\left[\frac{x-a}{h}\right]}(-1)^j\binom{\alpha}{j}\left(1+\frac{\alpha(3\alpha+1)}{24}\delta_x^2\right)^{-1}(u(x-jh)+u(x-(j-\alpha)h))\\ &+\sum_{j=0}^{\left[\frac{b-x}{h}\right]}(-1)^j\binom{\alpha}{j}\left(1+\frac{\alpha(3\alpha+1)}{24}\delta_x^2\right)^{-1}(u(x+jh)+u(x+(j-\alpha)h))\Bigg]+O(h^4).\end{aligned} \tag{2.121}$$

Finally, we derive another kind of *fourth-order numerical method for the Riesz derivative* which is presented in the following theorem.

Theorem 6 *Let $u(x)$ lie in $C^7(\mathbb{R})$ whose partial derivatives up to order seven belong to $L_1(\mathbb{R})$. Set*

$$\mathcal{L}_\theta u(x) = \sum_{k=-\infty}^{\infty} g_k^{(\alpha)} u(x-(k+\theta)h), \ \theta = -1, 0, 1,$$

in which

$$g_k^{(\alpha)} = \frac{(-1)^k\Gamma(\alpha+1)}{\Gamma\left(\frac{\alpha}{2}-k+1\right)\Gamma\left(\frac{\alpha}{2}+k+1\right)},$$

then we have

$$\frac{\partial^\alpha u(x)}{\partial|x|^\alpha} = \frac{1}{h^\alpha}\left[\frac{\alpha}{24}\mathcal{L}_{-1}u(x) - \left(1+\frac{\alpha}{12}\right)\mathcal{L}_0 u(x) + \frac{\alpha}{24}\mathcal{L}_1 u(x)\right] + O(h^4).$$

Proof. Here, we use the Fourier transform method to prove it. From [119], we know that the generating function with coefficients $g_k^{(\alpha)}$ satisfies

$$\left|2\sin\left(\frac{x}{2}\right)\right|^\alpha = \sum_{k=-\infty}^{\infty} g_k^{(\alpha)} \exp(ikx). \tag{2.122}$$

From (2.106) and Lemma 2.4.1, we get the Fourier transform of the Riesz derivative as follows

$$\begin{aligned}\mathcal{F}\left\{\frac{\partial^\alpha u(x)}{\partial|x|^\alpha};\omega\right\} &= -\Psi_\alpha\left[(i\omega)^\alpha + (-i\omega)^\alpha\right]\hat{u}(\omega)\\ &= -|\omega|^\alpha\hat{u}(\omega).\end{aligned} \tag{2.123}$$

Applying the Fourier transform to the difference operator

$$\frac{1}{h^\alpha}\left[\frac{\alpha}{24}\mathcal{L}_{-1}u(x) - \left(1+\frac{\alpha}{12}\right)\mathcal{L}_0 u(x) + \frac{\alpha}{24}\mathcal{L}_1 u(x)\right]$$

and using equation (2.122), gives

$$\begin{aligned}&\mathcal{F}\left\{\frac{1}{h^\alpha}\left[\frac{\alpha}{24}\mathcal{L}_{-1}u(x) - \left(1+\frac{\alpha}{12}\right)\mathcal{L}_0 u(x) + \frac{\alpha}{24}\mathcal{L}_1 u(x)\right];\omega\right\}\\ =\ &\frac{1}{h^\alpha}\left[\frac{\alpha}{24}\sum_{k=-\infty}^{\infty} g_k^{(\alpha)}\exp(-i(k-1)\omega h)\hat{u}(\omega)\right.\\ &-\left(1+\frac{\alpha}{12}\right)\sum_{k=-\infty}^{\infty} g_k^{(\alpha)}\exp(-ik\omega h)\hat{u}(\omega)\\ &\left.+\frac{\alpha}{24}\sum_{k=-\infty}^{\infty} g_k^{(\alpha)}\exp(-i(k+1)\omega h)\hat{u}(\omega)\right]\\ =\ &-\frac{1}{h^\alpha}\left[1+\frac{\alpha}{12}(1-\cos(\omega h))\right]\left|2\sin\left(\frac{\omega h}{2}\right)\right|^\alpha \hat{u}(\omega).\end{aligned}$$

Set

$$\begin{aligned}&|\omega|^\alpha\hat{u}(\omega) =\\ &\hat{C}(h,\omega) - \mathcal{F}\left\{\frac{1}{h^\alpha}\left[\frac{\alpha}{24}\mathcal{L}_{-1}u(x) - \left(1+\frac{\alpha}{12}\right)\mathcal{L}_0 u(x) + \frac{\alpha}{24}\mathcal{L}_1 u(x)\right];\omega\right\},\end{aligned} \tag{2.124}$$

then

$$\begin{aligned}\hat{C}(h,\omega) &= |\omega|^\alpha \left\{1 - \left[1 + \frac{\alpha}{12}(1-\cos(\omega h))\right]\left|\frac{2\sin\left(\frac{\omega h}{2}\right)}{\omega h}\right|^\alpha\right\}\hat{u}(\omega) \\ &= |\omega|^\alpha\left\{1 - \left[1 + \frac{\alpha}{24}(\omega h)^2 - \frac{\alpha}{288}(\omega h)^4 + O(\omega h)^6\right]\right. \\ &\quad \left.\cdot\left[1 - \frac{\alpha}{24}(\omega h)^2 + \alpha\left(\frac{1}{1920} + \frac{\alpha-1}{1152}\right)(\omega h)^4 + O(\omega h)^6\right]\right\}\hat{u}(\omega) \\ &= -|\omega|^\alpha\left\{\alpha\left(\frac{\alpha}{1152} + \frac{11}{2880}\right)(\omega h)^4 - O(\omega h)^6\right\}\hat{u}(\omega).\end{aligned} \tag{2.125}$$

Since $u(x) \in C^7(\mathbb{R})$ and its partial derivatives up to order seven belong to $L_1(\mathbb{R})$, there exists a positive constant $\tilde{C}_1$ such that

$$|\hat{u}(\omega)| \le \tilde{C}_1 (1 + |\omega|)^{-7}. \tag{2.126}$$

So, using (2.125) and (2.126) leads to

$$\begin{aligned}\left|\hat{C}(h,\omega)\right| \le \tilde{C}_2 h^4 |\omega|^{4+\alpha} |\hat{u}(\omega)| &\le \tilde{C}_2 h^4 (1 + |\omega|)^{4+\alpha} |\hat{u}(\omega)| \\ &\le \tilde{C}_3 h^4 (1 + |\omega|)^{\alpha-3},\end{aligned} \tag{2.127}$$

where $\tilde{C}_3 = \tilde{C}_1\tilde{C}_2$.

At this moment, taking the inverse Fourier transformation on both sides of (2.124) and noting (2.123) gives

$$\frac{\partial^\alpha u(x)}{\partial |x|^\alpha} = \frac{1}{h^\alpha}\left[\frac{\alpha}{24}\mathcal{L}_{-1}u(x) - \left(1 + \frac{\alpha}{12}\right)\mathcal{L}_0 u(x) + \frac{\alpha}{24}\mathcal{L}_1 u(x)\right] - C(h,\omega).$$

In view of (2.127), we have

$$\begin{aligned}|C(h,x)| &= \frac{1}{2\pi}\left|\int_{\mathbb{R}} \hat{C}(h,\omega)\exp(i\omega x)d\omega\right| \\ &\le \frac{1}{2\pi}\int_{\mathbb{R}} \left|\hat{C}(h,\omega)\right| d\omega \\ &\le \frac{\tilde{C}_3}{2\pi}\left(\int_{\mathbb{R}} (1 + |\omega|)^{\alpha-3} d\omega\right) h^4 \\ &= \tilde{C}h^4,\end{aligned}$$

where $\tilde{C} = \dfrac{\tilde{C}_3}{(2-\alpha)\pi}$, that is to say

$$\frac{\partial^\alpha u(x)}{\partial |x|^\alpha} = \frac{1}{h^\alpha}\left[\frac{\alpha}{24}\mathcal{L}_{-1}u(x) - \left(1 + \frac{\alpha}{12}\right)\mathcal{L}_0 u(x) + \frac{\alpha}{24}\mathcal{L}_1 u(x)\right] + O(h^4). \tag{2.128}$$

This finishes the proof. □

Furthermore, equation (2.128) can be rewritten as

$$\begin{aligned}\frac{\partial^\alpha u(x)}{\partial|x|^\alpha} &= \frac{\alpha}{24h^\alpha}\sum_{k=-\infty}^{\infty} g_k^{(\alpha)}u(x-(k+1)h)-\left(1+\frac{\alpha}{12}\right)\frac{1}{h^\alpha}\sum_{k=-\infty}^{\infty} g_k^{(\alpha)}u(x-kh)\\ &\quad+\frac{\alpha}{24h^\alpha}\sum_{k=-\infty}^{\infty} g_k^{(\alpha)}u(x-(k-1)h)+O(h^4).\end{aligned}$$

Combining (2.119) one can get

$$\begin{aligned}\frac{\partial^\alpha u(x_m)}{\partial|x|^\alpha} &= \frac{\alpha}{24h^\alpha}\sum_{k=-M+m+1}^{m-1} g_k^{(\alpha)}u(x_{m-(k+1)})\\ &\quad+\frac{\alpha}{24h^\alpha}\sum_{k=-M+m+1}^{m-1} g_k^{(\alpha)}u(x_{m-(k-1)})\\ &\quad-\left(1+\frac{\alpha}{12}\right)\frac{1}{h^\alpha}\sum_{k=-M+m+1}^{m-1} g_k^{(\alpha)}u(x_{m-k})+O(h^4).\end{aligned} \tag{2.129}$$

2.4.2 High-Order Algorithms (II)

In the above subsection, a new kind of second-order scheme and two classes of fourth-order schemes were established. In this subsection, we continue to construct much higher-order schemes. Next, we show how to build much *higher-order numerical schemes for the Riesz derivative*.

Define

$$\mathcal{H}_\theta u(x)=\sum_{k=-\infty}^{\infty} g_k^{(\alpha)}u(x-(k+\theta)h),\ \theta\in\chi=\{0,\pm1,\pm2,\pm3,\ldots\},$$

where

$$g_k^{(\alpha)}=\frac{(-1)^k\Gamma(\alpha+1)}{\Gamma\left(\frac{\alpha}{2}-k+1\right)\Gamma\left(\frac{\alpha}{2}+k+1\right)}.$$

Let

$$\frac{\partial^\alpha u(x)}{\partial|x|^\alpha}=\frac{1}{h^\alpha}\sum_{\theta\in\chi} Z_{\theta,p}\mathcal{H}_\theta u(x)+O(h^p),\quad p\geq 2,$$

where $Z_{\theta,p}$ are coefficients determined by the Fourier transform method. Obviously, in view of the above equation, we can obtain arbitrary order difference schemes by choosing various combination of θ values.

In this subsection, we give two different high-order difference schemes via the following theorem:

Theorem 7 *Suppose that $u(x)\in C^{11}(\mathbb{R})$ and all the derivatives of $u(x)$ up to order 11 belong to $L_1(\mathbb{R})$, then we have*

$$\frac{\partial^\alpha u(x)}{\partial|x|^\alpha}=\frac{D_s^\alpha u(x)}{h^\alpha}+O(h^6), \tag{2.130}$$

and

$$\frac{\partial^\alpha u(x)}{\partial |x|^\alpha} = \frac{D_e^\alpha u(x)}{h^\alpha} + O(h^8), \tag{2.131}$$

where

$$\begin{aligned} D_s^\alpha u(x) = \; & \mathcal{A}_1 \mathcal{H}_{-2} u(x) + \mathcal{A}_2 \mathcal{H}_{-1} u(x) + \mathcal{A}_3 \mathcal{H}_0 u(x) \\ & + \mathcal{A}_2 \mathcal{H}_1 u(x) + \mathcal{A}_1 \mathcal{H}_2 u(x), \\ D_e^\alpha u(x) = \; & \mathcal{B}_1 \mathcal{H}_{-3} u(x) + \mathcal{B}_2 \mathcal{H}_{-2} u(x) + \mathcal{B}_3 \mathcal{H}_{-1} u(x) + \mathcal{B}_4 \mathcal{H}_0 u(x) \\ & + \mathcal{B}_3 \mathcal{H}_1 u(x) + \mathcal{B}_2 \mathcal{H}_2 u(x) + \mathcal{B}_1 \mathcal{H}_3 u(x), \end{aligned}$$

$$\mathcal{A}_1 = -\left(\frac{\alpha}{1152} + \frac{11}{2880}\right)\alpha,\ \mathcal{A}_2 = \left(\frac{\alpha}{288} + \frac{41}{720}\right)\alpha,\ \mathcal{A}_3 = -\left(\frac{\alpha^2}{192} + \frac{17\alpha}{160} + 1\right),$$

$$\mathcal{B}_1 = \left(\frac{\alpha^2}{82944} + \frac{11\alpha}{69120} + \frac{191}{362880}\right)\alpha,\ \mathcal{B}_2 = -\left(\frac{\alpha^2}{13824} + \frac{7\alpha}{3840} + \frac{211}{30240}\right)\alpha,$$

$$\mathcal{B}_3 = \left(\frac{5\alpha^2}{27648} + \frac{3\alpha}{512} + \frac{7843}{120960}\right)\alpha,\ \mathcal{B}_4 = -\left(\frac{5\alpha^3}{20736} + \frac{29\alpha^2}{3456} + \frac{5297\alpha}{45360} + 1\right).$$

Proof. Applying the Fourier transform to the difference operator $\frac{D_s^\alpha u(x,t)}{h^\alpha}$ with respect to x yields

$$\begin{aligned} & \mathcal{F}\left\{\frac{D_s^\alpha u(x)}{h^\alpha}; \omega\right\} \\ = & \frac{1}{h^\alpha}\left[\mathcal{A}_1 \sum_{k=-\infty}^{\infty} g_k^{(\alpha)} e^{-i(k-2)\omega h} + \mathcal{A}_2 \sum_{k=-\infty}^{\infty} g_k^{(\alpha)} e^{-i(k-1)\omega h}\right. \\ & + \mathcal{A}_3 \sum_{k=-\infty}^{\infty} g_k^{(\alpha)} e^{-ik\omega h} + \mathcal{A}_2 \sum_{k=-\infty}^{\infty} g_k^{(\alpha)} e^{-i(k+1)\omega h} \\ & \left. + \mathcal{A}_1 \sum_{k=-\infty}^{\infty} g_k^{(\alpha)} e^{-i(k+2)\omega h}\right] \widehat{u}(\omega) \\ = & \frac{1}{h^\alpha}\left[2\mathcal{A}_1 \cos(2\omega h) + 2\mathcal{A}_2 \cos(\omega h) + \mathcal{A}_3\right]\left(\sum_{k=-\infty}^{\infty} g_k^{(\alpha)} e^{ik\omega h}\right)\widehat{u}(\omega). \end{aligned} \tag{2.132}$$

Note that the generating function of the coefficients $g_k^{(\alpha)}$ is $\left|2\sin\left(\frac{\omega h}{2}\right)\right|^\alpha$, that is to say [119]

$$\left|2\sin\left(\frac{\omega h}{2}\right)\right|^\alpha = \sum_{k=-\infty}^{\infty} g_k^{(\alpha)} e^{ik\omega h}.$$

According to Euler's formula $\exp(ix) = \cos(x) + i\sin(x)$, we easily obtain

$$\begin{aligned} |\xi|^{\alpha} &= \frac{1}{2\cos\left(\frac{\pi\alpha}{2}\right)}\left[|\xi|^{\alpha}\exp\left(i\frac{\pi\alpha}{2}sign(\xi)\right) + |\xi|^{\alpha}\exp\left(-i\frac{\pi\alpha}{2}sign(\xi)\right)\right] \\ &= \frac{1}{2\cos\left(\frac{\pi\alpha}{2}\right)}\left[(i\xi)^{\alpha} + (-i\xi)^{\alpha}\right]. \end{aligned}$$

Let $\xi = \frac{2\sin\left(\frac{\omega h}{2}\right)}{\omega h}$, $\omega \in \mathbb{R}$, then the above equation becomes

$$\begin{aligned} \left|\frac{2\sin\left(\frac{\omega h}{2}\right)}{\omega h}\right|^{\alpha} &= \frac{1}{2\cos\left(\frac{\pi\alpha}{2}\right)}\left[\left(\frac{2\sin\left(\frac{\omega h}{2}\right)}{\omega h}i\right)^{\alpha} + \left(-\frac{2\sin\left(\frac{\omega h}{2}\right)}{\omega h}i\right)^{\alpha}\right] \\ &= \frac{i^{\alpha} + (-i)^{\alpha}}{2\cos\left(\frac{\pi\alpha}{2}\right)}\left[1 - \frac{\alpha}{24}(\omega h)^2 + \left(\frac{1}{1920} + \frac{\alpha - 1}{1152}\right)\alpha(\omega h)^4\right. \\ &\quad \left. - \left(\frac{1}{322560} + \frac{\alpha - 1}{46080} + \frac{(\alpha - 1)(\alpha - 2)}{82944}\right)\alpha(\omega h)^6 + O(\omega h)^8\right] \\ &= \left[1 - \frac{\alpha}{24}(\omega h)^2 + \left(\frac{1}{1920} + \frac{\alpha - 1}{1152}\right)\alpha(\omega h)^4\right. \\ &\quad \left. - \left(\frac{1}{322560} + \frac{\alpha - 1}{46080} + \frac{(\alpha - 1)(\alpha - 2)}{82944}\right)\alpha(\omega h)^6 + O(\omega h)^8\right]. \end{aligned}$$

At this moment, one can rewrite (2.132) as

$$\begin{aligned} \mathcal{F}\left\{\frac{D_s^{\alpha}u(x)}{h^{\alpha}};\omega\right\} &= \frac{1}{h^{\alpha}}\left[2\mathcal{A}_1\cos(2\omega h) + 2\mathcal{A}_2\cos(\omega h) + \mathcal{A}_3\right]\left|2\sin\left(\frac{\omega h}{2}\right)\right|^{\alpha}\widehat{u}(\omega) \\ &= |\omega|^{\alpha}\left[2\mathcal{A}_1\cos(2\omega h) + 2\mathcal{A}_2\cos(\omega h) + \mathcal{A}_3\right]\left|\frac{2\sin\left(\frac{\omega h}{2}\right)}{\omega h}\right|^{\alpha}\widehat{u}(\omega) \\ &= |\omega|^{\alpha}\left(-1 + O(\omega h)^6\right)\widehat{u}(\omega). \end{aligned}$$

From [49], one has

$$\mathcal{F}\left\{\frac{\partial^{\alpha}u(x)}{\partial|x|^{\alpha}};\omega\right\} = -|\omega|^{\alpha}\widehat{u}(\omega).$$

Let

$$\widehat{\delta}(\omega, h) = \mathcal{F}\left\{\frac{D_s^{\alpha}u(x)}{h^{\alpha}};\omega\right\} - \mathcal{F}\left\{\frac{\partial^{\alpha}u(x)}{\partial|x|^{\alpha}};\omega\right\}. \tag{2.133}$$

Since $u(x) \in C^{11}(\mathbb{R})$ and its partial derivatives up to order eleven belong to $L_1(\mathbb{R})$, there exists a positive constant $\widehat{C}_1$ such that

$$\left|\widehat{u}(\omega)\right| \le \widehat{C_1}\,(1+|\omega|)^{-11}. \tag{2.134}$$

So, using (2.132) and (2.133) leads to

$$\begin{aligned}\left|\widehat{\delta}(\omega,h)\right| \le \widehat{C_2}h^6\,|\omega|^{6+\alpha}\left|\widehat{u}(\omega)\right| &\le \widehat{C_2}h^6\,(1+|\omega|)^{6+\alpha}\left|\widehat{u}(\omega)\right| \\ &\le \widehat{C_3}h^6\,(1+|\omega|)^{\alpha-5},\end{aligned} \tag{2.135}$$

where $\widehat{C_3} = \widehat{C_1}\widehat{C_2}$.

Furthermore, taking the inverse Fourier transform in both sides of (2.133) and combining with (2.135) give

$$\begin{aligned}\left|\frac{\partial^\alpha u(x)}{\partial|x|^\alpha} - \frac{D_s^\alpha u(x)}{h^\alpha}\right| &= |\delta(\omega,h)| = \frac{1}{2\pi}\left|\int_{\mathbb{R}} \widehat{\delta}(\omega,h)e^{i\omega h}d\omega\right| \\ &\le \frac{1}{2\pi}\int_{\mathbb{R}}\left|\widehat{\delta}(\omega,h)\right|d\omega \le \frac{\widehat{C_3}}{2\pi}\left(\int_{\mathbb{R}}(1+|\omega|)^{\alpha-5}\,d\omega\right)h^6 \\ &= \widehat{C}h^6,\end{aligned}$$

i.e.,

$$\frac{\partial^\alpha u(x)}{\partial|x|^\alpha} = \frac{D_s^\alpha u(x)}{h^\alpha} + O(h^6),$$

where $\widehat{C} = \dfrac{\widehat{C_3}}{(4-\alpha)\pi}$.

Let

$$\bar{\delta}(\omega,h) = \mathcal{F}\left\{\frac{D_e^\alpha u(x)}{h^\alpha};\omega\right\} - \mathcal{F}\left\{\frac{\partial^\alpha u(x)}{\partial|x|^\alpha};\omega\right\}.$$

Similarly, we can obtain

$$\begin{aligned}\bar{\delta}(\omega,h) &= |\omega|^\alpha\Big\{\Big[2(\mathcal{B}_1+\mathcal{B}_2+\mathcal{B}_3)+\mathcal{B}_4-(9\mathcal{B}_1+4\mathcal{B}_2+\mathcal{B}_3)(\omega h)^2 \\ &\quad+\left(\frac{27}{4}\mathcal{B}_1+\frac{4}{3}\mathcal{B}_2+\frac{1}{12}\mathcal{B}_3\right)(\omega h)^4-\left(\frac{81}{40}\mathcal{B}_1+\frac{8}{45}\mathcal{B}_2+\frac{1}{360}\mathcal{B}_3\right)(\omega h)^6 \\ &\quad+O(\omega h)^8\Big]\left|\frac{2\sin\left(\frac{\omega h}{2}\right)}{\omega h}\right|^\alpha+1\Big\}\bar{u}(\omega) \\ &= \left(|\omega|^{\alpha+8}\bar{u}(\omega)\right)O(h^8).\end{aligned}$$

It immediately follows that

$$\frac{\partial^\alpha u(x)}{\partial|x|^\alpha} = \frac{D_e^\alpha u(x)}{h^\alpha} + O(h^8).$$

Thus all this finishes the proof. □

If u^* is defined by

$$u^*(x) = \begin{cases} u(x), & x \in [a,b], \\ 0, & x \notin [a,b]. \end{cases}$$

such that $u^* \in C^{11}(\mathbb{R})$, and all derivatives up to order 11 belong to $L_1(\mathbb{R})$, then (2.130) and (2.131) at point (x_m, t_n) can be rewritten as

$$\begin{aligned} &\frac{\partial^\alpha u(x_m)}{\partial |x|^\alpha} = \\ &\frac{1}{h^\alpha}\Bigg\{\mathcal{A}_1 \sum_{k=-M+m+2}^{m-2} g_k^{(\alpha)} u(x_{m-(k-2)}) + \mathcal{A}_2 \sum_{k=-M+m+2}^{m-2} g_k^{(\alpha)} u(x_{m-(k-1)}) \\ &+ \mathcal{A}_3 \sum_{k=-M+m+2}^{m-2} g_k^{(\alpha)} u(x_{m-k}) + \mathcal{A}_2 \sum_{k=-M+m+2}^{m-2} g_k^{(\alpha)} u(x_{m-(k+1)}) \\ &+ \mathcal{A}_1 \sum_{k=-M+m+2}^{m-2} g_k^{(\alpha)} u(x_{m-(k+2)})\Bigg\} + O(h^6), \end{aligned} \tag{2.136}$$

and

$$\begin{aligned} &\frac{\partial^\alpha u(x_m)}{\partial |x|^\alpha} = \\ &\frac{1}{h^\alpha}\Bigg\{\mathcal{B}_1 \sum_{k=-M+m+3}^{m-3} g_k^{(\alpha)} u(x_{m-(k-3)}) + \mathcal{B}_2 \sum_{k=-M+m+3}^{m-3} g_k^{(\alpha)} u(x_{m-(k-2)}) \\ &+ \mathcal{B}_3 \sum_{k=-M+m+3}^{m-3} g_k^{(\alpha)} u(x_{m-(k-1)}) + \mathcal{B}_4 \sum_{k=-M+m+3}^{m-3} g_k^{(\alpha)} u(x_{m-k}, t_n) \\ &+ \mathcal{B}_3 \sum_{k=-M+m+3}^{m-3} g_k^{(\alpha)} u(x_{m-(k+1)}) + \mathcal{B}_2 \sum_{k=-M+m+3}^{m-3} g_k^{(\alpha)} u(x_{m-(k+2)}) \\ &+ \mathcal{B}_1 \sum_{k=-M+m+3}^{m-3} g_k^{(\alpha)} u(x_{m-(k+3)})\Bigg\} + O(h^8), \end{aligned} \tag{2.137}$$

respectively.

Remark 2.4.1 *In fact, we can use almost the same method to construct much higher-order difference schemes for the Riesz derivative, such as, 10th-order, 12th-order schemes, …, and so on; for more details see [39].*

2.4.3 High-Order Algorithms (III)

Although higher-order schemes can be constructed as above, we still think that it is necessary to reconsider the pth order schemes ($p = 2, 3, \cdots, 6$) with new methods due to the fact that such cases are mostly noticed. In [104], the asymptotical properties of the coefficients of the higher-order schemes for the Riemann–Liouville integrals (also derivatives) were given. In this subsection, we not only explicitly express these coefficients of the higher-order methods for Riesz derivatives, but also study their monotonicity.

If $f^{(k)}(a+) = 0$ $(k = 0, 1, \ldots, p-1)$, then it follows from [104] that the left Riemann–Liouville derivative has the following approximations

$$ {}_{RL}\mathrm{D}_{a,x}^{\alpha} f(x) = \frac{1}{h^{\alpha}} \sum_{\ell=0}^{\infty} \varpi_{p,\ell}^{(\alpha)} f(x - \ell h) + O(h^{p}), \tag{2.138} $$

in which h is the stepsize. Here we only show interests in $p = 2, 3, 4, 5, 6$.

The convolution (or weight) coefficients $\varpi_{p,\ell}^{(\alpha)}$ in the above equations are those of the Taylor series expansions of the corresponding *generating functions* $W_p^{(\alpha)}(z)$,

$$ W_p^{(\alpha)}(z) = \sum_{\ell=0}^{\infty} \varpi_{p,\ell}^{(\alpha)} z^{\ell}, \ \alpha \in (0, 2), $$

where

$$ \begin{aligned} W_2^{(\alpha)}(z) &= \left(\frac{3}{2} - 2z + \frac{1}{2} z^2 \right)^{\alpha}, \\ W_3^{(\alpha)}(z) &= \left(\frac{11}{6} - 3z + \frac{3}{2} z^2 - \frac{1}{3} z^3 \right)^{\alpha}, \\ W_4^{(\alpha)}(z) &= \left(\frac{25}{12} - 4z + 3z^2 - \frac{4}{3} z^3 + \frac{1}{4} z^4 \right)^{\alpha}, \\ W_5^{(\alpha)}(z) &= \left(\frac{137}{60} - 5z + 5z^2 - \frac{10}{3} z^3 + \frac{5}{4} z^4 - \frac{1}{5} z^5 \right)^{\alpha}, \\ W_6^{(\alpha)}(z) &= \left(\frac{147}{60} - 6z + \frac{15}{2} z^2 - \frac{20}{3} z^3 + \frac{15}{4} z^4 - \frac{6}{5} z^5 + \frac{1}{6} z^6 \right)^{\alpha}. \end{aligned} $$

By tedious but direct calculations, one has

$$ \varpi_{2,\ell}^{(\alpha)} = \left(\frac{3}{2} \right)^{\alpha} \sum_{\ell_1=0}^{\ell} \left(\frac{1}{3} \right)^{\ell_1} \varpi_{1,\ell_1}^{(\alpha)} \varpi_{1,\ell-\ell_1}^{(\alpha)}, $$

$$ \varpi_{3,\ell}^{(\alpha)} = \left(\frac{11}{6} \right)^{\alpha} \sum_{\ell_1=0}^{\ell} \sum_{\ell_2=0}^{\left[\frac{1}{2}\ell_1\right]} (-1)^{\ell_2} \left(\frac{7}{11} \right)^{\ell_1-\ell_2} \left(\frac{2}{7} \right)^{\ell_2} \frac{(\ell_1-\ell_2)!}{\ell_2!(\ell_1-2\ell_2)!} \varpi_{1,\ell-\ell_1}^{(\alpha)} \varpi_{1,\ell_1-\ell_2}^{(\alpha)}, $$

$$ \begin{aligned} \varpi_{4,\ell}^{(\alpha)} = {} & \left(\frac{25}{12} \right)^{\alpha} \sum_{\ell_1=0}^{\ell} \sum_{\ell_2=0}^{\left[\frac{2}{3}\ell_1\right]} \sum_{\ell_3=\max\{0, 2\ell_2-\ell_1\}}^{\left[\frac{1}{2}\ell_2\right]} (-1)^{\ell_2} \left(\frac{23}{25} \right)^{\ell_1-\ell_2} \left(\frac{13}{23} \right)^{\ell_2-\ell_3} \left(\frac{3}{13} \right)^{\ell_3} \times \\ & \frac{(\ell_1-\ell_2)!}{\ell_3!(\ell_2-2\ell_3)!(\ell_1+\ell_3-2\ell_2)!} \varpi_{1,\ell-\ell_1}^{(\alpha)} \varpi_{1,\ell_1-\ell_2}^{(\alpha)}, \end{aligned} $$

$$\varpi_{5,\ell}^{(\alpha)}=\left(\frac{137}{60}\right)^{\alpha}\sum_{\ell_1=0}^{\ell}\sum_{\ell_2=0}^{[\frac{3}{4}\ell_1]}\sum_{\ell_3=\max\{0,2\ell_2-\ell_1\}}^{[\frac{2}{3}\ell_2]}\sum_{\ell_4=\max\{0,2\ell_3-\ell_2\}}^{[\frac{1}{2}\ell_3]}(-1)^{\ell_2}\left(\frac{163}{137}\right)^{\ell_1-\ell_2}\left(\frac{137}{163}\right)^{\ell_2-\ell_3}$$
$$\times\left(\frac{63}{137}\right)^{\ell_3-\ell_4}\left(\frac{4}{21}\right)^{\ell_4}\frac{(\ell_1-\ell_2)!}{\ell_4!(\ell_3-2\ell_4)!(\ell_1+\ell_3-2\ell_2)!(\ell_2+\ell_4-2\ell_3)!}\varpi_{1,\ell-\ell_1}^{(\alpha)}\varpi_{1,\ell_1-\ell_2}^{(\alpha)},$$

and

$$\varpi_{6,\ell}^{(\alpha)}=\left(\frac{147}{60}\right)^{\alpha}\sum_{\ell_1=0}^{\ell}\sum_{\ell_2=0}^{[\frac{4}{5}\ell_1]}\sum_{\ell_3=\max\{0,2\ell_2-\ell_1\}}^{[\frac{3}{4}\ell_2]}\sum_{\ell_4=\max\{0,2\ell_3-\ell_2\}}^{[\frac{2}{3}\ell_3]}\sum_{\ell_5=\max\{0,2\ell_4-\ell_3\}}^{[\frac{1}{2}\ell_4]}$$
$$(-1)^{\ell_2}\left(\frac{213}{147}\right)^{\ell_1-\ell_2}\left(\frac{237}{213}\right)^{\ell_2-\ell_3}\left(\frac{163}{237}\right)^{\ell_3-\ell_4}\left(\frac{62}{163}\right)^{\ell_4-\ell_5}\left(\frac{5}{31}\right)^{\ell_5}\times$$
$$\frac{(\ell_1-\ell_2)!}{\ell_5!(\ell_4-2\ell_5)!(\ell_1+\ell_3-2\ell_2)!(\ell_2+\ell_4-2\ell_3)!(\ell_3+\ell_5-2\ell_4)!}\varpi_{1,\ell-\ell_1}^{(\alpha)}\varpi_{1,\ell_1-\ell_2}^{(\alpha)},$$
$$\ell=0,1,\ldots.$$

Here $\varpi_{1,j}^{(\alpha)}$ is the *first order coefficients* defined by $\varpi_{1,j}^{(\alpha)}=(-1)^j\frac{\Gamma(1+\alpha)}{\Gamma(j+1)\Gamma(1+\alpha-j)}$, $j=0,1,\cdots$. If $j\geq 2$, then $\varpi_{1,j-1}^{(\alpha)}\leq\varpi_{1,j}^{(\alpha)}$ for $\alpha\in(0,1)$ whilst $\varpi_{1,j-1}^{(\alpha)}\geq\varpi_{1,j}^{(\alpha)}$ for $\alpha\in(1,2)$. See [78] for more details.

On the other hand, if $f^{(k)}(b-)=0$ $(k=0,1,\ldots,p-1)$, then one has the approximations below,

$$_{RL}D_{x,b}^{\alpha}f(x)=\frac{1}{h^{\alpha}}\sum_{\ell=0}^{\infty}\varpi_{p,\ell}^{(\alpha)}f(x+\ell h)+O(h^p),\ \ p=2,\cdots,6, \tag{2.139}$$

where h is also the stepsize.

Based on (2.138) and (2.139), if $f(x)$, together with its derivatives, has homogeneous boundary value conditions, one easily gets

$$\frac{\partial^{\alpha}f(x)}{\partial|x|^{\alpha}}=-\frac{1}{2\cos(\pi\alpha/2)h^{\alpha}}\sum_{\ell=0}^{\infty}\varpi_{p,\ell}^{(\alpha)}(f(x-\ell h)+f(x+\ell h))+O(h^p). \tag{2.140}$$

Since $\alpha\in(0,2)$ is commonly used, we limit our interests in $\alpha\in(0,1)$ and $\alpha\in(1,2)$. When $\alpha=1$, we often set $\frac{\partial^{\alpha}f(x)}{\partial|x|^{\alpha}}=f'(x)$ which is the trivial case, so is omitted here.

The *second-order coefficients* have interesting properties some of which (for $\alpha\in(0,1)$) have been studied in [78]. Here we have the further results.

Theorem 8 *The second-order coefficients* $\varpi_{2,\ell}^{(\alpha)}$ $(\ell=0,1,\ldots)$ *satisfy*

$$\begin{aligned}(1)\quad &\varpi_{2,0}^{(\alpha)}=\left(\frac{3}{2}\right)^{\alpha}>0,\ \varpi_{2,1}^{(\alpha)}=-\frac{4\alpha}{3}\left(\frac{3}{2}\right)^{\alpha}<0,\\ &\varpi_{2,2}^{(\alpha)}=\frac{\alpha(8\alpha-3)}{9}\left(\frac{3}{2}\right)^{\alpha}>0,\ \varpi_{2,3}^{(\alpha)}=-\frac{4\alpha(\alpha-1)(8\alpha-7)}{81}\left(\frac{3}{2}\right)^{\alpha}<0,\\ &\varpi_{2,4}^{(\alpha)}=\frac{\alpha(\alpha-1)\left(64\alpha^2-176\alpha+123\right)}{486}\left(\frac{3}{2}\right)^{\alpha},\\ &\qquad\vdots\end{aligned}$$

(2) *When* $0<\alpha<1$, $\varpi_{2,\ell}^{(\alpha)}<0$ *and* $\varpi_{2,\ell}^{(\alpha)}<\varpi_{2,\ell+1}^{(\alpha)}$ *for* $\ell\geq 4$,

(3) *When* $1<\alpha<2$, $\varpi_{2,\ell}^{(\alpha)}>0$ *and* $\varpi_{2,\ell}^{(\alpha)}>\varpi_{2,\ell+1}^{(\alpha)}$ *for* $\ell\geq 5$.

Proof. (1) Direct calculations can finish it, so we omit the proof details.

(2) See [78] for details.

(3) Now we show the case $\alpha\in(1,2)$. We firstly show that $\varpi_{2,j}^{(\alpha)}>0$ for $\ell\geq 5$. For convenience, denote $\alpha=1+\gamma$, where $0<\gamma<1$. Lengthy calculations give

$$\begin{aligned}\varpi_{2,\ell}^{(\alpha)} &= \left(\frac{3}{2}\right)^{1+\gamma}\sum_{m=0}^{\ell}\left(\frac{1}{3}\right)^{m}\varpi_{1,m}^{(1+\gamma)}\varpi_{1,\ell-m}^{(1+\gamma)}\\ &= \left(\frac{3}{2}\right)^{1+\gamma}\Bigg[\varpi_{1,0}^{(1+\gamma)}\varpi_{1,\ell}^{(1+\gamma)}+\frac{1}{3}\varpi_{1,1}^{(1+\gamma)}\varpi_{1,\ell-1}^{(1+\gamma)}+\frac{1}{9}\varpi_{1,2}^{(1+\alpha)}\varpi_{1,\ell-2}^{(1+\gamma)}\\ &\quad+\left(\frac{1}{3}\right)^{\ell}\varpi_{1,0}^{(1+\gamma)}\varpi_{1,\ell}^{(1+\gamma)}+\left(\frac{1}{3}\right)^{\ell-1}\varpi_{1,1}^{(1+\gamma)}\varpi_{1,\ell-1}^{(1+\gamma)}\Bigg]\\ &\quad+\left(\frac{3}{2}\right)^{1+\gamma}\sum_{m=3}^{\ell-2}\left(\frac{1}{3}\right)^{m}\varpi_{1,m}^{(1+\gamma)}\varpi_{1,\ell-m}^{(1+\gamma)}\\ &= \left(\frac{3}{2}\right)^{1+\gamma}\Bigg[1-\frac{\gamma+1}{3}\frac{\ell}{(\ell-2-\gamma)}+\frac{\gamma(\gamma+1)}{18}\frac{\ell(\ell-1)}{(\ell-2-\gamma)(\ell-3-\gamma)}\\ &\quad+\left(\frac{1}{3}\right)^{\ell}\left(1-\frac{3\ell(\gamma+1)}{(\ell-2-\gamma)}\right)\Bigg]\varpi_{1,\ell}^{(1+\gamma)}+\left(\frac{3}{2}\right)^{1+\gamma}\sum_{m=3}^{\ell-2}\left(\frac{1}{3}\right)^{m}\varpi_{1,m}^{(1+\gamma)}\varpi_{1,\ell-m}^{(1+\gamma)}\\ &\geq \left(\frac{3}{2}\right)^{1-\alpha}\left[\frac{244}{243}-\frac{28(\gamma+1)}{81}\frac{\ell}{(\ell-2-\gamma)}+\frac{\gamma(\gamma+1)}{18}\frac{\ell(\ell-1)}{(\ell-2-\gamma)^2}\right]\varpi_{1,\ell}^{(1+\gamma)}\\ &\quad+\left(\frac{3}{2}\right)^{1+\gamma}\sum_{m=3}^{\ell-2}\left(\frac{1}{3}\right)^{m}\varpi_{1,m}^{(1+\gamma)}\varpi_{1,\ell-m}^{(1+\gamma)},\quad \ell\geq 5.\end{aligned}$$

Let

$$F(x,\gamma)=\frac{244}{243}-\frac{28(\gamma+1)}{81}\frac{x}{(x-2-\gamma)}+\frac{\gamma(\gamma+1)}{18}\frac{x(x-1)}{(x-2-\gamma)^2},\ x\geq 5,$$

and

$$\begin{aligned}G(x,\gamma) &= 486(x-2-\gamma)^2F(x,\gamma)\\ &= 488(x-2-\gamma)^2-168(\gamma+1)x(x-2-\gamma)+27\gamma(\gamma+1)x(x-1),\ x\geq 5.\end{aligned}$$

Then

$$G_x(x,\gamma) = 976(x-2-\gamma) - 168(\gamma+1)(2x-2-\gamma) + 27\gamma(\gamma+1)(2x-1)$$

and

$$G_{xx}(x,\gamma) = 54\gamma^2 - 282\gamma + 640.$$

Obviously,

$$G_{xx}(x,\gamma) \geq 0 \ \text{ for } \ 0 < \gamma < 1,$$

it immediately follows that $G_x(x,\gamma)$ is an increasing function and $G_x(x,\gamma) \geq G_x(5,\gamma)$ for $x \geq 5$.

Note that

$$G_x(5,\gamma) = 411\gamma^2 - 1909\gamma + 1585 > 0, \ \ 0 < \gamma < 1.$$

Hence $G(x,\gamma)$ is an increasing function too, and $G(x,\gamma) \geq G(5,\gamma)$ if $x \geq 5$. Simple calculations yields

$$G(5,\gamma) = 1868\gamma^2 - 3068\gamma + 1872 \geq G_{\min}(5,\gamma) = g\left(5, \frac{767}{934}\right) = 612\frac{131}{467},$$

so, $G(x,\gamma) \geq 0$. Therefore, the following inequality holds

$$F(x,\gamma) = \frac{G(x,\gamma)}{486(x-2-\gamma)^2} \geq 0,$$

which means $\varpi_{2,j}^{(1+\gamma)} > 0$ for $\ell \geq 5$.

Next, we show that $\varpi_{2,\ell}^{(\alpha)} > \varpi_{2,\ell+1}^{(\alpha)}$ for $\ell \geq 5$. Note that

$$\begin{aligned}
\varpi_{2,\ell}^{(\alpha)} - \varpi_{2,\ell+1}^{(\alpha)} &= (2+\gamma)\left(\frac{3}{2}\right)^{1+\gamma}\left(\frac{1}{3}\right)^{\ell}\sum_{\ell_1=0}^{\ell}\frac{3^{\ell_1}}{\ell_1+1}\varpi_{1,\ell_1}^{(1+\gamma)}\varpi_{1,\ell-\ell_1}^{(1+\gamma)} \\
&\quad - \left(\frac{3}{2}\right)^{1+\gamma}\left(\frac{1}{3}\right)^{\ell+1}\left(1-\frac{2+\gamma}{\ell+1}\right)\varpi_{1,\ell}^{(1+\gamma)} \\
&= \left(\frac{3}{2}\right)^{1+\gamma}\left(\frac{1}{3}\right)^{\ell}\left[(2+\gamma)\sum_{\ell_1=0}^{\ell}\frac{3^{\ell_1}}{\ell_1+1}\varpi_{1,\ell_1}^{(1+\gamma)}\varpi_{1,\ell-\ell_1}^{(1+\gamma)} - \frac{1}{3}\varpi_{1,\ell}^{(1+\gamma)}\right] \\
&\quad + \frac{2+\gamma}{\ell+1}\left(\frac{3}{2}\right)^{1+\gamma}\left(\frac{1}{3}\right)^{\ell+1}\varpi_{1,\ell}^{(1+\gamma)} \\
&\geq (2+\gamma)\left(\frac{3}{2}\right)^{1+\gamma}\left(\frac{1}{3}\right)^{\ell}\left[\sum_{\ell_1=0}^{\ell}\frac{3^{\ell_1}}{\ell_1+1}\varpi_{1,\ell_1}^{(1+\gamma)}\varpi_{1,\ell-\ell_1}^{(1+\gamma)} - \frac{1}{6}\varpi_{1,\ell}^{(1+\gamma)}\right] \\
&\quad + \frac{2+\gamma}{\ell+1}\left(\frac{3}{2}\right)^{1+\gamma}\left(\frac{1}{3}\right)^{\ell+1}\varpi_{1,\ell}^{(1+\gamma)} \\
&= (2+\gamma)\left(\frac{3}{2}\right)^{1+\gamma}\left(\frac{1}{3}\right)^{\ell}P(\ell,\gamma)\varpi_{1,\ell}^{(1+\gamma)} + \frac{2+\gamma}{\ell+1}\left(\frac{3}{2}\right)^{1+\gamma}\left(\frac{1}{3}\right)^{\ell+1}\varpi_{1,\ell}^{(1+\gamma)} \\
&\quad + (2+\gamma)\left(\frac{3}{2}\right)^{1+\gamma}\left(\frac{1}{3}\right)^{\ell}\sum_{\ell_1=3}^{\ell-3}\frac{3^{\ell_1}}{\ell_1+1}\varpi_{1,\ell_1}^{(1+\gamma)}\varpi_{1,\ell-\ell_1}^{(1+\gamma)}.
\end{aligned}$$

Here

$$\begin{aligned} P(\ell,\gamma) &= \left[\frac{5}{6}-\frac{3\ell(\gamma+1)}{2(\ell-2-\gamma)}+\frac{3\gamma(\gamma+1)\ell(\ell-1)}{2(\ell-2-\gamma)(\ell-3-\gamma)}\right] \\ &\quad +\frac{3^\ell}{\ell+1}\left[1-\frac{(\ell+1)(\gamma+1)}{3(\ell-2-\gamma)}+\frac{\gamma(\gamma+1)\ell(\ell+1)}{18(\ell-2-\gamma)(\ell-3-\gamma)}\right]. \end{aligned}$$

Obviously, the last two terms in the right-hand side of the last equality are both nonnegative, so we only need to prove that the factor $P(\ell,\gamma)$ in the first term is positive.

Let

$$P_1(\ell,\gamma)=\frac{5}{6}-\frac{3\ell(\gamma+1)}{2(\ell-2-\gamma)}+\frac{3\gamma(\gamma+1)\ell(\ell-1)}{2(\ell-2-\gamma)(\ell-3-\gamma)},$$

$$P_2(\ell,\gamma)=1-\frac{(\ell+1)(\gamma+1)}{3(\ell-2-\gamma)}+\frac{\gamma(\gamma+1)\ell(\ell+1)}{18(\ell-2-\gamma)(\ell-3-\gamma)},$$

then

$$P(\ell,\gamma)=P_1(\ell,\gamma)+\frac{3^\ell}{\ell+1}P_2(\ell,\gamma).$$

If $\ell=5$, then

$$P_2(5,\gamma)=1-\frac{2(\gamma+1)}{(3-\gamma)}+\frac{5\gamma(\gamma+1)}{3(3-\gamma)(2-\gamma)}>0.$$

Now we consider the case $\ell\geq 6$. Let

$$Q(x,\gamma)=18(x-2-\gamma)(x-2.5-\gamma)P_3(x,\gamma),\ \ x\in[6,\infty),$$

where

$$P_3(x,\gamma)=1-\frac{(x+1)(\gamma+1)}{3(x-2-\gamma)}+\frac{\gamma(\gamma+1)x(x+1)}{18(x-2-\gamma)(x-2.5-\gamma)}.$$

Then

$$Q_{xx}(x,\gamma)=2\gamma^2-10\gamma+24>0,\ \ 0<\gamma<1.$$

So $Q_x(x,\gamma)$ is an increasing function and

$$Q_x(x,\gamma)\geq Q_x(6,\gamma)=19\gamma^2-80\gamma+72>0,\ x\geq 6,\ 0<\gamma<1.$$

It immediately follows that $Q(x,\gamma)$ is an increasing function with respect to x and

$$Q(x,\gamma)\geq Q(6,\gamma)=102\gamma^2-198\gamma+105>0,\ \ 0<\gamma<1,$$

i.e., $P_3(x,\gamma)>0$. Noticing $P_2(\ell,\gamma)>P_3(\ell,\gamma)$ yields

$$P_2(\ell,\gamma)>0,\ \ \ell\in[5,\infty).$$

Therefore,

$$\begin{aligned} P(\ell,\gamma) &= P_1(\ell,\gamma)+\frac{3^\ell}{\ell+1}P_2(\ell,\gamma)\geq P_1(\ell,\gamma)+\frac{81}{2}P_2(\ell,\gamma)\\ &= \frac{124}{3}-\frac{3(\gamma+1)(10\ell+9)}{2(\ell-2-\gamma)}+\frac{3\ell\gamma(\gamma+1)(5\ell+1)}{4(\ell-2-\gamma)(\ell-3-\gamma)}\\ &> \frac{124}{3}-\frac{3(\gamma+1)(10\ell+9)}{2(\ell-2-\gamma)}+\frac{3\ell\gamma(\gamma+1)(5\ell+1)}{4(\ell-2-\gamma)^2}. \end{aligned}$$

When $\ell=5$, we easily know that $P(\ell,\gamma)>0$ by direct calculation. Next, we discuss the case $\ell\geq 6$. Let

$$P_4(\ell,\gamma)=\frac{124}{3}-\frac{3(\gamma+1)(10\ell+9)}{2(\ell-2-\gamma)}+\frac{3\ell\gamma(\gamma+1)(5\ell+1)}{4(\ell-2-\gamma)^2},$$

and

$$R(x,\gamma)=12(x-2-\gamma)^2P_4(x,\gamma),\ x\geq 6.$$

Differentiating twice with respect to x gives

$$R_{xx}(x,\gamma)=90\gamma^2-270\gamma+632,$$

which is positive when $\gamma\in(0,1)$.

So, $R_x(x,\gamma)$ is an increasing function and

$$R_x(x,\gamma)>R_x(6,\gamma)=729\gamma^2-2225\gamma+2006>0.$$

Furthermore, $R(x,\gamma)$ is an increasing function as well and

$$R(x,\gamma)>R(6,\gamma)=3412\gamma^2-6020\gamma+2968>0,\ \ x\in[6,\infty).$$

So, $P_4(\ell,\gamma)>0$ implies $P(\ell,\gamma)>0$. It follows that $\varpi_{2,\ell}^{(\alpha)}\geq\varpi_{2,\ell+1}^{(\alpha)}$ for $\ell\geq 5$. The proof is thus finished. □

The monotonicity of the *second-order coefficients* $\varpi_{2,\ell}^{(\alpha)}$ is often used to prove the stability and convergence of the constructed algorithms for the time fractional differential equations. For the space fractional differential equations, we use the following theorem instead of the monotonicity of the coefficients $\varpi_{p,\ell}^{(\alpha)}$ to show the stability and convergence for the derived algorithms. Now we establish the following theorem. Here we focus on studying the case $\alpha\in(0,1)$.

Theorem 9 *For $0<\alpha<1$, then the following relation holds:*

$$\sum_{\ell=0}^{\infty}\varpi_{p,\ell}^{(\alpha)}\cos(\ell\theta)\geq 0,\ \theta\in[-\pi,\pi],\ p=2,3,5,6.$$

Proof. We only prove $p=2$, the other cases can be almost similarly shown so are left out here. Let

$$f_1(\alpha,\theta)=\sum_{\ell=0}^{\infty}\varpi_{2,\ell}^{(\alpha)}\cos(\ell\theta),$$

which can be expanded as

$$
\begin{aligned}
f_1(\alpha,\theta) &= \sum_{\ell=0}^{\infty} \varpi_{2,\ell}^{(\alpha)} \cos(\ell\theta) = \frac{1}{2}\sum_{\ell=0}^{\infty} \varpi_{2,\ell}^{(\alpha)} (\exp(i\ell\theta)+\exp(-i\ell\theta)) \\
&= \frac{1}{2}\left[(1-\exp(i\theta))^{\alpha}\left(\frac{3}{2}-\frac{1}{2}\exp(i\theta)\right)^{\alpha} + (1-\exp(-i\theta))^{\alpha}\left(\frac{3}{2}-\frac{1}{2}\exp(-i\theta)\right)^{\alpha}\right].
\end{aligned}
$$

Note that $f_1(\alpha,\theta)$ is a real-value and even function, so we need only consider $\theta \in [0,\pi]$. Using the following equations

$$
(1-\exp(\pm i\theta))^{\alpha} = \left(2\sin\frac{\theta}{2}\right)^{\alpha} \exp\left(\pm i\alpha\left(\frac{\theta-\pi}{2}\right)\right)
$$

and

$$
(x-yi)^{\alpha} = \left(x^2+y^2\right)^{\frac{\alpha}{2}} \exp(i\alpha\phi), \quad \phi = -\arctan\frac{y}{x}.
$$

Now we can rewrite $f_1(\alpha,\theta)$ as

$$
f_1(\alpha,\theta) = \left(2\sin\frac{\theta}{2}\right)^{\alpha} \left(\lambda_1^2(\theta)+\mu_1^2(\theta)\right)^{\frac{\alpha}{2}} \cos\alpha\left(\frac{\theta-\pi}{2}+\phi_1\right),
$$

where

$$
\lambda_1(\theta) = 3-\cos\theta, \quad \mu_1(\theta) = \sin\theta, \quad \phi_1 = -\arctan\frac{\mu_1(\theta)}{\lambda_1(\theta)}.
$$

Let

$$
z(\theta) = \frac{\theta-\pi}{2}+\phi_1, \quad 0 \le \theta \le \pi.
$$

Then

$$
z'(\theta) = \left(\frac{\theta-\pi}{2}+\phi_1\right)' = \frac{3\sin^2\left(\frac{\theta}{2}\right)}{1+3\sin^2\left(\frac{\theta}{2}\right)} \ge 0.
$$

Hence $z(\theta)$ is an increasing function in $[0,\pi]$ and

$$
z_{\min}(\theta) = z(0) = -\frac{\pi}{2}, \quad z_{\max}(\theta) = z(\pi) = 0.
$$

It is simple to see that $\alpha \in (0,1)$ and $\theta \in [0,\pi]$ imply $\cos\alpha\left(\frac{\theta-\pi}{2}+\phi_1\right) \ge 0$, so one has

$$
f_1(\alpha,\theta) = \left(2\sin\frac{\theta}{2}\right)^{\alpha} \left(\lambda_1^2(\theta)+\mu_1^2(\theta)\right)^{\frac{\alpha}{2}} \cos\alpha\left(\frac{\theta-\pi}{2}+\phi_1\right) \ge 0.
$$

All this ends the proof. □

For $p = 4$, α can not be very close to 1. But we have the following theorem.

Theorem 10 *If* $0 < \alpha \le \dfrac{\pi}{\pi - \arccos\frac{1}{5} + 2\arctan\frac{191\sqrt{6}}{317}} \approx 0.8439$, *then the following relation holds:*

$$
\sum_{\ell=0}^{\infty} \varpi_{4,\ell}^{(\alpha)} \cos(\ell\theta) \ge 0, \; \theta \in [-\pi,\pi].
$$

Proof. Let $f_2(\alpha,\theta) = \sum\limits_{\ell=0}^{\infty} \varpi_{4,\ell}^{(\alpha)} \cos(\ell\theta)$. By almost the same reasoning as that of Theorem 9, we can get

$$f_2(\alpha,\theta) = \left(2\sin\frac{\theta}{2}\right)^{\alpha} \left(\lambda_2^2(\theta)+\mu_2^2(\theta)\right)^{\frac{\alpha}{2}} \cos\alpha\left(\frac{\theta-\pi}{2}+\phi_2\right),$$

where

$$\lambda_2(\theta) = 25 - 23\cos\theta + 13\cos 2\theta - 3\cos 3\theta,$$
$$\mu_2(\theta) = 23\sin\theta - 13\sin 2\theta + 3\sin 3\theta, \quad \phi_2 = -\arctan\frac{\mu_2(\theta)}{\lambda_2(\theta)}.$$

Since

$$\lambda_2(\theta) = 14\left(\cos(\theta)-\frac{1}{2}\right)^2 + 24\cos^2(\theta)\sin^2\left(\frac{\theta}{2}\right) + \frac{17}{2} > 0,$$

and

$$\mu_2(\theta) = \left[12\left(\cos(\theta)-\frac{13}{12}\right)^2 + \frac{71}{24}\right]\sin(\theta) \geq 0,$$

so $\phi_2 \in [-\frac{\pi}{2}, 0]$. We need only consider $0 \leq \theta \leq \pi$, therefore $-\pi \leq \alpha\left(\frac{\theta-\pi}{2}+\phi_2\right) \leq 0$.

Obviously, if $\cos\alpha\left(\frac{\theta-\pi}{2}+\phi_2\right) \geq 0$, then $f_2(\alpha,\theta) \geq 0$. A sufficient condition for $\cos\alpha\left(\frac{\theta-\pi}{2}+\phi_2\right) \geq 0$ is

$$-\frac{\pi}{2} \leq \min_{\theta\in[0,\pi]} \alpha\left(\frac{\theta-\pi}{2}+\phi_2\right) \leq 0,$$

i.e.,

$$0 < \alpha \leq \min_{\theta\in[0,\pi]} \left\{\frac{\pi}{\pi-\theta-2\phi_2}\right\}.$$

Let $y(\theta) = \pi - \theta - 2\phi_2$, then

$$y'(\theta) = \frac{1920(5\cos\theta-1)\sin^4\left(\frac{\theta}{2}\right)}{a_2^2(\theta)+b_2^2(\theta)}.$$

It is clear that $\theta = \arccos\frac{1}{5}$ is a unique maximum point of $y(\theta)$ when $\theta \in [0,\pi]$, so

$$y_{\max}(\theta) = y_{\max}\left(\arccos\frac{1}{5}\right) = \pi - \arccos\frac{1}{5} + 2\arctan\frac{191\sqrt{6}}{317},$$

it follows that

$$\min_{\theta\in[0,\pi]} \left\{\frac{\pi}{\pi-\theta-2\phi_2}\right\} = \frac{\pi}{\pi - \arccos\frac{1}{5} + 2\arctan\frac{191\sqrt{6}}{317}} \approx 0.8439,$$

i.e.,

$$0 < \alpha \leq 0.8439.$$

This finishes the proof. □

Now we again return to discuss the properties of the other *high-order coefficients*.

Remark 2.4.2 *The monotonicity of the coefficients $\varpi_{2,\ell}^{(\alpha)}$ with respect to ℓ (see Theorems 8) are often used for stability and convergence analysis for the time fractional differential equations.*

Although it is not facile to prove the monotonicity of the coefficients $\varpi_{p,\ell}^{(\alpha)}$, $p = 3,\cdots,6$, we can explicitly write their expressions; see the beginning part of this subsection for more details. The coefficients are explicitly expressed which are beneficial for numerical calculations. Besides, through numerical simulations, one can find the monotonicity of the coefficients $\varpi_{p,\ell}^{(\alpha)}$, $p = 3, 4, 5$.

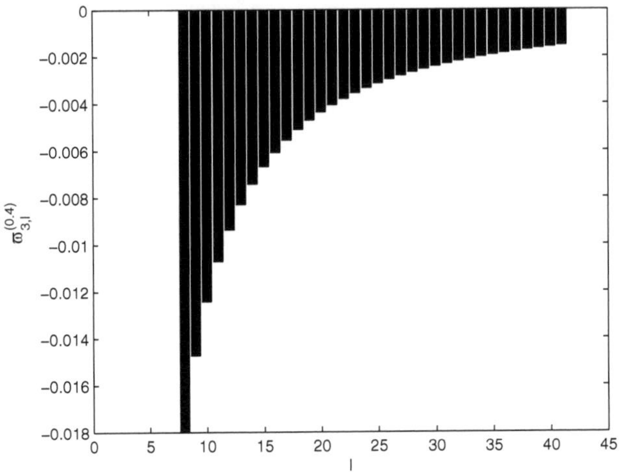

FIGURE 2.1: The values of coefficient $\varpi_{3,\ell}^{(\alpha)}$ $(\ell = 4,5,\cdots)$ for $\alpha = 0.4$.

Figs. 2.1 and 2.2 show the monotonicity of the coefficients $\varpi_{3,\ell}^{(\alpha)}$ for $\alpha \in (0,1)$, Figs. 2.3 and 2.4 for $\alpha \in (1,2)$. Figs. 2.5 and 2.6 display the monotonicity of the coefficients $\varpi_{4,\ell}^{(\alpha)}$ for $\alpha \in (0,1)$, Figs. 2.7 and 2.8 for $\alpha \in (1,2)$. Figs. 2.9 and 2.10 present the monotonicity of the coefficients $\varpi_{5,\ell}^{(\alpha)}$ for $\alpha \in (0,1)$, and Figs. 2.11 and 2.12 for $\alpha \in (1,2)$. But through the numerical simulations, $\varpi_{6,\ell}^{(\alpha)}$, $\alpha \in (0,1)$ and $\alpha \in (1,2)$ seem not to have the monotonicity.

In the following, we provide a conjecture which is seemingly primary but is hard to prove.

Conjecture 2.4.1

(1) *If* $0 < \alpha < 1$, *then* $\varpi_{3,\ell}^{(\alpha)} \leq \varpi_{3,\ell+1}^{(\alpha)}$ *for* $\ell \geq 4$, $\varpi_{4,\ell}^{(\alpha)} \leq \varpi_{4,\ell+1}^{(\alpha)}$ *for* $\ell \geq 7$, *and* $\varpi_{5,\ell}^{(\alpha)} \leq \varpi_{5,\ell+1}^{(\alpha)}$ *for* $\ell \geq 12$.

(2) *If* $1 < \alpha < 2$, *then* $\varpi_{3,\ell}^{(\alpha)} \geq \varpi_{3,\ell+1}^{(\alpha)}$ *for* $\ell \geq 7$, $\varpi_{4,\ell}^{(\alpha)} \geq \varpi_{4,\ell+1}^{(\alpha)}$ *for* $\ell \geq 12$, *and* $\varpi_{5,\ell}^{(\alpha)} \geq \varpi_{5,\ell+1}^{(\alpha)}$ *for* $\ell \geq 16$.

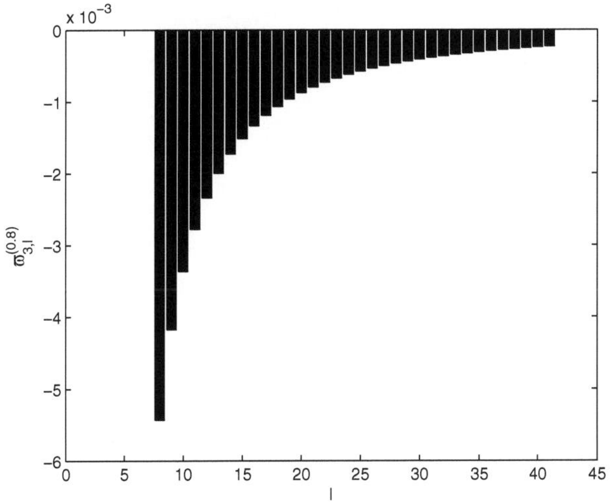

FIGURE 2.2: The values of coefficient $\varpi_{3,\ell}^{(\alpha)}$ ($\ell = 4, 5, \cdots$) for $\alpha = 0.8$.

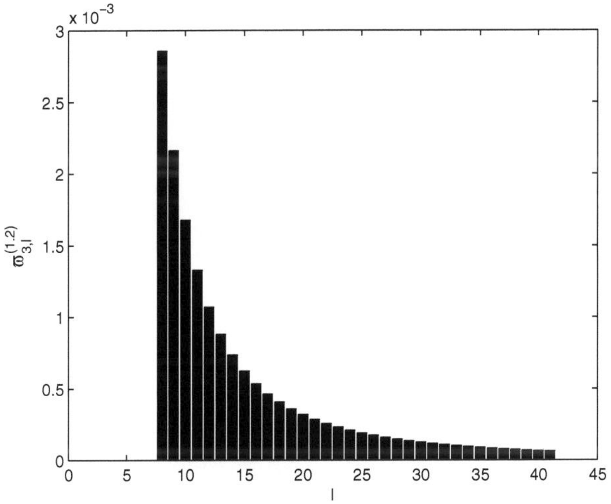

FIGURE 2.3: The values of coefficient $\varpi_{3,\ell}^{(\alpha)}$ ($\ell = 7, 8, \cdots$) for $\alpha = 1.2$.

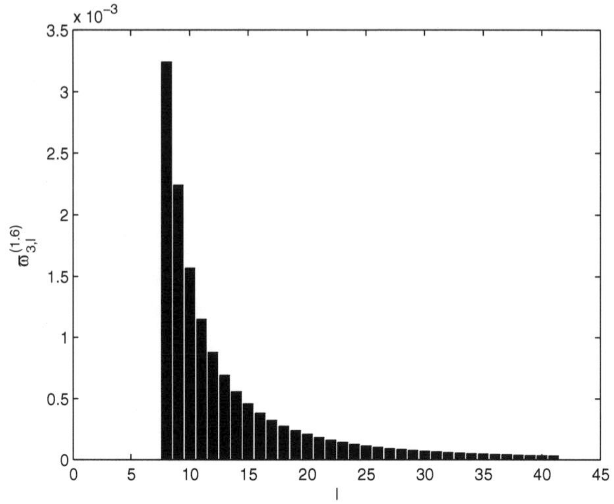

FIGURE 2.4: The values of coefficient $\varpi_{3,\ell}^{(\alpha)}$ ($\ell = 7, 8, \cdots$) for $\alpha = 1.6$.

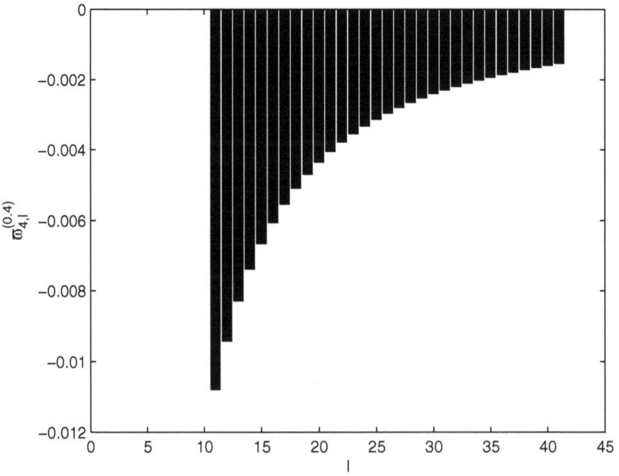

FIGURE 2.5: The values of coefficient $\varpi_{4,\ell}^{(\alpha)}$ ($\ell = 7, 8, \cdots$) for $\alpha = 0.4$.

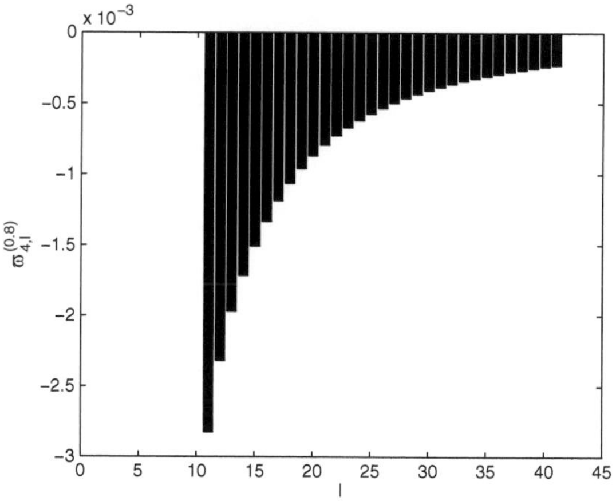

FIGURE 2.6: The values of coefficient $\varpi_{4,\ell}^{(\alpha)}$ $(\ell = 7, 8, \cdots)$ for $\alpha = 0.8$.

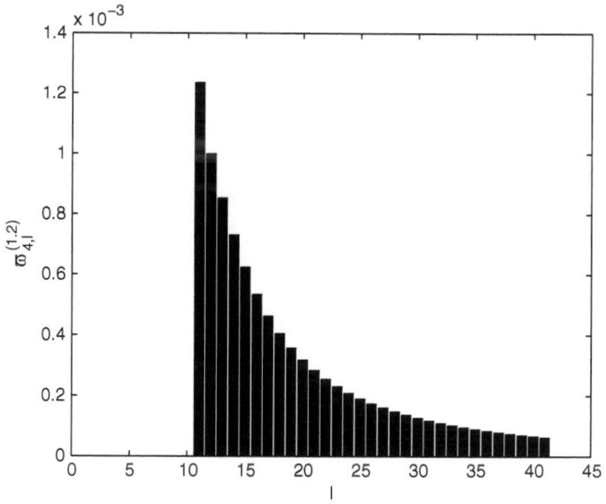

FIGURE 2.7: The values of coefficient $\varpi_{4,\ell}^{(\alpha)}$ $(\ell = 10, 11, \cdots)$ for $\alpha = 1.2$.

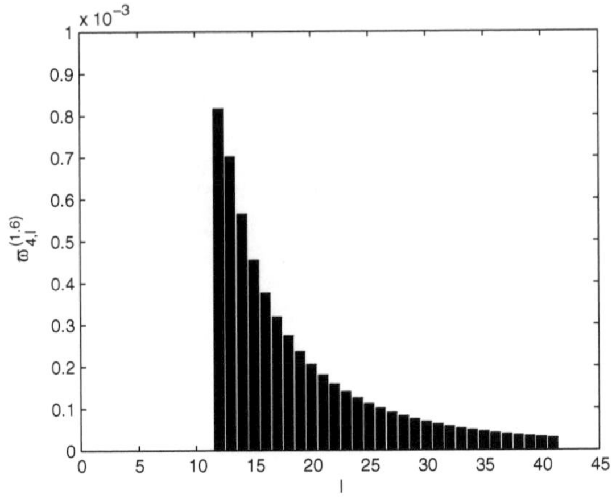

FIGURE 2.8: The values of coefficient $\varpi_{4,\ell}^{(\alpha)}$ ($\ell = 10, 11, \cdots$) for $\alpha = 1.6$.

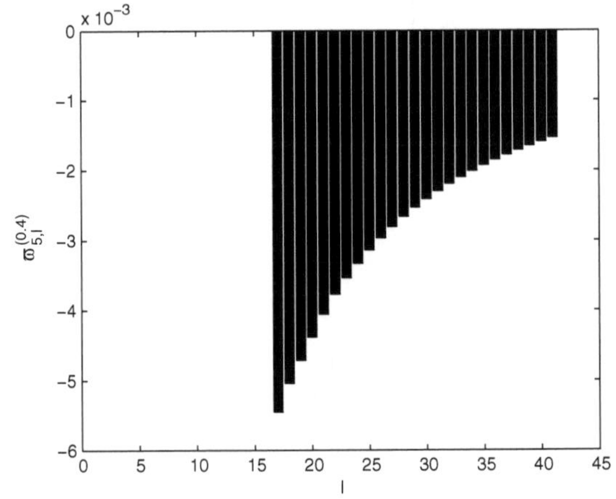

FIGURE 2.9: The values of coefficient $\varpi_{5,\ell}^{(\alpha)}$ ($\ell = 12, 13, \cdots$) for $\alpha = 0.4$.

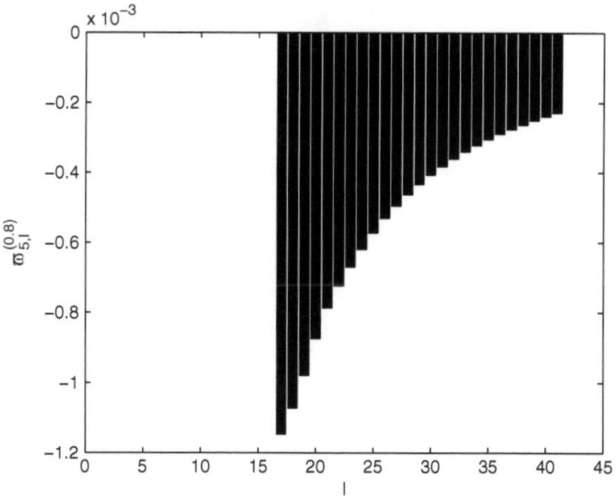

FIGURE 2.10: The values of coefficient $\varpi_{5,\ell}^{(\alpha)}$ $(\ell = 12, 13, \cdots)$ for $\alpha = 0.8$.

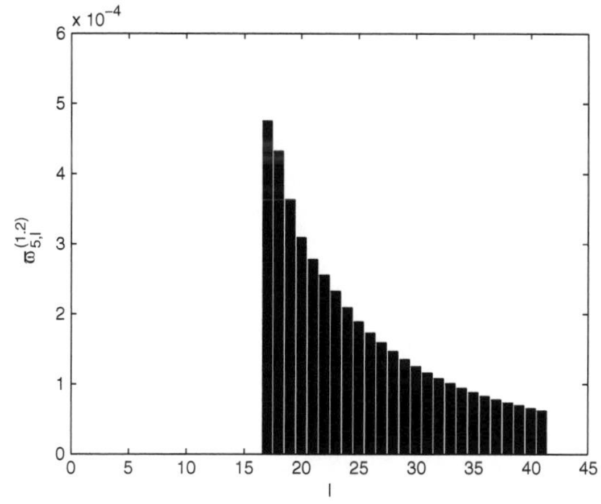

FIGURE 2.11: The values of coefficient $\varpi_{5,\ell}^{(\alpha)}$ $(\ell = 16, 17, \cdots)$ for $\alpha = 1.2$.

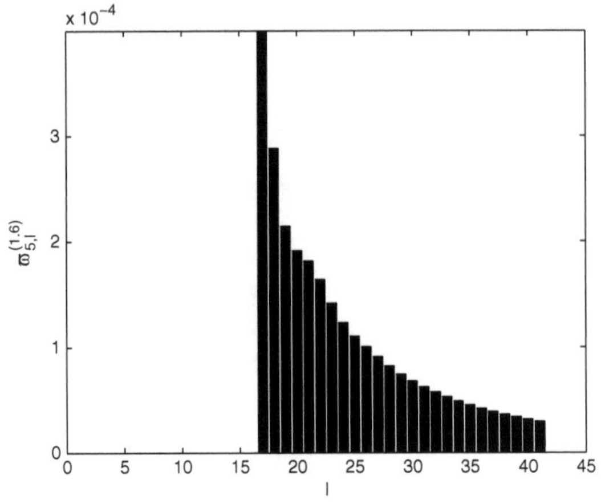

FIGURE 2.12: The values of coefficient $\varpi_{5,\ell}^{(\alpha)}$ ($\ell = 16, 17, \cdots$) for $\alpha = 1.6$.

These properties of *high-order coefficients* are generally suitable for stability and convergence analysis of the designed algorithms for the time-fractional partial differential equations.

2.4.4 Numerical Examples

Now we test the higher-order schemes for Riesz derivatives.

Example 1 *Consider the function $f_p(x) = x^p(1-x)^p$, $x \in [0,1]$, $p = 2,3,4,5,6$.*
The Riesz derivative of the above function is analytically expressed as

$$\frac{\partial^\alpha f_p(x)}{\partial|x|^\alpha} = -\frac{1}{2\cos(\pi\alpha/2)}\sum_{\ell=0}^{p}(-1)^\ell \frac{p!(p+l)!}{\ell!(p-l)!\Gamma(p+\ell+1-\alpha)}\left[x^{p+\ell-\alpha} + (1-x)^{p+\ell-\alpha}\right].$$

We first verify the convergence orders for numerical scheme (2.120) and (2.121). The computational results are showed in Tables 2.1–2.2. These computational results confirm the second-order and fourth-order of the numerical formulas (2.120) and (2.121), respectively.

Then, we numerically solve the Riesz derivative of $f_p(x)$ by using numerical scheme (2.140). The numerical results are presented in Tables 2.3–2.7. From these tables, the experimental orders are in line with the theoretical orders p ($p = 2,3,4,5,6$).

TABLE 2.1: The absolute error, convergence order of Example 1 by numerical scheme (2.120) with $p = 2$.

α	$1/h$	L^∞-error	Order
0.2	10	3.630966e−03	
	20	9.120270e−04	1.9932
	40	2.285315e−04	1.9967
	80	5.719787e−05	1.9984
0.4	10	5.124542e−03	
	20	1.289681e−03	1.9904
	40	3.234889e−04	1.9952
	80	8.100606e−05	1.9976
0.6	10	4.629914e−03	
	20	1.164707e−03	1.9910
	40	2.920982e−04	1.9954
	80	7.314118e−05	1.9977
0.8	10	2.652282e−03	
	20	6.653815e−04	1.9950
	40	1.666617e−04	1.9973
	80	4.170691e−05	1.9986
1.2	10	2.371107e−03	
	20	5.878877e−04	2.0119
	40	1.464631e−04	2.0050
	80	3.655831e−05	2.0023
1.4	10	3.614913e−03	
	20	8.891952e−04	2.0234
	40	2.207491e−04	2.0101
	80	5.500869e−05	2.0047
1.6	10	3.311900e−03	
	20	8.088550e−04	2.0337
	40	2.001866e−04	2.0145
	80	4.981273e−05	2.0068
1.8	10	1.748268e−03	
	20	4.265789e−04	2.0350
	40	1.055454e−04	2.0149
	80	2.625954e−05	2.0069

TABLE 2.2: The absolute error, convergence order of Example 1 by numerical scheme (2.121) with $p = 4$.

α	$1/h$	L^∞-error	Order
0.2	20	1.571776e−08	
	40	9.923695e−10	3.9854
	80	6.218021e−11	3.9963
	160	3.888658e−12	3.9991
0.4	20	6.255711e−08	
	40	3.955450e−09	3.9833
	80	2.479333e−10	3.9958
	160	1.550707e−11	3.9990
0.6	20	1.660570e−07	
	40	1.051535e−08	3.9811
	80	6.593628e−10	3.9953
	160	4.124394e−11	3.9988
0.8	20	3.501076e−07	
	40	2.219986e−08	3.9792
	80	1.392502e−09	3.9948
	160	8.711096e−11	3.9987
1.2	20	7.029869e−07	
	40	4.446255e−08	3.9828
	80	2.787163e−09	3.9957
	160	1.743236e−10	3.9990
1.4	20	4.372243e−08	
	40	3.985260e−09	3.4556
	80	2.690303e−10	3.8888
	160	1.712778e−11	3.9734
1.6	20	3.751536e−06	
	40	2.440706e−07	3.9421
	80	1.540687e−08	3.9857
	160	9.653215e−10	3.9964
1.8	20	1.592788e−05	
	40	1.034107e−06	3.9451
	80	6.524548e−08	3.9864
	160	4.087469e−09	3.9966

TABLE 2.3: The absolute error, convergence order of Example 1 by numerical scheme (2.140) with $p = 2$.

α	$1/h$	L^∞-error	Order
0.1	20	9.799438e−05	
	40	2.468338e−05	1.9892
	80	6.158789e−06	2.0028
	160	1.536039e−06	2.0034
	320	3.834192e−07	2.0022
0.5	20	1.093549e−03	
	40	2.557319e−04	2.0963
	80	6.143944e−05	2.0574
	160	1.503151e−05	2.0312
	320	3.715825e−06	2.0162
0.9	20	6.874862e−03	
	40	1.269052e−03	2.4376
	80	2.603644e−04	2.2851
	160	5.793868e−05	2.1679
	320	1.358819e−05	2.0922

TABLE 2.4: The absolute error, convergence order of Example 1 by numerical scheme (2.140) with $p = 3$.

α	$1/h$	L^∞-error	Order
0.2	40	3.146678e−07	
	60	1.576085e−07	1.7052
	80	7.991483e−08	2.3608
	100	4.501080e−08	2.5726
	120	2.761993e−08	2.6786
0.5	40	1.015756e−05	
	60	3.330915e−06	2.7499
	80	1.470163e−06	2.8430
	100	7.721886e−07	2.8856
	120	4.542595e−07	2.9100
0.9	40	3.110547e−04	
	60	9.436636e−05	2.9418
	80	4.020332e−05	2.9659
	100	2.069282e−05	2.9764
	120	1.201409e−05	2.9821

TABLE 2.5: The absolute error, convergence order of Example 1 by numerical scheme (2.140) with $p = 4$.

α	$1/h$	L^∞-error	Order
0.2	20	3.254967e−06	
	25	1.421712e−06	3.7121
	30	7.061638e−07	3.8381
	35	3.863551e−07	3.9123
	40	2.279637e−07	3.9509
0.5	20	2.352556e−05	
	25	9.541773e−06	4.0441
	30	4.506623e−06	4.1143
	35	2.379569e−06	4.1429
	40	1.366626e−06	4.1531
0.8	20	1.466022e−04	
	25	5.460563e−05	4.4258
	30	2.420431e−05	4.4625
	35	1.216174e−05	4.4647
	40	6.707129e−06	4.4568

TABLE 2.6: The absolute error, convergence order of Example 1 by numerical scheme (2.140) with $p = 5$.

α	$1/h$	L^∞-error	Order
0.3	80	2.498729e−10	
	100	1.091359e−10	3.7122
	120	5.108016e−11	4.1641
	140	2.599689e−11	4.3816
	160	1.423638e−11	4.5096
0.6	80	5.385482e−09	
	100	1.900398e−09	4.6680
	120	7.985621e−10	4.7554
	140	3.806168e−10	4.8071
	160	1.994064e−10	4.8412
0.9	80	9.315443e−08	
	100	3.135150e−08	4.8803
	120	1.279618e−08	4.9150
	140	5.979648e−09	4.9353
	160	3.088216e−09	4.9484

TABLE 2.7: The absolute error, convergence order of Example 1 by numerical scheme (2.140) with $p = 6$.

α	$1/h$	L^∞-error	Order
0.3	20	1.224183e−07	
	40	5.170505e−09	4.5654
	80	8.866763e−11	5.8658
	160	1.345871e−12	6.0418
	320	2.028938e−14	6.0517
0.6	20	1.564617e−06	
	40	3.647148e−08	5.4229
	80	5.062291e−10	6.1708
	160	6.799919e−12	6.2181
	320	9.539537e−14	6.1555
0.9	20	2.577647e−05	
	40	3.796038e−07	6.0854
	80	3.895063e−09	6.6067
	160	4.007447e−11	6.6028
	320	4.600099e−13	6.4449

2.5 Matrix Approach

This approach is based on using a triangular strip matrix to discretize the differentiation and integration operators with arbitrary order (integer and noninteger) [125, 126]. Using this technique one can obtain all the numerical solutions at the mesh grids at once, avoiding the traditional step-by-step method by moving from the previous time layer to the next one. The *matrix approach* is quite simple to put into implementation.

The matrices with the special structure such as the triangular strip type are introduced in order to describe this approach. The following two kinds of triangular strip matrices are needed:

Lower triangular strip matrices

$$L_N = \begin{pmatrix} \omega_0 & 0 & 0 & 0 & \cdots & 0 \\ \omega_1 & \omega_0 & 0 & 0 & \dots & 0 \\ \omega_2 & \omega_1 & \omega_0 & 0 & \dots & 0 \\ \ddots & \ddots & \ddots & \ddots & \dots & \dots \\ \omega_{N-1} & \ddots & \omega_2 & \omega_1 & \omega_0 & 0 \\ \omega_N & \omega_{N-1} & \ddots & \omega_2 & \omega_1 & \omega_0 \end{pmatrix}, \tag{2.141}$$

and *upper triangular strip matrices*

$$U_N = \begin{pmatrix} \omega_0 & \omega_1 & \omega_2 & \dots & \omega_{N-1} & \omega_N \\ 0 & \omega_0 & \omega_1 & \ddots & \ddots & \omega_{N-1} \\ 0 & 0 & \omega_0 & \ddots & \dots & \omega_{N-2} \\ 0 & 0 & \ddots & \ddots & \ddots & \dots \\ \dots & \dots & \dots & \dots & \omega_0 & \omega_1 \\ 0 & 0 & 0 & \dots & 0 & \omega_0 \end{pmatrix}. \tag{2.142}$$

Denote by $\mathbf{u}_n^{(\alpha)} = \left(\left[{}_{RL}\mathrm{D}_{0,t}^{\alpha} f(t)\right]_{t=t_n}, \left[{}_{RL}\mathrm{D}_{0,t}^{\alpha} f(t)\right]_{t=t_{n-1}}, \cdots, \left[{}_{RL}\mathrm{D}_{0,t}^{\alpha} f(t)\right]_{t=t_0}\right)^T$, and let $\omega_k = (-1)^k\binom{\alpha}{k}$ in (2.141), we get

$$\mathbf{u}_n^{(\alpha)} \approx B_N^{(\alpha)}[u(t_0), \cdots, u(t_{N-1}), u(t_N)]^T, \tag{2.143}$$

where $B_N^{(\alpha)} = \frac{L_N}{\Delta t^{\alpha}}$. (2.143) is just another representation of the Grünwald–Letnikov formula for the left Riemann–Liouville derivative. In fact, $B_n^{(\alpha)}$ can be seen as a kind of *fractional differential matrix* . If $\alpha = 1$, $B_n^{(\alpha)}$ is just the *differential matrix* for the classical first-order derivative corresponding to the first-order backward difference.

The right Riemann–Liouville (or Caputo) derivative can be approximated in a similar way, where the upper triangular strip matrix (2.142) is used accordingly. The symmetric Riesz derivative can also be approximated by this approach, while the fractional differential matrix is different from the Riemann–Liouville or Caputo case; refer to [126] for details.

In [125, 126], this approach is successfully adopted to solve the classical differential equations such as classical diffusion equations, and the FDEs such as diffusion equations with time-fractional derivatives, diffusion equations with spatial derivatives in the Riesz sense, general diffusion equations with time-space fractional derivatives and fractional diffusion equations with delay. This method can also be extended to the cases of nonlinear problems, see [125, 126] for more details.

2.6 Short Memory Principle

Unlike the classical differential operator, the fractional differential operator is not a local one. From the definitions of the fractional Grünwald–Letnikov, Riemann–Liouville and Caputo derivatives, one can easily find that these fractional derivatives of a given function $f(t)$ depend on the whole interval $(0,t)$, which means the fractional derivatives of $f(t)$ depend on the "historical" behavior of the function $f(t)$ [124]. However, it follows from the expressions of the coefficients $\{\omega_j^{(\alpha)}\}$ in the Grünwald–Letnikov definition (1.3) that for large j, $\omega_j^{(\alpha)}$ is reduced to zero (In fact,

TABLE 2.8: Variation of Grünwaldian expansion coefficients with different α

j	$\alpha = 0.5$	$\alpha = 1.5$	$\alpha = 2.5$
1	1.0000e+00	1.0000e+00	1.0000e+00
2	-5.0000e−01	-1.5000e+00	-2.5000e+00
3	-1.2500e−01	3.7500e−01	1.8750e+00
4	-6.2500e−02	6.2500e−02	-3.1250e−01
5	-3.9063e−02	2.3438e−02	-3.9063e−02
10	-1.0910e−02	2.1820e−03	-8.3923e−04
50	-8.2880e−04	2.6172e−05	-1.4071e−06
100	-2.8747e−04	4.4226e−06	-1.1458e−07
500	-2.5326e−05	7.6361e−08	-3.8449e−10
1000	-8.9374e−06	1.3440e−08	-3.3717e−11

$\left\{|\omega_j^{(\alpha)}|\right\}$ is a monotone decreasing sequence for $j \geq J_\alpha$, J_α is a positive integer only dependent on α). Table 2.8 gives the coefficients $\omega_j^{(\alpha)}$ with different α. We can find that $|\omega_j^{(\alpha)}|$ decreases rapidly when α increases.

Actually, $\omega_j^{(\alpha)} = O(j^{-\alpha-1})$ [104] for any $\alpha > 0$, which means that for large j, the behavior of function $f(t)$ near the lower terminal ($t = a$ in (1.3)) can be neglected under certain conditions. Those observations lead to the formulation of the *"short memory" principle*, which takes into account the behavior of $f(t)$ only in the "recent past." For the Grünwald–Letnikov definition (1.3), it means that there exists a positive integer N_α, such that [120]

$$\left[{}_{GL}\mathrm{D}_{0,t}^{\alpha} f(t)\right]_{t=t_N} \approx \Delta t^{-\alpha} \sum_{j=0}^{N_\alpha} \omega_j^{(\alpha)} f(t_{N-j}). \tag{2.144}$$

Denote by

$$R(N,N_\alpha) = \Delta t^{-\alpha} \sum_{j=0}^{N} \omega_j^{(\alpha)} f(t_{N-j}) - \Delta t^{-\alpha} \sum_{j=0}^{N_\alpha} \omega_j^{(\alpha)} f(t_{N-j}) = \Delta t^{-\alpha} \sum_{j=N_\alpha+1}^{N} \omega_j^{(\alpha)} f(t_{N-j}).$$

Noticing that $\omega_j^{(\alpha)} = O(j^{-\alpha-1})$, i.e. $|\omega_j^{(\alpha)}| \leq C j^{-\alpha-1}$, we have

$$\begin{aligned} |R(N,N_\alpha)| &= \left|\Delta t^{-\alpha} \sum_{j=N_\alpha+1}^{N} \omega_j^{(\alpha)} f(t_{N-j})\right| \leq \Delta t^{-\alpha} \sum_{j=N_\alpha+1}^{N} \left|\omega_j^{(\alpha)}\right| \left|f(t_{N-j})\right| \\ &\leq C\Delta t^{-\alpha} \max_{N_\alpha+1 \leq j \leq N} \left|f(t_{N-j})\right| \sum_{j=N_\alpha+1}^{N} j^{-\alpha-1}. \end{aligned} \tag{2.145}$$

For the fixed Δt, when N_α is big enough, $|R(N,N_\alpha)|$ can be small enough for $\alpha > 0$.

Next, we discuss the "short memory" principle for the Riemann–Liouville derivative, which gives the theoretical standard to determine N_α in (2.144). The Riemann–Liouville derivative can be written in the form of a finite-part integral (see [124] p.194)

$$ {}_{RL}\mathrm{D}^{\alpha}_{a,t}f(t) = \frac{1}{\Gamma(-\alpha)} p.f. \int_a^t (t-s)^{-\alpha-1} f(s)\,\mathrm{d}s. \tag{2.146} $$

Let L be a positive number, denote by

$$ R(t,L) = {}_{RL}\mathrm{D}^{\alpha}_{a,t}f(t) - {}_{RL}\mathrm{D}^{\alpha}_{t-L,t}f(t). $$

Using (2.146) leads to

$$ |R(t,L)| = |{}_{RL}\mathrm{D}^{\alpha}_{a,t}f(t) - {}_{RL}\mathrm{D}^{\alpha}_{t-L,t}f(t)| \le \frac{ML^{-\alpha}}{|\Gamma(1-\alpha)|}, \quad a+L \le t \le b, \tag{2.147} $$

where $|f(t)| \le M, M > 0$. For any $\epsilon > 0$, letting

$$ |R(t,L)| \le \epsilon, $$

one has

$$ L \ge \left(\frac{M}{\epsilon |\Gamma(1-\alpha)|} \right)^{1/\alpha}, $$

which means that

$$ {}_{RL}\mathrm{D}^{\alpha}_{t-L,t}f(t) = {}_{RL}\mathrm{D}^{\alpha}_{a,t}f(t) - R(t,L) $$

with $|R(t,L)| \le \epsilon$ when $L \ge \left(\frac{M}{\epsilon|\Gamma(1-\alpha)|} \right)^{1/\alpha}$. Here, L is also called the "*memory length*." Therefore, N_α can be chosen as

$$ N_\alpha = \left\lceil \frac{L}{\Delta t} \right\rceil, \quad L \ge \left(\frac{M}{\epsilon |\Gamma(1-\alpha)|} \right)^{1/\alpha}. $$

Deng [22] considered this approach to solve time fractional differential equations. Up to now, the short memory principal has not been thoroughly studied so is seldom used in the real applications.

2.7 Other Approaches

In [159], Yuan and Agrawal proposed a method to calculate the Caputo derivative of order α $(0 < \alpha < 1)$, in which the weakly singular kernel is removed. Using the definition of the gamma function and the formula of complement variable below

$$ \Gamma(\alpha)\Gamma(1-\alpha) = \frac{\pi}{\sin(\pi\alpha)}, $$

they transformed the Caputo derivative into the following equivalent form

$$ {}_{C}\mathrm{D}^{\alpha}_{a,t}f(t) = \frac{2\sin(\pi\alpha)}{\pi} \int_0^{\infty} y^{2\alpha-1} \left(\int_0^t e^{-(t-s)y^2} f'(s)\,\mathrm{d}s \right) \mathrm{d}y. \tag{2.148} $$

Denote by

$$\phi(y,t) = y^{2\alpha-1}\left(\int_0^t e^{-(t-s)y^2} f'(s)\,\mathrm{d}s\right).$$

Then, the integral $\int_0^\infty \phi(y,t)\,\mathrm{d}y$ is approximated by the *Laguerre integral formula* [21]

$$\int_0^\infty \phi(y,t)\,\mathrm{d}y \approx \sum_{i=1}^{N} w_i e^{y_i}\phi(y_i,t),$$

where w_i and y_i, $i=1,2,\cdots,N\in\mathbb{N}$ are Laguerre weights and node points, respectively. Therefore,

$$_C\mathrm{D}_{a,t}^{\alpha} f(t) \approx \frac{2\sin(\pi\alpha)}{\pi}\sum_{i=1}^{N} w_i e^{y_i}\phi(y_i,t). \tag{2.149}$$

Obviously, $\phi(y_i,t)$ can be integrated by the standard integration method, and the numerical calculation for $_C\mathrm{D}_{a,t}^{\alpha} f(t)$ is accomplished.

Remark 2.7.1 *The case of* $1<\alpha<2$ *is considered in [147], where the equivalent form of the* α*th Caputo derivative is given by*

$$_C\mathrm{D}_{a,t}^{\alpha} f(t) = \frac{2\sin(\pi(\alpha-1))}{\pi}\int_0^\infty y^{2\alpha-3}\left(\int_0^t e^{-(t-s)y^2} f''(s)\,\mathrm{d}s\right)\mathrm{d}y. \tag{2.150}$$

There exist other methods to approximate the fractional integrals and derivatives. The algorithm based on *Haar wavelet approximation* theory for the fractional integrals was proposed in [57]. The definitions of *variable-order derivatives* and their numerical approximations were investigated in [143, 149, 177]. The Grünwald formula for *vector fractional derivatives* was developed in [110].

Chapter 3

Numerical Methods for Fractional Ordinary Differential Equations

In the previous chapter, various kinds of numerical methods for fractional integrals and fractional derivatives are displayed. In the present chapter, we focus on introducing numerical methods for fractional ordinary differential equations. Since Riemann–Liouville derivatives can be changed into the Caputo ones (see Eq. (1.11)) under suitable conditions, we study only the Caputo-type ordinary differential equations.

3.1 Introduction

In this chapter, we study the numerical methods for the typical initial-value problem below.

$$\begin{cases} {}_C\mathrm{D}_{0,t}^{\alpha}u(t) = f(t,u(t)),\ m-1<\alpha<m \in Z^+, \\ u^{(j)}(0) = u_0^j,\ j=0,1,\cdots,m-1. \end{cases} \tag{3.1}$$

The following Theorems 11 and 12 of *existence and uniqueness for initial-value problems* (3.1) can be found in [29].

Theorem 11 (existence) *Assume that* $\mathscr{D} := [0,\chi^*]\times[u_0^0-\delta, u_0^0+\delta]$ *with some* $\chi^*>0$ *and some* $\delta>0$, *and let the function* $f:\mathscr{D}\to\mathbb{R}$ *be continuous. Furthermore, define* $\chi := \min\{\chi^*, (\delta\Gamma(\alpha+1)/\|f\|_\infty)^{1/\alpha}\}$. *Then, there exists a function* $u:[0,\chi]\to\mathbb{R}$, *solving the initial value problem* (3.1).

Theorem 12 (uniqueness) *Assume* $\mathscr{D} := [0,\chi^*]\times[u_0^0-\delta, u_0^0+\delta]$ *with* $\chi^*>0$ *and* $\delta>0$. *Furthermore, let the function* $f:\mathscr{D}\to\mathbb{R}$ *be bounded on* $\mathscr{D}$ *and fulfill a Lipschitz condition with respect to the second variable, i.e.,*

$$|f(t,x)-f(t,y)| \le L|x-y| \tag{3.2}$$

with constant $L>0$ *independent of* t,x *and* y. *Then, denoting* χ *as in Theorem 11, there exists at most one function* $u:[0,\chi]\to\mathbb{R}$ *solving the initial value problem* (3.1).

If the initial value problem (3.1) has a unique solution $u(t)$, and $f(t,u(t))$ satisfies some smooth conditions, then one can obtain the following properties of the solution $u(t)$, see [32].

Theorem 13 *(a) Assume $f \in C^2(G)$. Define $\hat{\nu} = \lceil 1/\alpha \rceil - 1$. Then there exist functions $\psi \in C^1[0,T]$ and $c_1, \cdots, c_{\hat{\nu}} \in \mathbb{R}$ such that the solution $u(t)$ of (3.1) can be expressed in the form*

$$u(t) = \psi(t) + \sum_{\nu=1}^{\hat{\nu}} c_\nu t^{\nu\alpha}.$$

(b) Assume that $f \in C^3(G)$. Define $\hat{\nu} = \lceil 2/\alpha \rceil - 1$ and $\tilde{\nu} = \lceil 1/\alpha \rceil - 1$. Then there exists a function $\psi \in C^2[0,T]$, $c_1, \cdots, c_{\hat{\nu}} \in \mathbb{R}$, and $d_1, \cdots, d_{\tilde{\nu}} \in \mathbb{R}$ such that the solution $u(t)$ of (3.1) can be expressed in the form

$$u(t) = \psi(t) + \sum_{\nu=1}^{\hat{\nu}} c_\nu t^{\nu\alpha} + \sum_{\nu=1}^{\tilde{\nu}} d_\nu t^{1+\nu\alpha}.$$

There are several ways to discretize equation (3.1); the most often used two techniques are based on the following ideas:

- Discretizing the Caputo derivative directly to get the numerical schemes.
- Transforming the original fractional equation (3.1) into the fractional integral equation, then applying the corresponding numerical methods to discretize the fractional integral to get the numerical schemes.

In the following, we introduce the typical numerical methods for equation (3.1).

3.2 Direct Methods

In the section, we discretize the Caputo derivative operator directly to get the numerical schemes for equation (3.1). Obviously, we can use the numerical methods for Caputo derivative operator developed in Section 2.4. Next, we just list some of numerical methods, and we do not give the stability and convergence analysis, which will be discussed in the following section.

- **L1 Method**

For $0 < \alpha < 1$, L1 method (2.81) is often used to discreteize the Caputo derivative. L1 method for the initial value problem (3.1) is

$$\sum_{j=0}^{n-1} b_{n-j-1}(u_{j+1} - u_j) = f(t_n, u_n), \tag{3.3}$$

where u_n is the approximate solution of $u(t_n)$, and $b_j = \frac{\Delta t^{-\alpha}}{\Gamma(2-\alpha)}[(j+1)^{1-\alpha} - j^{1-\alpha}]$, see (2.81) for more details.

- **Product Trapezoidal Method**

This approach is based on the fact that the Riemann–Liouville derivative is equivalent to the *Hadamard finite-part integral* [26, 48, 124], i.e.,

$$ {}_{RL}\mathrm{D}_{0,t}^{\alpha}u(t) = \frac{1}{\Gamma(-\alpha)}\ \text{p.f.}\int_0^t \frac{u(s)}{(t-s)^{\alpha+1}}\,\mathrm{d}s, \quad \alpha \neq 0,1,2,\cdots. \tag{3.4}$$

For $0<\alpha<1$, the above quadrature is approximated by the *first-degree compound quadrature formula* [25, 26, 34], which is given by

$$\frac{1}{\Gamma(-\alpha)}\ \text{p.f.}\int_0^{t_n} \frac{u(s)}{(t_n-s)^{\alpha+1}}\,\mathrm{d}s \approx \sum_{j=0}^{n} a_{j,n}u(t_{n-j}), \tag{3.5}$$

where

$$a_{j,n} = \frac{\Delta t^{-\alpha}}{\Gamma(2-\alpha)}\begin{cases} 1, & j=0,\\ (j+1)^{1-\alpha}-2j^{1-\alpha}+(j-1)^{1-\alpha}, & 0<j<n,\\ (1-\alpha)n^{-\alpha}-n^{1-\alpha}+(n-1)^{1-\alpha}, & j=n. \end{cases} \tag{3.6}$$

Using the relationship

$$ {}_{RL}\mathrm{D}_{0,t}^{\alpha}[u(t)-u(0)] = {}_{C}\mathrm{D}_{0,t}^{\alpha}u(t), \quad 0<\alpha<1,$$

we get the numerical scheme for (3.1) as follows

$$\sum_{j=0}^{n} a_{j,n}(u_{n-j}-u_0) = f(t_n,u_n), \tag{3.7}$$

where u_n is the approximate solution of $u(t_n)$.

In [25], when $f(t,u)=\beta u(t)+f(t), \beta\le 0$, the error estimate for the above method (3.7) is given by

$$|u(t_n)-u_n| \le C\Delta t^{2-\alpha}. \tag{3.8}$$

- **Grünwald–Letnikov Formula**

One knows that the Riemann–Liouville derivative can be approximated by the Grünwald–Letnikov formula (2.51)

$$ {}_{RL}\mathrm{D}_{0,t}^{\alpha}u(t)|_{t=t_n} \approx \frac{1}{\Delta t^{\alpha}}\sum_{j=0}^{n}\omega_j^{(\alpha)}u(t_{n-j}), \quad \omega_j^{(\alpha)} = (-1)^j\binom{\alpha}{j}.$$

By the relationship

$$ {}_{RL}\mathrm{D}_{0,t}^{\alpha}\Big[u(t)-\sum_{k=0}^{m}\frac{t^k}{k!}u^{(k)}(0)\Big] = {}_{C}\mathrm{D}_{0,t}^{\alpha}u(t),$$

one gets the following method for the initial value problem (3.1) as

$$\frac{1}{\Delta t^\alpha}\sum_{j=0}^{n}\omega_{n-j}^{(\alpha)}\Big[u_j-\sum_{k=0}^{m}\frac{u_0^k}{k!}t_j^k\Big]=f(t_n,u_n). \tag{3.9}$$

Of course, one can use the shifted Grünwald–Letnikov formula (2.52) to approximate the Caputo derivative.

The above method (3.9) is valid for any $\alpha>0$. One can construct high-order (pth-order, $p=1,2,\cdots,6$) methods of the form (3.9), where $\{\omega_j^{(\alpha)}\}$ are coefficients of the Taylor series expansions of the following generating functions [28, 87]

$$w_p^{(\alpha)}(z)=\left[\sum_{j=1}^{p}\frac{1}{j}(z-1)^j\right]^\alpha.$$

3.3 Integration Methods

Consider the *classical ordinary differential equation* (ODE)

$$\begin{cases} u'=f(t,u), \\ u(0)=\eta. \end{cases} \tag{3.10}$$

If we integrate (3.10) on the interval $[t_n,t_{n+1}]$, we can get

$$u(t_{n+1})-u(t_n)=\int_{t_n}^{t_{n+1}}f(s,u(s))\,\mathrm{d}s. \tag{3.11}$$

Therefore, the classical numerical methods (such as the rectangular formula, the trapezoidal rule, and Simpson's formula, etc.) for the integral $\int_{t_n}^{t_{n+1}}f(s,y(s))\,\mathrm{d}s$ can be used to derive the corresponding numerical methods for the ODE (3.10).

This idea can be adopted for the numerical solution of the FODE (3.1). Similarly, if we apply $\mathrm{D}_{0,t}^{-\alpha}$ on the both sides of (3.1), we can obtain the following equivalent Volterra integral equation [32]

$$u(t)=\sum_{j=0}^{m-1}\frac{t^j}{j!}u_0^{(j)}+\frac{1}{\Gamma(\alpha)}\int_0^t(t-s)^{\alpha-1}f(s,u(s))\,\mathrm{d}s=\sum_{j=0}^{m-1}\frac{t^j}{j!}u_0^{(j)}+\mathrm{D}_{0,t}^{-\alpha}f(t,u(t)). \tag{3.12}$$

Next, we adopt the numerical methods developed in Section 2.2 for the fractional operator $\mathrm{D}_{0,t}^{-\alpha}$ to derive the numerical methods for (3.1) or (3.12).

- **Fractional Euler Methods**

(1) ***Fractional forward Euler method:*** $\left[D_{0,t}^{-\alpha}f(t,u(t))\right]_{t=t_{n+1}}$ is approximated by the left fractional rectangular formula (2.6)

$$u_{n+1} = \sum_{j=0}^{m-1} \frac{t_{n+1}^j}{j!} u_0^{(j)} + \Delta t^\alpha \sum_{j=0}^{n} b_{j,n+1} f(t_j, u_j), \tag{3.13}$$

where

$$b_{j,n+1} = \frac{1}{\Gamma(\alpha+1)}[(n-j+1)^\alpha - (n-j)^\alpha]. \tag{3.14}$$

(2) ***Fractional backward Euler method:*** $\left[D_{0,t}^{-\alpha}f(t,u(t))\right]_{t=t_{n+1}}$ is approximated by the right fractional rectangular formula (2.8)

$$u_{n+1} = \sum_{j=0}^{m-1} \frac{t_{n+1}^j}{j!} y_0^{(j)} + \Delta t^\alpha \sum_{j=0}^{n} b_{j,n+1} f(t_{j+1}, u_{j+1}), \tag{3.15}$$

where $b_{j,n+1}$ is defined by (3.14).

(3) ***Fractional weighted difference method:*** $\left[D_{0,t}^{-\alpha}f(t,u(t))\right]_{t=t_{n+1}}$ is approximated by the weight fractional rectangular formula (2.9)

$$u_{n+1} = \sum_{j=0}^{m-1} \frac{t_{n+1}^j}{j!} u_0^{(j)} + \Delta t^\alpha \sum_{j=0}^{n} b_{j,n+1}[\theta f(t_j, u_j) + (1-\theta) f(t_{j+1}, u_{j+1})], \tag{3.16}$$

where $b_{j,n+1}$ is defined by (3.14).

Remark 3.3.1 *If* $\alpha = 1$, *the methods* (3.13), (3.15), *and* (3.16) *with* $\theta = 1/2$ *are reduced to the classical forward Euler method, the backward Euler method, and the trapezoidal formula for the classical ODE* (3.10), *respectively.*

Next, we investigate the stability and convergence of the methods (3.13), (3.15) and (3.16). The following *generalized discretized Gronwall's inequality* [42, 87] plays a crucial role in the stability and convergence analysis for numerical methods for the FODEs.

Lemma 3.3.1 ([87]) *Suppose that* $b_{j,n} = (n-j)^{\alpha-1}$ $(j = 1, 2, \cdots, n-1)$ *and* $b_{j,n} = 0$ *for* $j \geq n$, $\alpha, \Delta t, M, T > 0$, $k\Delta t \leq T$ *and k is a positive integer. Let* $\sum_{j=m}^{n} b_{j,n}|e_j| = 0$ *for* $m > n \geq 1$. *If*

$$|e_n| \leq M\Delta t^\alpha \sum_{j=1}^{n-1} b_{j,n}|e_j| + |\eta_0|, \quad n = 1, 2, \cdots, k, \tag{3.17}$$

then

$$|e_k| \leq C|\eta_0|, \quad k = 1, 2, \cdots. \tag{3.18}$$

where C is a positive constant independent of Δt *and k.*

Remark 3.3.2 *If $\alpha \geq 1$, Lemma 3.3.1 is reduced to the commonly discretized Gronwall inequality, and the bound in* (3.18) *can be determined by* $C = \exp(MT^{\alpha})$.

Now, we consider the stability of the fractional forward Euler method (3.13).

The stability here means that if there are perturbations in the initial conditions, then the small changes would not cause large errors in the numerical solutions.

For instance, the fractional forward Euler method (3.13) is stable if u_j and $v_j, (j = 1,2,..,n)$ are two solutions of the method (3.13); then there exists a positive constant C independent of Δt and n, such that

$$|u_n - v_n| \leq C|u_0 - v_0|.$$

Suppose that $u_0^{(i)}$ $(i = 0,1,\cdots,m-1)$ and $u_j (j = 0,1,\cdots,k+1)$ have perturbations $\tilde{u}_0^{(i)}$ and $\tilde{u}_j$, respectively. Denote by $\eta_0 = \max\limits_{0 \leq k \leq N} \left\{ \sum_{j=0}^{m-1} \frac{T^j}{j!} |\tilde{u}_0^{(j)}| + \frac{L\Delta t^{\alpha} b_k}{\Gamma(\alpha+1)} |\tilde{u}_0| \right\}$. Then we get the perturbation equation as follows

$$u_{n+1} + \tilde{u}_{n+1} = \sum_{j=0}^{m-1} \frac{t_{n+1}^j}{j!} (u_0^{(j)} + \tilde{u}_0^{(j)}) + \Delta t^{\alpha} \sum_{j=0}^{n} b_{n+1,j} f(t_j, u_j + \tilde{u}_j), \tag{3.19}$$

We also suppose that $f(t,u)$ satisfies the following Lipschitz condition

$$|f(t,u) - f(t,v)| \leq L|u - v|, \quad L > 0. \tag{3.20}$$

By (3.13), (3.19), and (3.20), one has

$$\begin{aligned}
|\tilde{u}_{n+1}| &= |\sum_{j=0}^{m-1} \frac{t_{n+1}^j}{j!} \tilde{u}_0^{(j)} + \Delta t^{\alpha} \sum_{j=0}^{n} b_{j,n+1} (f(t_j, u_j + \tilde{u}_j) - f(t_j, u_j))| \\
&\leq \eta_0 + \Delta t^{\alpha} \sum_{j=1}^{n} b_{j,n+1} |f(t_j, u_j + \tilde{u}_j) - f(t_j, u_j)| \\
&\leq \eta_0 + L\Delta t^{\alpha} \sum_{j=1}^{n} b_{j,n+1} |\tilde{u}_j|.
\end{aligned} \tag{3.21}$$

Applying Lemma 3.3.1 yields

$$|\tilde{u}_{k+1}| \leq C\eta_0.$$

Therefore, we get the theorem below.

Theorem 14 *Suppose that u_j $(j = 1,2,\cdots,n+1)$ are the solutions of the fractional forward Euler method* (3.13)*, $f(t,y)$ satisfies the Lipschitz condition with respect to the second argument u with a Lipschitz constant L on the existing interval of its unique solution. Then the fractional forward Euler method* (3.13) *is stable.*

By almost the same reasoning, we can prove that the methods (3.15) and (3.16) are stable too.

Remark 3.3.3 *In Theorem 14, we have supposed that f satisfies the global Lipschitz condition, which is a little strong. In fact, if $f(t,u)$ satisfies the local Lipschitz condition with a Lipschitz constant L (or $|\frac{\partial f}{\partial u}|\leq L$) on the suitable domain, then Theorem 14 still holds. One can refer to [87] for more details.*

Next, we investigate the convergence of the methods (3.13), (3.15) and (3.16). We first consider the fractional forward Euler method (3.15).

Theorem 15 *Assume that $u(t)$ is the solution of* (3.12), *$f(t,u)$ satisfies the Lipschitz condition with respect to u with a Lipschitz constant L, and $f(t,u(t)), u(t)\in C^1[0,T]$, u_j $(1\leq j\leq N)$ are the solutions of the fractional forward Euler method* (3.13). *Then we have*

$$|u(t_{k+1})-u_{k+1}|\leq C\Delta t,\ k=0,1,\cdots,N-1, \tag{3.22}$$

where C is a positive constant independent of Δt and k.

Proof. Denote by $e_n = u(t_n)-u^n\,(n=0,1,\cdots,k,k+1)$. By (2.20) (see also Theorem 2.4 in [32]), we get

$$\left|\Delta t^{\alpha}\sum_{j=0}^{n} b_{j,n+1}f(t_j,u(t_j))-\frac{1}{\Gamma(\alpha)}\int_0^{t_{n+1}}(t_{n+1}-s)^{\alpha-1}f(s,u(s))\,\mathrm{d}s\right|\leq C\Delta t. \tag{3.23}$$

By (3.12) and (3.13), we get the error equation

$$u(t_{n+1})-u_{n+1}=\frac{1}{\Gamma(\alpha)}\int_0^{t_{n+1}}(t_{n+1}-t)^{\alpha-1}f(t,u(t))\,\mathrm{d}t-\Delta t^{\alpha}\sum_{j=0}^{n} b_{j,n+1}f(t_j,y_j).$$

Therefore

$$\begin{aligned}
|e_{n+1}| &=\left|\frac{1}{\Gamma(\alpha)}\int_0^{t_{n+1}}(t_{n+1}-t)^{\alpha-1}f(t,u(t))\,\mathrm{d}t-\Delta t^{\alpha}\sum_{j=0}^{k} b_{j,k+1}f(t_j,u_j)\right|\\
&\leq\left|\frac{1}{\Gamma(\alpha)}\int_0^{t_{n+1}}(t_{n+1}-t)^{\alpha-1}f(t,u(t))\,\mathrm{d}t-\Delta t^{\alpha}\sum_{j=0}^{n} b_{j,n+1}f(t_j,u(t_j))\right|\\
&\quad+\Delta t^{\alpha}\sum_{j=1}^{n} b_{j,n+1}|f(t_j,u(t_j))-f(t_j,u_j)| \qquad (3.24)\\
&\leq C\Delta t+L\Delta t^{\alpha}\sum_{j=1}^{n} b_{j,n+1}|e_j|\\
&\leq C\Delta t+C\Delta t^{\alpha}\sum_{j=1}^{n}(n+1-j)^{\alpha-1}|e_j|.
\end{aligned}$$

Applying Lemma 3.3.1 leads to the desired result. The proof is completed. □

By similar reasoning, one can prove that the methods (3.15) and (3.16) are convergent of order one.

- **Fractional Adams Method**

If $\left[\mathrm{D}_{0,t}^{-\alpha} f(t)\right]_{t=t_{n+1}}$ is approximated by the fractional trapezoidal formula (2.12), the following ***fractional trapezoidal rule*** is derived

$$u_{n+1} = \sum_{j=0}^{m-1} \frac{t_{n+1}^{j}}{j!} u_0^{(j)} + \sum_{j=0}^{n+1} a_{j,n+1} f(t_j, u_j), \tag{3.25}$$

where

$$a_{j,n+1} = \frac{\Delta t^{\alpha}}{\Gamma(\alpha+2)} \begin{cases} n^{\alpha+1} - (n-\alpha)(n+1)^{\alpha}, & j=0, \\ (n-j+2)^{\alpha+1} - 2(n-j+1)^{\alpha+1} + (n-j)^{\alpha+1}, & 1 \le j \le n, \\ 1, & j=n+1. \end{cases} \tag{3.26}$$

The above scheme (3.25) is implicit, which needs much more computation. On one hand, we can use an approximation method to get u_{n+1} in (3.25), such as the *Newton iterative method*, the *Adomian decomposition method* [83], and so on. On the other hand, similar to the *predictor-corrector method* for the ODEs, we first use (3.13) to get u_{n+1}^P (predictor), then we use (3.25) to get u_{n+1} (corrector) by replacing u_{n+1} with u_{n+1}^P on the right-hand side of (3.25), which leads to the *fractional Adams method* [27, 31, 32, 33, 82]

$$\begin{cases} u_{n+1}^P = \sum_{j=0}^{m-1} \frac{t_{n+1}^{j}}{j!} u_0^{(j)} + \sum_{j=0}^{n} b_{j,n+1} f(t_j, u_j), \\ u_{n+1} = \sum_{j=0}^{m-1} \frac{t_{n+1}^{j}}{j!} u_0^{(j)} + \sum_{j=0}^{n} a_{j,n+1} f(t_j, u_j) + a_{n+1,n+1} f(t_{n+1}, u_{n+1}^P). \end{cases} \tag{3.27}$$

Remark 3.3.4 *If $\alpha = 1$, the fractional Adams method* (3.27) *is reduced to the classical predictor-corrector method for* (3.10)

$$\begin{cases} u_{n+1}^P = u_n + \Delta t f(t_n, u_n), \\ u_{n+1} = \frac{\Delta t}{2}\Big(f(t_n, u_n) + f(t_{n+1}, u_{n+1}^P)\Big). \end{cases} \tag{3.28}$$

Remark 3.3.5 *In [117], a predictor-corrector algorithm was presented based on the generalized Taylor's formula, which is similar to* (3.27)*, except that the predictor is chosen as*

$$u_{n+1}^P = u_n + \frac{\Delta t^{\alpha}}{\Gamma(\alpha+1)} f(t_n, u_n).$$

Remark 3.3.6 *The detailed error analysis of the fractional Adams method is investigated in [30] and [82], where the error estimates were proved by the mathematical induction method for the small enough T. Using Lemma 3.3.1, one can easily get the error bounds for the fractional Adams method* (3.27)*. The stability of the method* (3.27) *can be also proved similarly to that of the fractional Euler method, see [87] for more information.*

- **High Order Methods**

Based on the polynomial interpolation, one can theoretically construct high order methods for FODEs (3.1) [13, 69, 73, 165].

Consider the discretization of the fractional integral

$$\mathrm{D}_{0,t}^{-\alpha} f(t) = \int_0^t (t-s)^{\alpha-1} f(s)\,\mathrm{d}s. \tag{3.29}$$

We first let $t = t_{n+1}$ in (3.29), which yields from (2.16)

$$\left[\mathrm{D}_{0,t}^{-\alpha} f(t)\right]_{t=t_{n+1}} = \sum_{k=0}^{n+1} c_{k,n+1} f(t_k) + \sum_{k=0}^{n} \hat{c}_{k,n+1} f(t_{k+\frac{1}{2}}) + O(\Delta t^3), \tag{3.30}$$

where $c_{k,n+1}$ and $\hat{c}_{k,n+1}$ are defined by (2.17) and (2.18), respectively.

Then, we choose $t = t_{k+\frac{1}{2}}$ in (3.29), which leads to

$$\begin{aligned}
\left[D_{0,t}^{-\alpha} f(t)\right]_{t=t_{n+\frac{1}{2}}} &= \frac{1}{\Gamma(\alpha)} \int_0^{t_{n+\frac{1}{2}}} (t_{n+\frac{1}{2}} - s)^{\alpha-1} f(s)\,\mathrm{d}s, \\
&= \frac{1}{\Gamma(\alpha)} \sum_{j=0}^{n-1} \int_{t_j}^{t_{j+1}} (t_{n+\frac{1}{2}} - s)^{\alpha-1} f(s)\,\mathrm{d}s \\
&\quad + \frac{1}{\Gamma(\alpha)} \int_{t_n}^{t_{n+\frac{1}{2}}} (t_{n+\frac{1}{2}} - s)^{\alpha-1} f(s)\,\mathrm{d}s.
\end{aligned} \tag{3.31}$$

On each subinterval $[t_j, t_{j+1}], j = 0, 1, \cdots, n-1$, $f(t)$ is approximated by the quadric polynomials defined on the nodes $\{t_j, t_{j+\frac{1}{2}}, t_{j+1}\}$. On the interval $[t_n, t_{n+\frac{1}{2}}]$, $f(t)$ is approximated by the quadric polynomial defined on the nodes $\{t_n, t_{n+\frac{1}{4}}, t_{n+\frac{1}{2}}\}$ with $t_{n+\frac{1}{4}} = (t_n + t_{n+\frac{1}{2}})/2$. Let $S_1^j = \{j, j+\frac{1}{2}, j+1\}$ and $S_2^n = \{n, n+\frac{1}{4}, n+\frac{1}{2}\}$, we obtain

$$f(t) \approx \tilde{f}(t) = \begin{cases} \sum\limits_{k\in S_1^j} f(t_k)\varphi_{j,k}(t), & t \in [t_j, t_{j+1}] \\ \sum\limits_{k\in S_2^n} f(t_k)\psi_{n,k}(t), & t \in [t_n, t_{n+\frac{1}{2}}] \end{cases} \tag{3.32}$$

where

$$\varphi_{j,k}(t) = \prod_{i\in S_1^j, i\neq k} \frac{t - t_i}{t_k - t_i}, \quad k \in S_1^j,$$

and

$$\psi_{n,k}(t) = \prod_{i\in S_2^n, i\neq k} \frac{t - t_i}{t_k - t_i}, \quad k \in S_2^n.$$

Substituting $f(t)$ for $\tilde{f}(t)$ in (3.31), we have

$$\left[\mathrm{D}_{0,t}^{-\alpha} f(t)\right]_{t=t_{n+\frac{1}{2}}} = \sum_{j=0}^{n-1} \sum_{k\in S_1^j} d_{j,k}^n f(t_k) + \sum_{k\in S_2^n} \hat{d}_k^n f(t_k) + O(\Delta t^3), \tag{3.33}$$

in which

$$d_{j,k}^{n}=\frac{1}{\Gamma(\alpha)}\int_{t_j}^{t_{j+1}}(t_{n+\frac{1}{2}}-s)^{\alpha-1}\varphi_{j,k}(s)\,\mathrm{d}s,$$

and

$$\hat{d}_{k}^{n}=\frac{1}{\Gamma(\alpha)}\int_{t_n}^{t_{n+\frac{1}{2}}}(t_{n+\frac{1}{2}}-s)^{\alpha-1}\psi_{n,k}(s)\,\mathrm{d}s.$$

Note that we have used the value $f(t_{n+\frac{1}{4}})$. We find that

$$f(t_{n+\frac{1}{4}})=\frac{3}{8}f(t_n)+\frac{3}{4}f(t_{n+\frac{1}{2}})-\frac{1}{8}f(t_{n+1})+O(\Delta t^3).$$

Therefore, we get the following approximation of $\left[D_{0,t}^{-\alpha}f(t)\right]_{t=t_{n+\frac{1}{2}}}$

$$\begin{aligned}\left[\mathrm{D}_{0,t}^{-\alpha}f(t)\right]_{t=t_{n+\frac{1}{2}}}=&\sum_{j=0}^{n-1}\sum_{k\in S_1^j}d_{j,k}^{n}f(t_k)+\hat{d}_{n}^{n}f(t_n)+\hat{d}_{n+\frac{1}{2}}^{n}f(t_{n+\frac{1}{2}})\\&+\hat{d}_{n+\frac{1}{4}}^{n}\left[\frac{3}{8}f(t_n)+\frac{3}{4}f(t_{n+\frac{1}{2}})-\frac{1}{8}f(t_{n+1})\right]+O(\Delta t^3).\end{aligned}\tag{3.34}$$

Suppose that $u_j, j=0,1,\cdots,n$ and $u_{j+\frac{1}{2}}, j=0,1,\cdots,n-1$ are known. Then, by (3.30), (3.12), and (3.34), we derive the following high order numerical scheme

$$\begin{cases}u_{n+\frac{1}{2}}=\displaystyle\sum_{j=0}^{m-1}\frac{t_{n+\frac{1}{2}}^{j}}{j!}u_0^{(j)}+\sum_{j=0}^{n-1}\sum_{k\in S_1^j}d_{j,k}^{n}f(t_k)+\hat{d}_{n}^{n}f(t_n,u_n)+\hat{d}_{n+\frac{1}{2}}^{n}f(t_{n+\frac{1}{2}},u_{n+\frac{1}{2}})\\ \qquad+\hat{d}_{n+\frac{1}{4}}^{n}\left[\dfrac{3}{8}f(t_n,u_n)+\dfrac{3}{4}f(t_{n+\frac{1}{2}},u_{n+\frac{1}{2}})-\dfrac{1}{8}f(t_{n+1},u_{n+1})\right],\\ u_{n+1}=\displaystyle\sum_{j=0}^{m-1}\frac{t_{n+1}^{j}}{j!}u_0^{(j)}+\sum_{k=0}^{n+1}c_{k,n+1}f(t_k,u_k)+\sum_{k=0}^{n}\hat{c}_{k,n+1}f(t_{k+\frac{1}{2}},u_{k+\frac{1}{2}}).\end{cases}\tag{3.35}$$

Obviously, the method (3.35) is implicit and nonlinear, which can be solved by the iteration method. It needs more computational time to get u_{n+1} and $u_{n+\frac{1}{2}}$ from the nonlinear system (3.35). The predictor-corrector method is a good approach to present the explicit high order method to solve the nonlinear equation as (3.12), which costs less computational time.

One simple way is to use the fractional Adams method to obtain u_{n+1}^{P} as a predictor of u_{n+1} based on the grid points $\{t_0,t_1,\cdots,t_n,t_{n+1}\}$. The predictor $u_{n+\frac{1}{2}}^{P}$ of $u_{n+\frac{1}{2}}$ can be similarly derived by the fractional Adams method based on the grid points $\{t_0,t_{\frac{1}{2}},\cdots,t_{n-\frac{1}{2}},t_{n+\frac{1}{2}}\}$. So, u_{n+1} and $u_{n+\frac{1}{2}}$ can be calculated from (3.35) by replacing u_{n+1} and $u_{n+\frac{1}{2}}$ on the right-hand side of (3.35) with u_{n+1}^{P} and $u_{n+\frac{1}{2}}^{P}$.

Next, we introduce another *predictor-corrector method* [73] based on the **fractional Simpson's formula** (2.16) for (3.1) or (3.12).

Let $t = t_{n+1}$ in (3.12). By (2.16), one gets the following implicit method

$$u_{n+1} = \sum_{j=0}^{m-1} \frac{t_{n+1}^j}{j!} u_0^{(j)} + \sum_{k=0}^{n+1} c_{k,n+1} f(t_k, u_k) + \sum_{k=0}^{n} \hat{c}_{k,n+1} f(t_{k+\frac{1}{2}}, u_{k+\frac{1}{2}}), \tag{3.36}$$

where $c_{k,n+1}$ and $\hat{c}_{k,n+1}$ are defined by (2.17) and (2.18), respectively.

In order to get u_{n+1} from (3.29), the nonlinear equation about u_{n+1} needs to be solved. One way to solve this problem is to use the fractional rectangular formula (2.8) to get the predictor of u_{n+1}, which is given by

$$u_{n+1}^P = \sum_{j=0}^{m-1} \frac{t_{n+1}^j}{j!} u_0^{(j)} + \sum_{k=0}^{n} b_{n-k} f(t_k, u_k), \tag{3.37}$$

where $b_k = \frac{\Delta t^\alpha}{\Gamma(\alpha+1)}[(k+1)^\alpha - k^\alpha]$ is defined by (2.5).

Noticing that $u_{n+\frac{1}{2}}$ is unknown, one can also use the fractional rectangular formula (2.8) to approximate $u_{n+\frac{1}{2}}$, which reads as

$$u_{n+\frac{1}{2}} = \sum_{j=0}^{m-1} \frac{t_{n+\frac{1}{2}}^j}{j!} u_0^{(j)} + \sum_{k=0}^{n} \hat{b}_{n-k} f(t_k, u_k), \tag{3.38}$$

where $\hat{b}_{n-k} = \frac{\Delta t^\alpha}{\Gamma(\alpha+1)}[(n+\frac{1}{2}-k)^\alpha - (n-\frac{1}{2}-k)^\alpha]$.

Hence, we get the initial *predictor-corrector method* developed in [73] as follows

$$u_{n+1} = \sum_{j=0}^{m-1} \frac{t_{n+1}^j}{j!} u_0^{(j)} + \sum_{k=0}^{n} c_{k,n+1} f(t_k, u_k) + c_{n+1,n+1} f(t_{n+1}, u_{n+1}^P) + \sum_{k=0}^{n} \hat{c}_{k,n+1} f(t_{k+\frac{1}{2}}, u_{k+\frac{1}{2}}), \tag{3.39}$$

$$u_{n+1}^P = \sum_{j=0}^{m-1} \frac{t_{n+1}^j}{j!} u_0^{(j)} + \sum_{k=0}^{n} b_{n-k} f(t_k, u_k), \tag{3.40}$$

$$u_{n+\frac{1}{2}} = \sum_{j=0}^{m-1} \frac{t_{n+\frac{1}{2}}^j}{j!} u_0^{(j)} + \sum_{k=0}^{n} \hat{b}_{n-k} f(t_k, u_k). \tag{3.41}$$

The error estimate of the above method (3.39)–(3.41) is similar to that of the fractional Adams method. In fact, the implicit method (3.36) has higher order accuracy due to the quadric interpolation of f if f is smooth enough. If the accuracy of the predictors u_{n+1}^P and $u_{n+\frac{1}{2}}$ in the method (3.39)–(3.41) is improved, then one can get the higher order method. It is easy to prove that the predictor u_{n+1}^P and $u_{n+\frac{1}{2}}$ in (3.39)–(3.41) have first-order accuracy. An improved algorithm with high-order predictors is described below.

Improved algorithm

Step 1: Use the fractional Simpson's formula to get the implicit method (3.36) based on the nodes $\{t_k, t_{k+\frac{1}{2}}, k=0,1,\cdots,n-1\} \cup \{t_{n+\frac{1}{2}}\}$. Replace u_{n+1} on the right-hand side of (3.36) by the predictor u^P_{n+1}, which leads to (3.39).

Step 2: Use the fractional Adams method (3.27) based on the grid nodes $\{t_k, k = 0,1,\cdots,n+1\}$ to get the approximate value of u_{n+1}, which is denoted by u^P_{n+1} that is given below

$$\begin{cases} v^P_{n+1} = \sum_{j=0}^{m-1} \frac{t^j_{n+1}}{j!} u_0^{(j)} + \sum_{j=0}^{n} b_{j,n+1} f(t_j, u_j), \\ u^P_{n+1} = \sum_{j=0}^{m-1} \frac{t^j_{n+1}}{j!} u_0^{(j)} + \sum_{j=0}^{n} a_{j,n+1} f(t_j, u_j) + a_{n+1,n+1} f(t_{n+1}, v^P_{n+1}). \end{cases} \tag{3.42}$$

Step 3: Use the fractional Adams method (3.27) based on the grid nodes $\{t_k, k = 0,1,\cdots,n\} \cup \{t_{n+\frac{1}{2}}\}$ to get the approximate value of $u_{n+\frac{1}{2}}$, which is given by

$$\begin{cases} u_{n+\frac{1}{2}} = \sum_{j=0}^{m-1} \frac{t^j_{n+\frac{1}{2}}}{j!} u_0^{(j)} + \sum_{k=0}^{n} e_{k,n+1} f(t_k, u_k) + e_{n+1,n+1} f(t_{n+\frac{1}{2}}, u^P_{n+\frac{1}{2}}), \\ u^P_{n+\frac{1}{2}} = \sum_{j=0}^{m-1} \frac{t^j_{n+\frac{1}{2}}}{j!} u_0^{(j)} + \sum_{k=0}^{n} \hat{b}_{n-k} f(t_k, u_k), \end{cases} \tag{3.43}$$

where

$$e_{k,n+1} = \frac{\Delta t^\alpha}{\Gamma(\alpha+2)} \begin{cases} \left(n+\frac{1}{2}\right)^{\alpha+1} - \left(n+\frac{1}{2}\right)^{\alpha}\left(n-\frac{1}{2}-\alpha\right), & k=0, \\ \left(n-k+\frac{3}{2}\right)^{\alpha+1} - 2\left(n-k+\frac{1}{2}\right)^{\alpha+1} + \left(n-k-\frac{1}{2}\right)^{\alpha+1}, & 1 \le k < n, \\ \left(\frac{3}{2}\right)^{\alpha+1} - 3\left(\frac{1}{2}\right)^{\alpha+1}, & k=n, \\ \left(\frac{1}{2}\right)^{\alpha}, & k=n+1. \end{cases} \tag{3.44}$$

Therefore, (3.39), (3.42) and (3.43) give the improved algorithm.

If $f(t,u)$ satisfies the Lipschitz condition (3.20) and $f(t,u(t))$ is suitably smooth, one can easily prove that the predictor-corrector method (3.39)–(3.41) is convergent of order $O(\Delta t^{1+\sigma(\alpha)})$, where $\sigma(\alpha) = 1$ for $\alpha \ge 1$, and $\sigma(\alpha) = \alpha$ for $0 < \alpha < 1$. The **Improved Algorithm** is convergent of order $O(\Delta t^{1+2\sigma(\alpha)})$ for the suitably smooth $f(t,u(t))$, and the high order method (3.35) is convergent of order $O(\Delta t^3)$ for suitably smooth $f(t,u(t))$.

Remark 3.3.7 *According to Lemma 3.3.1, if we want to prove the stability of* (3.39)–(3.41)*, we just need to prove*

$$\max_{0\le k\le n} \left\{|c_{k,n+1}|, |\hat{c}_{k,n+1}|, |b_{n-k}|, |\hat{b}_{n-k}|\right\} \le C(n+1-k)^{\alpha-1}, \tag{3.45}$$

TABLE 3.1: The absolute errors at $t = 1$ for Example 2 by the fractional forward Euler method (3.13).

$1/\Delta t$	$\alpha = 0.1$	$\alpha = 0.3$	$\alpha = 0.5$	$\alpha = 0.7$	$\alpha = 0.9$
10	6.9223e−02	6.2373e−02	5.6086e−02	5.1208e−02	4.7865e−02
20	3.7760e−02	3.1940e−02	2.8002e−02	2.5684e−02	2.4474e−02
40	1.9692e−02	1.5838e−02	1.3749e−02	1.2762e−02	1.2359e−02
80	9.9614e−03	7.7267e−03	6.7202e−03	6.3292e−03	6.2062e−03
160	4.9498e−03	3.7476e−03	3.2891e−03	3.1418e−03	3.1085e−03
320	2.4367e−03	1.8175e−03	1.6155e−03	1.5622e−03	1.5552e−03
640	1.1944e−03	8.8363e−04	7.9654e−04	7.7798e−04	7.7776e−04
EOC	1.0287	1.0404	1.0202	1.0057	0.9997

where C is independent of n and k. Obviously, (3.45) *holds by the simple calculation. The stability of the* ***Improved Algorithm*** *((3.39), (3.42) and (3.43)) can be proved similarly. Of course, this is also true for the high order method* (3.35).

3.3.1 Numerical Examples

This subsection gives some numerical results for the numerical methods in this section.

Example 2 *Consider the following nonlinear FODE*

$$ {}_{C}D_{0,t}^{\alpha}y(t) + y^2(t) = f(t), \quad 0 < \alpha < 2, \ t > 0, \tag{3.46}$$

where

$$f(t) = \frac{\Gamma(6)}{\Gamma(6-\alpha)}t^{5-\alpha} - \frac{3\Gamma(5)}{\Gamma(5-\alpha)}t^{3-\alpha} + \frac{\Gamma(5)}{\Gamma(4-\alpha)}t^{3-\alpha} + (t^5 - 3t^4 + 2t^3)^2.$$

The exact solution is $y(t) = t^5 - 3t^4 + 2t^3$ *with the following initial conditions:*

$$0 < \alpha < 1, \ y(0) = 0.$$

In this example, we test the fractional forward Euler method (3.13) and the Adams method (3.27), respectively. The results are shown in Tables 3.1–3.2 . We can find that the experimental order of convergence (EOC) of the fractional forward Euler method (3.13) and the fractional Adams method (3.27) are 1 and 2 respectively, which are in line with the theoretical analysis. The EOC here is computed by the formula: $\log_2 \frac{E(\Delta t,T)}{E(\Delta t/2,T)}$, where $E(\Delta t, T) = |y(T) - y_{T/\Delta t}|$.

Table 3.3 displays the long-term integration of the two methods (3.13) and (3.27), where the step size $\Delta t = 1e-5$ (10000 steps). We can see that the satisfactory and reliable results are obtained for the three methods in this example.

TABLE 3.2: The absolute errors at $t = 1$ for Example 2 by the fractional Adams method (3.27).

$1/\Delta t$	$\alpha = 0.1$	$\alpha = 0.3$	$\alpha = 0.5$	$\alpha = 0.7$	$\alpha = 0.9$
10	5.4721e−03	4.1744e−03	3.6407e−03	3.4025e−03	3.2901e−03
20	1.4456e−03	9.7613e−04	8.5514e−04	8.1615e−04	8.0180e−04
40	3.4634e−04	2.2288e−04	2.0346e−04	1.9890e−04	1.9779e−04
80	7.7788e−05	5.0421e−05	4.8993e−05	4.8957e−05	4.9099e−05
160	1.6044e−05	1.1214e−05	1.1877e−05	1.2118e−05	1.2228e−05
320	2.7132e−06	2.3899e−06	2.8839e−06	3.0090e−06	3.0508e−06
640	1.5675e−07	4.5897e−07	6.9794e−07	7.4833e−07	7.6184e−07
EOC	4.1135	2.3805	2.0468	2.0075	2.0016

TABLE 3.3: The absolute errors at $t = 1$ for Example 2 with $\Delta t = 1e-5$.

Methods	$\alpha = 0.2$	$\alpha = 0.5$	$\alpha = 0.8$
Forward Euler method (3.13)	5.7566e-05	4.9565e-05	4.9614e-05
Adams method (3.27)	1.7998e-08	2.1195e-09	3.0764e-09

Example 3 *Consider the following fractional differential equation*

$$ {}_{C}D^{\alpha}_{0,t}y(t) = -y(t) + \frac{t^{4-\alpha}}{\Gamma(5-\alpha)}, \quad 0 < \alpha < 1, \quad y(0) = 0,\ t > 0. $$

Its exact solution is

$$ y(t) = t^4 E_{\alpha,5}(-t^{\alpha}), $$

where $E_{\alpha,\beta}(z) = \sum_{k=0}^{\infty} z^k/\Gamma(\alpha k + \beta)$ *is the two-parameter Mittag–Leffler function.*

We also apply the fractional forward Euler method (3.13) and the fractional Adams method (3.27) to get the numerical solutions; the results are shown in Tables 3.4–3.5. The numerical results show good agreement with the exact solution.

The long-term integration (10000 steps) is still tested in this example; the results are shown in Table 3.6, which shows good agreement with the analytical solutions.

3.4 Fractional Linear Multistep Methods

The fractional linear multistep method (FLMM) for fractional calculus was first studied by Lubich [102, 103, 104], which can be seen as the generalization of the linear multistep method (LMM) for classical calculus.

TABLE 3.4: The absolute errors at $t = 1$ for Example 3 by the fractional forward Euler method (3.13).

$1/\Delta t$	$\alpha = 0.1$	$\alpha = 0.3$	$\alpha = 0.5$	$\alpha = 0.7$	$\alpha = 0.9$
10	4.1205e−03	4.1397e−03	4.2821e−03	4.5571e−03	4.9388e−03
20	2.0192e−03	1.9788e−03	2.0515e−03	2.2207e−03	2.4550e−03
40	9.8557e−04	9.4634e−04	9.8894e−04	1.0892e−03	1.2222e−03
80	4.8052e−04	4.5427e−04	4.8032e−04	5.3731e−04	6.0937e−04
160	2.3428e−04	2.1912e−04	2.3490e−04	2.6625e−04	3.0416e−04
320	1.1429e−04	1.0623e−04	1.1553e−04	1.3235e−04	1.5192e−04
640	5.5794e−05	5.1741e−05	5.7071e−05	6.5923e−05	7.5914e−05

TABLE 3.5: The absolute errors at $t = 1$ for Example 3 by the fractional Adams method (3.27).

$1/\Delta t$	$\alpha = 0.1$	$\alpha = 0.3$	$\alpha = 0.5$	$\alpha = 0.7$	$\alpha = 0.9$
10	8.3901e−03	2.8610e−03	1.4115e−03	8.3648e−04	5.6869e−04
20	4.0702e−03	1.0375e−03	4.4345e−04	2.3463e−04	1.4543e−04
40	1.7756e−03	3.7406e−04	1.4182e−04	6.7031e−05	3.7535e−05
80	7.4735e−04	1.3656e−04	4.6334e−05	1.9468e−05	9.7498e−06
160	3.1280e−04	5.0689e−05	1.5430e−05	5.7295e−06	2.5432e−06
320	1.3159e−04	1.9114e−05	5.2187e−06	1.7035e−06	6.6531e−07
640	5.5820e−05	7.3069e−06	1.7863e−06	5.1041e−07	1.7440e−07

TABLE 3.6: The absolute errors at $t = 1$ for Example 3 with $\Delta t = 1e-5$.

Methods	$\alpha = 0.2$	$\alpha = 0.5$	$\alpha = 0.8$
Forward Euler method (3.13)	3.0345e-06	3.5706e-06	4.5247e-06
Adams method (3.27)	5.0991e-07	2.7219e-08	1.8964e-09

In this section, we introduce the FLMMs for the FODE (3.1). For simplicity, we consider the case of $0 < \alpha < 1$, i.e.,

$$ {}_C\mathrm{D}_{0,t}^{\alpha} u(t) = f(t, u(t)), \qquad u(0) = u_0. \tag{3.47} $$

The equivalent form of (3.47) reads

$$ u(t) = u_0 + \frac{1}{\Gamma(\alpha)} \int_0^t (t-s)^{\alpha-1} f(s, u(s))\,\mathrm{d}s. \tag{3.48} $$

If $\alpha = 1$, the FODE (3.47) is reduced to the classical ODE as

$$ u'(t) = f(t, u(t)), \qquad u(0) = u_0. \tag{3.49} $$

It is well known that the p-step LMM for (3.49) has the form

$$ \sum_{k=0}^{p} \alpha_k u_{n+k} = \Delta t \sum_{k=0}^{p} \beta_k f(t_{n+k}, u_{n+k}), \quad n = 0, 1, \cdots. \tag{3.50} $$

The first and second character polynomials of the LMM (3.50) read as

$$ \rho(\xi) = \sum_{k=0}^{p} \alpha_k \xi^k, \qquad \sigma(\xi) = \sum_{k=0}^{p} \beta_k \xi^k. $$

Denote by

$$ w(\xi) = \frac{\sigma(1/\xi)}{\rho(1/\xi)}. \tag{3.51} $$

It is also known that the equivalent form of (3.50) can be written as

$$ u_n - u_0 = \Delta t \sum_{j=0}^{n} \omega_{n-j}^{(1)} f(t_j, u_j) + \Delta t \sum_{j=0}^{s} w_{n,j}^{(1)} f(t_j, u_j), \tag{3.52} $$

where $\{\omega_j^{(1)}\}$ are the coefficients of the Taylor expansions of the generating function $\omega(\xi)$ defined by (3.51). In fact, $\Delta t \sum_{j=0}^{n} \omega_{n-j}^{(1)} f(t_j, u(t_j)) + \Delta t \sum_{j=0}^{s} w_{n,j}^{(1)} f(t_j, u(t_j))$ is just the pth-order approximation of $\int_0^{t_n} f(s, u(s)) = \left[D_{0,t}^{-1} f(t, u(t))\right]_{t=t_n}$.

One can also easily obtain

$$ \Delta t w(e^{-\Delta t}) = 1 + O(\Delta t^p), $$

which yields

$$ \left(\Delta t w(e^{-\Delta t})\right)^{\alpha} = (1 + O(\Delta t^p))^{\alpha} = 1 + O(\Delta t^p). \tag{3.53} $$

The following theorem states that the FLMM for (3.48) has a similar form as (3.52).

Theorem 16 ([104]) *Let $\rho(\xi)$ and $\sigma(\xi)$ denote the first and second character polynomials of the pth-order LMM* (3.50). *Assume that the zeros of $\sigma(\xi)$ have absolute value less than* 1. *Let $\{\omega_k^{(\alpha)}\}$ denote the Taylor expansions of the generating function $w^{(\alpha)}(\xi) = \left(\frac{\sigma(1/\xi)}{\rho(1/\xi)}\right)^{\alpha}$. Then the convolution quadrature*

$$\Delta t^{\alpha} \sum_{j=0}^{n} \omega_{n-j}^{(\alpha)} f(t_j) + \Delta t^{\alpha} \sum_{j=0}^{s} w_{n,j}^{(\alpha)} f(t_j)$$

is convergent of order p with respect to $D_{0,t}^{-\alpha} = {}_{RL}D_{0,t}^{-\alpha}$, i.e.,

$$\left[{}_{RL}\mathrm{D}_{0,t}^{-\alpha} f(t)\right]_{t=t_n} = \left[\mathrm{D}_{0,t}^{-\alpha} f(t)\right]_{t=t_n} = \Delta t^{\alpha} \sum_{j=0}^{n} \omega_{n-j}^{(\alpha)} f(t_j) + \Delta t^{\alpha} \sum_{j=0}^{s} w_{n,j}^{(\alpha)} f(t_j) + O(\Delta t^{p}), \tag{3.54}$$

where $w_{n,j}^{(\alpha)}$ are starting weights such that the above equation is exact for $f(t) = t^{\mu}, \mu < p$.

By Theorem 16 one can construct the pth-order FLMMs for (3.48) as follows

$$u_n - u_0 = \Delta t^{\alpha} \sum_{j=0}^{n} \omega_{n-j}^{(\alpha)} f(t_j, u_j) + \Delta t^{\alpha} \sum_{j=0}^{s} w_{n,j}^{(\alpha)} f(t_j, u_j), \tag{3.55}$$

where $w_{n,j}^{(\alpha)}$ are the starting weights that are chosen such that the asymptotic behavior of the function $f(t, u(t))$ near the origin is taken into account [28, 104], and $\omega_j^{(\alpha)}$ can be the coefficients of the Taylor expansion of the following generating functions

$$w^{(\alpha)}(\xi) = \left(\frac{\sigma(1/\xi)}{\rho(1/\xi)}\right)^{\alpha}. \tag{3.56}$$

In fact, $w^{(\alpha)}(z)$ can be also other generating functions; we just list some often used generating functions (see also (2.43)–(2.45)) and their convergence orders as follows

$$w^{(\alpha)}(\xi) = \begin{cases} \left(\frac{1}{2}\frac{1+\xi}{1-\xi}\right)^{\alpha}, & \text{order } 2, \\ \left(\sum\limits_{j=1}^{p} \frac{1}{j}(1-\xi)^{j}\right)^{-\alpha}, & \text{order } p, \\ (1-\xi)^{-\alpha}\left(\gamma_0 + \gamma_1(1-\xi) + \cdots + \gamma_{p-1}(1-\xi)^{p-1}\right), & \text{order } p, \end{cases} \tag{3.57}$$

where γ_i satisfies

$$\sum_{i=0}^{\infty} \gamma_i (1-\xi)^i = \left(\frac{\ln \xi}{\xi - 1}\right)^{-\alpha}.$$

Remark 3.4.1 *The fractional order α in* (3.56) *and* (3.57) *can be negative. In such a case, the convolution quadrature* (3.54) *is just the pth-order approximation of the $(-\alpha)$th-order Riemann–Liouville fractional derivative operator.*

It is known that $\omega_n^{(\alpha)} = O(n^{\alpha-1})$ and $w_{n,j}^{(\alpha)} = O(n^{\alpha-1})$ when the generating functions (3.56) or (3.57) are used [104]. So the convergence of the FLMM (3.55) can be easily obtained by using the generalized Gronwall's inequality from Lemma 3.3.1. See [87, 95, 102] for more details.

Next, we investigate the stability properties of the FLMMs (3.55). For simplicity, we consider the following linear model problem

$$\begin{cases} {}_C\mathrm{D}_{0,t}^{\alpha} u(t) = \lambda u(t), & 0 < \alpha < 1, \\ u(0) = u_0. \end{cases} \tag{3.58}$$

The above FODE (3.58) is equivalent to the following Abel integral equation of the second kind

$$u(t) - u_0 = \frac{\lambda}{\Gamma(\alpha)} \int_0^t (t-s)^{\alpha-1} u(s)\,\mathrm{d}s, \quad 0 < \alpha < 1. \tag{3.59}$$

The more general form of the Abel integral equation of the second kind reads

$$u(t) = g(t) + \frac{\lambda}{\Gamma(\alpha)} \int_0^t (t-s)^{\alpha-1} u(s)\,\mathrm{d}s, \quad 0 < \alpha < 1. \tag{3.60}$$

The equation (3.59) can be seen as a special case of (3.60). Here we mainly study the FLMM for (3.60), which reads

$$u_n = g_n + \Delta t^{\alpha} \lambda \sum_{j=0}^{n} \omega_{n-j}^{(\alpha)} u_j, \quad n > s. \tag{3.61}$$

Here $\omega_{n-j}^{(\alpha)}$ are the Taylor expansions of the generating functions defined by (3.56) or (3.57), $g_n = g(t_n) + \Delta t^{\alpha} \lambda \sum_{j=0}^{s} w_{n,j}^{(\alpha)} u_j$, $u_0, u_1, \cdots, u_s$ are given starting values which are usually computed by a different method.

Theorem 17 ([103]) *Consider the integral equation* (3.60) *with* $g \in C[0,\infty)$ *and* $|\arg\lambda - \pi| < (1 - \frac{1}{2}\alpha)\pi$. *Then there exists a unique solution* $u \in C[0,\infty)$ *which satisfies*

(a) $u(t) \to 0$ *as* $t \to \infty$ *when* $g(t)$ *has a finite limit as* $t \to \infty$.

(b) $u(t)$ *is bounded on* $[0,\infty)$ *when* $g(t)$ *is bounded.*

In the following, we introduce several concepts of stability based on Theorem 17 which extend the classical stability concepts [103].

Definition 12 *The FLMM* (3.61) *is called* ***A-stable*** *if the numerical solution* u_n *given by* (3.61) *satisfies*

$$u_n \to 0 \text{ as } n \to \infty \text{ whenever } \{g_n\} \text{ has a finite limit}$$

for every stepsize Δt *and for all* λ *in* $|\arg\lambda - \pi| < (1 - \frac{1}{2}\alpha)\pi$, *the analytical stability region of* (3.60).

Definition 13 *The* ***stability region*** S *of the FLMM* (3.61) *is the set of all complex* $z = \Delta t^{\alpha}\lambda$ *for which the numerical solution* $\{u_n\}$ *given by* (3.61) *satisfies*

$$u_n \to 0 \text{ as } n \to \infty \text{ whenever } \{g_n\} \text{ has a finite limit.}$$

The method is called ***strongly stable****, if for any* λ *with* $|\arg\lambda - \pi| < (1 - \frac{1}{2}\alpha)\pi$ *there exists* $h_0(\lambda) > 0$ *such that* $\Delta t^{\alpha}\lambda$ *is constrained in* S *for all* $0 < \Delta t \le h_0(\lambda)$. *The method is called* ***A(θ)-stable*** *if* S *contains the sector* $|\arg\lambda - \pi| < \theta$.

Theorem 18 ([103]) *If* $\omega_n^{(\alpha)}$ *is the coefficients of the Taylor expansions of* $w^{(\alpha)}(\xi)$ *with*

$$\omega_n^{(\alpha)} = \frac{n^{\alpha-1}}{\Gamma(\alpha)} + v_n \quad (n \ge 1) \quad \text{with} \quad \sum_{n=1}^{\infty} |v_n| < \infty, \tag{3.62}$$

then the stability region of the FLMMs (3.61) *is*

$$\mathbb{C} \setminus \{1/w^{(\alpha)}(\xi) : |\xi| \le 1\}.$$

Proof. In order to prove this theorem, we need the following two properties.

(a) (**Wiener's inversion theorem**) If $\{a_n\}$ is in ℓ^1 and $a(\xi) = \sum_{n=0}^{\infty} a_n \xi^n \ne 0$ for $|\xi| \le 1$, then the coefficients of the Taylor expansions of $1/a(\xi)$ is again in ℓ^1.

(b) Assume that the coefficients of the Taylor expansions of $a(\xi)$ is in ℓ^1. Let $|\xi_0| \le 1$. Then the coefficients of the Taylor expansions of

$$b(\xi) = \frac{a(\xi) - a(\xi_0)}{\xi - \xi_0}$$

converge to zero.

Let $z = \Delta t^{\alpha}\lambda$. Since 0 is neither contained in the stability region S nor in $\tilde{S} = \mathbb{C} \setminus \{1/w^{(-\alpha)}(\xi) : |\xi| \le 1\}$, we can from now on assume $z \ne 0$. In terms of the corresponding power series we can rewrite (3.61) as

$$u(\xi) = g(\xi) + z w^{(\alpha)}(\xi) u(\xi),$$

or equivalently,

$$u(\xi) = \frac{g(\xi)}{1 - z w^{(\alpha)}(\xi)} = \frac{(1-\xi)^{\alpha} g(\xi)}{(1-\xi)^{\alpha}[1 - z w^{(\alpha)}(\xi)]}, \tag{3.63}$$

where

$$u(\xi) = \sum_{n=0}^{\infty} u_n \xi^n, \quad g(\xi) = \sum_{n=0}^{\infty} g_n \xi^n.$$

Next, we prove that $\tilde{S} \subset S$. By the property (a), the sequence of the coefficients of the Taylor expansions of $(1-\xi)^{\alpha}[1 - z w^{(\alpha)}(\xi)]$ is in ℓ^1. If $z \in \tilde{S}$, we have

$$1 - z w^{(\alpha)}(\xi) \ne 0 \quad \text{for} \quad |\xi| \le 1 \quad \text{with} \quad \xi \ne 1.$$

Since

$$(-1)^n\binom{-\alpha}{n} = \frac{n^{\alpha-1}}{\Gamma(\alpha)}[1+O(n^{-1})], \tag{3.64}$$

the condition (3.62) is equivalent to

$$\omega_n^{(\alpha)} = \frac{n^{\alpha-1}}{\Gamma(\alpha)} + v_n \quad (n\geq 0) \quad \text{with} \quad \sum_{n=1}^{\infty}|v_n| < \infty. \tag{3.65}$$

Hence

$$w^{(\alpha)}(\xi) = (1-\xi)^{-\alpha} + v(\xi), \quad v(\xi) = \sum_{n=0}^{\infty} v_n\xi^n. \tag{3.66}$$

Therefore, $(1-\xi)^\alpha[1-zw^{(\alpha)}(\xi)] = (1-\xi)^\alpha[1-zv(\xi)]-z$, which leads to

$$(1-\xi)^\alpha[1-zw^{(\alpha)}(\xi)] \neq 0 \quad \text{for} \quad \xi\neq 1.$$

Wiener's inversion theorem now yields that the sequence of the coefficients of the Taylor expansions of

$$\frac{1}{(1-\xi)^\alpha[1-zw^{(\alpha)}(\xi)]} \quad \text{is in} \quad \ell^1. \tag{3.67}$$

Let $\tilde{g}_n = g_n - g_\infty$, so that we can write

$$g(\xi) = \frac{g_\infty}{1-\xi} + \tilde{g}(\xi).$$

We now show that the coefficient sequence of

$$(1-\xi)^\alpha g(\xi) = (1-\xi)^{\alpha-1} f_\infty + (1-\xi)^\alpha \tilde{g}(\xi)$$

converges to zero. By (3.64), the coefficient sequence of $(1-\xi)^{\alpha-1}$ tends to zero. Also the coefficient sequence of $(1-\xi)^\alpha\tilde{g}(\xi)$ converges to zero, since the coefficient sequence of $(1-\xi)^\alpha$ is in ℓ^1 and

$$\ell^1 * \mathbb{C}_0 \subseteq \mathbb{C}_0, \tag{3.68}$$

where $*$ denotes convolution, and $\mathbb{C}_0$ is the space of sequences convergent to 0; formula (3.68) is a result of dominated convergence (see [121]): for $\{l_n\}$ and $\{d_n\}\in\mathbb{C}_0$ we have

$$\lim_{n\to\infty}\sum_{j=0}^{n} l_j d_{n-j} = 0.$$

Using (3.63), (3.67), and (3.68) we can finally conclude that the coefficient sequence $\{u_n\}$ of $u(\xi)$ tends to zero. Hence $z\in S$.

In order to prove that S is exhausted by $\tilde{S}$ we assume that

$$1-zw(\xi_0) = 0 \quad \text{for some} \quad |\xi_0|\leq 1. \tag{3.69}$$

By assumption (3.66) we have $\xi_0 \neq 1$. We show that $z \notin S$. We choose

$$u(\xi) = \frac{(1-\xi)^\alpha}{\xi - \xi_0} = \frac{(1-\xi)^\alpha - (1-\xi_0)^\alpha}{\xi - \xi_0} + \frac{(1-\xi_0)^\alpha}{\xi - \xi_0}. \tag{3.70}$$

The coefficient sequence of the first expression of the sum tends to zero by the property (b) whereas the coefficient sequence of

$$\frac{1}{\xi - \xi_0} = -\sum_{n=0}^{\infty} \xi_0^{-n-1} \xi^n$$

diverges. Hence $\{u_n\}$ also diverges. From (3.63), (3.69), and (3.70) we obtain

$$\begin{aligned} g(\xi) =&[1 - zw^{(\alpha)}(\xi)]u(\xi) = (1-\xi)^\alpha[1 - zw^{(\alpha)}(\xi)](1-\xi)^{-\alpha}u(\xi) \\ =& \frac{(1-\xi)^\alpha[1 - zw^{(\alpha)}(\xi)] - (1-\xi_0)^\alpha[1 - zw^{(\alpha)}(\xi_0)]}{\xi - \xi_0}. \end{aligned}$$

Now the property (b) yields $g_n \to 0$, but, as we have seen before, $u_n \nrightarrow 0$. Hence $z \notin S$. □

Corollary 3.4.1 *If the FLMM* (3.61) *with the condition* (3.62) *is used, and* $\Delta t^\alpha \lambda \in S$, *then*

$$\{u_n\} \textit{ is bounded whenever } \{g_n\} \textit{ is bounded.} \tag{3.71}$$

Conversely, if (3.71) *holds, then* $\Delta t^\alpha \lambda \in S$ *is contained in the closure of* S.

Corollary 3.4.2 *The FLMM* (3.61) *with the condition* (3.62) *is* ***strongly stable***.

In [103], Lubich proved that there is the order barrier for the A-**stable** method (3.55), which is the same as that of the LMM for ODEs [20]. The result is given in the following theorem.

Theorem 19 ([103]) *The order, p, of an* A-***stable*** *FLMM* (3.61) *with the condition* (3.62) *and* (3.53) *can not exceed* 2.

Remark 3.4.2 *Obviously, Theorems 18 and 19, Corollaries 3.4.1 and 3.4.2 hold for FLMM* (3.55) *when* $f(t,u) = \lambda u(t)$.

As is known, the backward Euler method and the trapezoidal rule are two A-**stable** numerical methods for the ODE $u'(t) = \lambda u(t)$, $Re(\lambda) < 0$, $u(0) = u_0$. Next, we study the corresponding methods for the FODE (3.58). Let us first introduce two lemmas.

Lemma 3.4.1 ([104]) *If* $y(t) = t^{\nu-1}, \nu > 0$, *then*

$$\mathrm{D}_{0,t}^{-\alpha} y(t)\Big|_{t=t_n} = \tau^\alpha \sum_{k=0}^{n} \omega_{n-k}^{(\alpha)} y(t_k) + O(\Delta t^p) + O(\Delta t^\nu),$$

where $\omega_k^{(\alpha)}$ *can be the coefficients of the Taylor series of the generating functions defined as* (3.56) *or* (3.57).

Lemma 3.4.2 ([168]) *Denote by*

$$u_n = \Delta t^\alpha \sum_{k=0}^{n} \omega_{n-k}^{(\alpha)} G(t_k, u_k), \tag{3.72}$$

where $\{\omega_k^{(\alpha)}\}$ *are the coefficients of Taylor expansions of the generating function* $w^{(\alpha)}(z)$ *defined by* (3.56) *or* (3.57). *Then,* (3.72) *is equivalent to the following equation*

$$\sum_{k=0}^{n} \omega_k u_{n-k} = \Delta t^\alpha \sum_{k=0}^{n} \theta_{n-k} G(t_k, u_k) \tag{3.73}$$

where ω_k *and* θ_k *are the coefficients of the Taylor expansions of* $\alpha(z)$ *and* $\theta(z)$ *satisfying* $w^{(\alpha)}(z) = \theta(z)/\omega(z)$.

Proof. We first rewrite $w^{(\alpha)}(z) = \theta(z)/\omega(z)$ into the following form

$$\Big(\sum_{k=0}^{\infty} \omega_k z^k\Big)\Big(\sum_{k=0}^{\infty} \omega_k^{(\alpha)} z^k\Big) = \sum_{k=0}^{\infty} \theta_k z^k,$$

which yields

$$\theta_m = \sum_{k=0}^{m} \omega_k^{(\alpha)} \omega_{m-k}, \quad m = 0, 1, \cdots, n. \tag{3.74}$$

By (3.72), one obtains $u_m = \Delta t^\alpha \sum\limits_{k=0}^{m} \omega_{m-k}^{(\alpha)} G(t_k, u_k)$. Hence, we have

$$\sum_{m=0}^{n} \omega_{n-m} u_m = \sum_{m=0}^{n} \omega_{n-m} \Big[\Delta t^\alpha \sum_{k=0}^{m} \omega_{m-k}^{(\alpha)} G(t_k, u_k)\Big]. \tag{3.75}$$

Rearranging the right-hand side of (3.75) and using (3.74) yield the desired result. The proof is completed. □

Using Lemma 3.4.2, the FLMM (3.61) can be written in the following equation

$$\sum_{k=0}^{n} \omega_{n-k} u_k = \Delta t^\alpha \lambda \sum_{j=0}^{n} \theta_{n-j} u_j + \Delta t^\alpha \lambda \sum_{k=0}^{n} \omega_{n-k} \sum_{j=0}^{s} w_{k,j}^{(\alpha)} u_j + \sum_{k=0}^{n} \omega_{n-k} g(t_k). \tag{3.76}$$

Denote by

$$w_1^{(\alpha)}(\xi) = (1-\xi)^{-\alpha}; \tag{3.77}$$

$$w_2^{(\alpha)}(\xi) = \left(\frac{1}{2}\frac{1+\xi}{1-\xi}\right)^{\alpha}; \tag{3.78}$$

$$w_3^{(\alpha)}(\xi) = (1-\xi)^{-\alpha}\Big[1 - \frac{\alpha}{2} + \frac{\alpha}{2}\xi\Big]. \tag{3.79}$$

Next, we study these three cases of generating functions $w_i^{(\alpha)}(z), i = 1, 2, 3$, which yields the absolute stable numerical methods for (3.58).

By Lemma 3.4.2, we obtain the equivalent form of (3.61) as follows

$$\sum_{k=0}^{n} \omega_k(u_{n-k} - g_{n-k}) = \lambda \Delta t^{\alpha} \sum_{k=0}^{n} \theta_{n-k} u_k. \tag{3.80}$$

Hence, one can obtain the FLMM for (3.58) as

$$\sum_{k=0}^{n} \omega_k(u_{n-k} - u_0) = \lambda \Delta t^{\alpha} \sum_{k=0}^{n} \theta_{n-k} u_k. \tag{3.81}$$

- **Case (1)**: If the generating function (3.77) is used in (3.61), then $\omega(\xi)$ and $\theta(\xi)$ in Lemma 3.4.2 are chosen as $\omega(\xi) = (1-\xi)^{\alpha}$ and $\theta(\xi) = 1$. Thus, ω_j and θ_j in (3.81) are given by $\omega_j = (-1)^j\binom{\alpha}{j}$ and $\theta_0 = 1, \theta_j = 0, j > 0$.
- **Case (2)**: If the generating function (3.78) is used in (3.61), then $\omega(\xi)$ and $\theta(\xi)$ in Lemma 3.4.2 are chosen as $\omega(\xi) = (1-\xi)^{\alpha}$ and $\theta(\xi) = 1$. Thus, ω_j and θ_j in (3.81) are given by $\omega_j = (-1)^j\binom{\alpha}{j}$ and $\theta_j = \frac{(-1)^j}{2^{\alpha}}\omega_j$.
- **Case (3)**: If the generating function (3.77) is used in (3.61), then $\omega(\xi)$ and $\theta(\xi)$ in Lemma 3.4.2 are chosen as $\omega(\xi) = (1-\xi)^{\alpha}$ and $\theta(\xi) = 1$. Thus, ω_j and θ_j in (3.81) are given by $\omega_j = (-1)^j\binom{\alpha}{j}$ and $\theta_0 = 1 - \frac{\alpha}{2}, \theta_1 = \frac{\alpha}{2}, \theta_j = 0, j > 1$.

Obviously, the method (3.81) is reduced to the classical Euler method for the classical ODE under the condition of **Case (1)** with $\alpha = 1$. And the method (3.81) is reduced to the classical trapezoidal rule under the condition of **Case (2)** or **Case (3)** with $\alpha = 1$. We can directly prove that the method (3.81) is absolutely stable under the condition in **Case (1)**, **Case (2)**, or **Case (3)**.

Lemma 3.4.3 ([55]) *Let $\omega_k = (-1)^k\binom{\alpha}{k}, 0 < \alpha < 1$. Then we have*

$$\begin{aligned}
&\omega_0 = 1,\ \omega_n < 0,\ |\omega_{n+1}| < |\omega_n|, \quad n = 1,2,\cdots,\\
&\omega_0 = -\sum_{k=1}^{\infty} \omega_k > -\sum_{k=1}^{n} \omega_k > 0, \quad n = 1,2,\cdots,\\
&b_{n-1} = \sum_{k=0}^{n-1} \omega_k = \frac{\Gamma(n-\alpha)}{\Gamma(1-\alpha)\Gamma(n)} = \frac{n^{-\alpha}}{\Gamma(1-\alpha)} + O(n^{-1-\alpha}), \quad n \text{ suitably large.}
\end{aligned} \tag{3.82}$$

Furthermore, $b_n - b_{n-1} = \omega_n < 0$ for $n > 0$, i.e., $b_n \le b_{n-1}$.

Theorem 20 *Let u_n be the solution to the method* (3.81) *under the condition in* ***Case (1)***, ***Case (2)*** *or* ***Case (3)***, *$g_k = u_0$, and $\lambda < 0$. Then*

$$|u_k| \le |u_0|. \tag{3.83}$$

Proof. We only prove $|u_k| \le |g|_{\infty}$ under the condition of **Case (2)**, the other two cases are very similar so are omitted here. Under the condition of **Case (2)** and Lemma 3.4.3, we can rewrite (3.81) as the following form

$$u_n = \frac{1}{1-\lambda\Delta t^\alpha 2^{-\alpha}}\left\{\sum_{k=1}^{n}(b_{k-1}-b_k)\left[1-\lambda\Delta t^\alpha(-1)^k 2^{-\alpha}\right]u_{n-k}+b_n u_0\right\}. \tag{3.84}$$

Note that if $\omega_k = b_k - b_{k-1} \le 0$ and $\lambda < 0$, one has

$$|u_n| \le \sum_{k=1}^{n}(b_{k-1}-b_k)|u_{n-k}| + b_n u_0. \tag{3.85}$$

Next, we use the mathematical induction method to prove (3.83). Let $n = 1$ in (3.85) yields $|u_1| \le (b_0 - b_1)|u_0| + b_1|u_0| = |u_0|$. Suppose that (3.83) holds for $0 < n < m$. Next, one needs to prove that (3.83) holds for $n = m$. From (3.85)

$$|u_m| \le \sum_{k=1}^{m}(b_{k-1}-b_k)|u_{m-k}| + b_n|u_0| \le \sum_{k=1}^{m}(b_{k-1}-b_k)|u_0| + b_m|u_0| = |u_0|. \tag{3.86}$$

Hence, (3.83) holds for any $n > 0$. The proof is completed. □

Next, we consider the convergence of the FLMM (3.81) applied to the FODE of the form

$$_C D_{0,t}^{\alpha} u(t) = f(t,u(t)) = \lambda u(t) + g(t), \qquad u(0) = u_0. \tag{3.87}$$

Assume that the analytical solution $u(t)$ to (3.87) is suitably smooth. We first transform the FODE (3.87) into the following integral equation

$$u(t) - u(0) = D_{0,t}^{-\alpha} f(t,u(t)) = \lambda D_{0,t}^{-\alpha} u(t) + D_{0,t}^{-\alpha} g(t). \tag{3.88}$$

For $u(t) \in C^2([0,T])$, one has $_C D_{0,t}^{\alpha} u(t) = \frac{u'(0)}{\Gamma(2-\alpha)} t^{1-\alpha} + D_{0,t}^{-(2-\alpha)} u''(t)$. Using Lemma 3.4.1 yields

$$u(t_n) - u(0) = \left[D_{0,t}^{-\alpha} f(t,u(t))\right]_{t=t_n} = \Delta t^\alpha \sum_{k=0}^{n} \omega_{n-k}^{(\alpha)} f(t_k, u(t_k)) + \hat{R}_n, \tag{3.89}$$

where $\hat{R}_n = O(t_n^{1-p}\Delta t^p) + O(t_n^{\alpha-1}\Delta t^{2-\alpha})$ for $p = 1$ with the generating function (3.77), and $p = 2$ with the generating function (3.78) or (3.79).

Applying Lemma 3.4.2 yields the equivalent form of (3.89) as

$$\frac{1}{\Delta t^\alpha}\sum_{k=0}^{n}\omega_{n-k}(u(t_k)-u(0)) = \sum_{k=0}^{n}\theta_{n-k} f(t_k,u(t_k)) + \frac{1}{\Delta t^\alpha}\sum_{k=0}^{n}\omega_{n-k}\hat{R}_k. \tag{3.90}$$

Whether or not the generating function (3.77), (3.78), or (3.79) is used, we always have $\omega_k = (-1)^k\binom{\alpha}{k}$, and

$$R_n = \frac{1}{\Delta t^\alpha}\sum_{k=0}^{n}\omega_{n-k}\hat{R}_k = \Delta t^{p-\alpha}\sum_{k=1}^{n}\omega_{n-k}O(k^{1-p}) + \Delta t\sum_{k=1}^{n}\omega_{n-k}O(k^{\alpha-1}). \tag{3.91}$$

Exactly speaking, R^n has the following expression

$$R_n = \begin{cases} O(n^{-\alpha}\Delta t^{1-\alpha}), & p = 1 \text{ with (3.77) used,} \\ O(n^{-1}\Delta t^{1-\alpha}), & p = 2 \text{ with (3.78) or (3.79) used,} \end{cases} \tag{3.92}$$

where we have used the following relation [104]

$$\sum_{k=1}^{n} \omega_{n-k} k^{\gamma-1} = O(n^{\gamma-1-\alpha}) + O(n^{-\alpha-1}), \quad \gamma \neq -1, -2, \cdots. \tag{3.93}$$

Let u_k be the approximate solution of $u(t_k)$. Then we can obtain the numerical methods for (3.87) as follows

$$\frac{1}{\Delta t^{\alpha}} \sum_{k=0}^{n} \omega_{n-k}(u_k - u_0) = \sum_{k=0}^{n} \theta_{n-k} f(t_k, u_k), \tag{3.94}$$

where $\omega_k = (-1)^k \binom{\alpha}{k}$, and

$$\theta_0 = 1, \theta_k = 0, k > 0, \quad p = 1 \text{ with (3.77) used,} \tag{3.95}$$

$$\theta_k = \frac{1}{2^{\alpha}}(-1)^k \omega_k, k \geq 0, \quad p = 2 \text{ with (3.78) used,} \tag{3.96}$$

$$\theta_0 = 1 - \frac{\alpha}{2}, \theta_1 = \frac{\alpha}{2}, \theta_k = 0, k > 1, \quad p = 2 \text{ with (3.79) used.} \tag{3.97}$$

Similar to the proof of Theorem 20, one can prove that the FLMM (3.94) is unconditionally stable under condition (3.95), (3.96), or (3.97) if $f(t, u(t)) = \lambda u(t) + g(t)$. And the global convergence rate is $O(\Delta t^{1-\alpha})$.

Theorem 21 *Let $u \in C^2([0,T])$ be the solution of* (3.87), *and $u_k, k > 0$ be the solution of* (3.94) *with condition* (3.95), (3.96), *or* (3.97). *Then there exists a positive constant C independent of n such that*

$$|u(t_n) - u_n| \leq C\Delta t^{1-\alpha}. \tag{3.98}$$

Proof. We only prove the error estimate for (3.94) with the condition (3.95); the other two cases are almost the same. Let $e_k = u(t_k) - u_k$. Then we can derive the error equation below

$$\sum_{k=0}^{n} \omega_{n-k} e_k = \lambda \Delta t^{\alpha} e_n + \Delta t^{\alpha} R_n, \tag{3.99}$$

By Lemma 3.4.3, we can rewrite the above equation into

$$e_n = \frac{1}{1 - \lambda \Delta t^{\alpha}} \sum_{k=1}^{n} (b_{k-1} - b_k) e_{n-k} + \frac{\Delta t^{\alpha}}{1 - \lambda \Delta t^{\alpha}} R_n. \tag{3.100}$$

Next, we use the mathematical induction method to prove that

$$|e_n| \leq C|R|_{\infty} = C \max_{0 < k \leq n_T} |R_k|. \tag{3.101}$$

For $n = 1$, by $e_0 = 0$ and $\lambda < 0$, one has from (3.100) $|e_1| \leq \Delta t^{\alpha}|R_1| \leq C|R|_{\infty}$. Suppose that (3.101) holds for $0 < n < m$. For $n = m$ one has

$$
\begin{aligned}
|e_m| &\leq \frac{1}{1-\lambda\Delta t^{\alpha}} \sum_{k=1}^{m} (b_{k-1} - b_k)|e_{m-k}| + \frac{\Delta t^{\alpha}}{1-\lambda\Delta t^{\alpha}}|R_m| \\
&\leq \sum_{k=1}^{m} (b_{k-1} - b_k)|e_{m-k}| + Cb_m|R|_{\infty} \\
&\leq \sum_{k=1}^{m} (b_{k-1} - b_k)C|R|_{\infty} + Cb_m|R|_{\infty} = C|R|_{\infty}.
\end{aligned}
\tag{3.102}
$$

Hence (3.101) holds for any $0 < n \leq n_T$. Hence $|e_n| \leq C \max_{0<k\leq n_T} |R_k| \leq C\Delta t^{1-\alpha}$. The proof is completed. □

It is known that the Euler method and the trapezoidal method for the classical ODE are of first-order and second-order accuracy for the smooth solutions. Next, we can construct the corresponding schemes for the FODE (3.87). We just need to make an improvement of (3.94) to get the desired schemes.

- **Improved algorithms I:**

From (3.88), we have

$$
u(t) - u(0) = \lambda \mathrm{D}_{0,t}^{-\alpha}(u(t) - u(0)) + \frac{\lambda u(0) t^{\alpha}}{\Gamma(\alpha+1)} + \mathrm{D}_{0,t}^{-\alpha} g(t). \tag{3.103}
$$

Applying Lemma 3.4.1 to (3.103) yields

$$
\begin{aligned}
u(t_n) - u(0) =& \lambda \left[\mathrm{D}_{0,t}^{-\alpha}(u(t) - u(0))\right]_{t=t_n} + \frac{\lambda u(0) t_n^{\alpha}}{\Gamma(\alpha+1)} + \left[D_{0,t}^{-\alpha} g(t)\right]_{t=t_n} \\
=& \lambda\Delta t^{\alpha} \sum_{k=0}^{n} \omega_{n-k}^{(\alpha)}(u(t_k) - u(0)) + \frac{\lambda u(0) t_n^{\alpha}}{\Gamma(\alpha+1)} + G_n + \hat{R}_n,
\end{aligned}
\tag{3.104}
$$

where $G_n = \left[\mathrm{D}_{0,t}^{-\alpha} g(t)\right]_{t=t_n}$ $\hat{R}_n = O(t_n^{1+\alpha-p}\Delta t^p) + O(t_n^{\alpha-1}\Delta t^2)$ for $p = 1$ if the generating function (3.77) is used, and $p = 2$ if the generating function (3.78) or (3.79) is used.

Applying Lemma 3.4.2 yields the equivalent form of (3.104) as

$$
\frac{1}{\Delta t^{\alpha}} \sum_{k=0}^{n} \omega_{n-k}(u(t_k) - u_0) = \lambda \sum_{k=0}^{n} \theta_{n-k} u(t_k) + \lambda B_n u_0 + \frac{1}{\Delta t^{\alpha}} \sum_{k=0}^{n} \omega_{n-k} G_k + R_n, \tag{3.105}
$$

in which $\omega_k = (-1)^k \binom{\alpha}{k}$, and θ_k is defined as in (3.95)–(3.97), B_n is defined by

$$
B_n = \frac{1}{\Gamma(1+\alpha)} \sum_{k=0}^{n} \omega_{n-k} k^{\alpha} - \sum_{k=0}^{n} \theta_k, \tag{3.106}
$$

and R^n is given by

$$R_n = \frac{1}{\Delta t^\alpha}\sum_{k=0}^{n}\omega_{n-k}\hat{R}_k = \Delta t\sum_{k=1}^{n}\omega_{n-k}O(k^{1+\alpha-p}) + \Delta t\sum_{k=1}^{n}\omega_{n-k}O(k^{\alpha-1}). \tag{3.107}$$

By (3.93), we can obtain

$$R_n = \begin{cases} O(\Delta t), & p = 1 \text{ with (3.77) used,} \\ O(n^{-1}\Delta t), & p = 2 \text{ with (3.78) or (3.79) used,} \end{cases} \tag{3.108}$$

From (3.105), we derive the improved algorithms for (3.87)

$$\frac{1}{\Delta t^\alpha}\sum_{k=0}^{n}\omega_{n-k}u_k = \lambda\sum_{k=0}^{n}\theta_{n-k}u_k + \lambda B_n u_0 + \frac{1}{\Delta t^\alpha}\sum_{k=0}^{n}\omega_{n-k}G^k. \tag{3.109}$$

Note that $B_n = O(n^{-1})$. Similar to the proofs of Theorems 20 and 21, we can prove that the improved algorithm (3.109) is unconditionally stable and convergent of order 1. In the real computations, one can find that the improved algorithm (3.109) can attain second-order accuracy if the generating function (3.78) or (3.79) is used, since the local truncation error $R_n = O(n^{-1}\Delta t)$ has second-order accuracy when n is big enough. One surprising finding is that the average error satisfies

$$\sqrt{\Delta t\sum_{k=0}^{n}|u(t_k) - u_k|^2} = O(\Delta t^{1.5})$$

when the generating function (3.78) or (3.79) is used.

- **Improved algorithms II:**

Now we introduce another two improved algorithms for (3.87) such that the convergence rate is of order two when the solutions are smooth enough.

We still consider the discretization of (3.103). We use the FLMM (3.54) to discretize $\mathrm{D}_{0,t}^{-\alpha}(u(t) - u(0))$ in (3.103), which gives

$$\left[\mathrm{D}_{0,t}^{-\alpha}(u(t) - u(0))\right]_{t=t_n} = \Delta t^\alpha\sum_{k=0}^{n}\omega_{n-k}^{(\alpha)}(u(t_k) - u(0)) + \Delta t^\alpha w_{n,1}^{(\alpha)}(u(t_1) - u(0)) + O(t_n^\alpha\Delta t^2), \tag{3.110}$$

where $\omega_k^{(\alpha)}$ are the coefficients of the Taylor expansion of the generating function (3.78) or (3.79), $w_{n,1}^{(\alpha)}$ is the starting weight such that (3.110) is exact for $u(t) = t$, which is given by

$$w_{n,1}^{(\alpha)} = \frac{\Gamma(2)}{\Gamma(2+\alpha)}n^{1+\alpha} - \sum_{k=1}^{n}\omega_{n-k}^{(\beta)}k = O(n^{\alpha-1}). \tag{3.111}$$

From (3.103) and (3.111), we can derive the following discretization

$$
\begin{aligned}
u(t_n)-u(0) =&\lambda\left[\mathrm{D}_{0,t}^{-\alpha}(u(t)-u(0))\right]_{t=t_n}+\frac{\lambda u(0)t_n^{\alpha}}{\Gamma(\alpha+1)}+\left[\mathrm{D}_{0,t}^{-\alpha}g(t)\right]_{t=t_n}\\
=&\lambda\Delta t^{\alpha}\sum_{k=0}^{n}\omega_{n-k}^{(\alpha)}(u(t_k)-u(0))+\lambda\Delta t^{\alpha}w_{n,1}^{(\alpha)}(u(t_1)-u(0))\\
&+\frac{\lambda u(0)t_n^{\alpha}}{\Gamma(\alpha+1)}+G_n+\hat{R}_n,
\end{aligned}
\tag{3.112}
$$

where $G_n=\left[\mathrm{D}_{0,t}^{-\alpha}g(t)\right]_{t=t_n}$, $\hat{R}_n=O(t_n^{\alpha}\Delta t^2)$, and the generating function (3.78) or (3.79) is utilized.

Applying Lemma 3.4.2 yields the equivalent form of (3.112)

$$
\begin{aligned}
\frac{1}{\Delta t^{\alpha}}\sum_{k=0}^{n}\omega_{n-k}(u(t_k)-u_0)=&\lambda\sum_{k=0}^{n}\theta_{n-k}u(t_k)+\lambda B_n u_0+\lambda C_n(u(t_1)-u_0)\\
&+\frac{1}{\Delta t^{\alpha}}\sum_{k=0}^{n}\omega_{n-k}G^k+R_n,
\end{aligned}
\tag{3.113}
$$

in which $\omega_k=(-1)^k\binom{\alpha}{k}$, and θ_k is defined as in (3.96)–(3.97), B_n is defined by (3.106), C_n is defined by

$$
C_n=\sum_{k=0}^{n}\omega_{n-k}w_{k,1}^{(\alpha)}=\frac{\Gamma(2)}{\Gamma(\alpha+2)}\sum_{k=0}^{n}\omega_{n-k}k^{\alpha+1}-\sum_{k=1}^{n}\theta_{n-k}k=O(n^{-1}),\tag{3.114}
$$

and R^n is given by

$$
R_n=\frac{1}{\Delta t^{\alpha}}\sum_{k=0}^{n}\omega_{n-k}\hat{R}_k=\Delta t^2\sum_{k=1}^{n}\omega_{n-k}O(k^{\alpha})=O(\Delta t^2).\tag{3.115}
$$

From (3.113), we obtain the improved algorithm for (3.87) below

$$
\begin{aligned}
\frac{1}{\Delta t^{\alpha}}\sum_{k=0}^{n}\omega_{n-k}(u_k-u_0)=&\lambda\sum_{k=0}^{n}\theta_{n-k}u_k+\lambda B_n u_0+\lambda C_n(u_1-u_0)\\
&+\frac{1}{\Delta t^{\alpha}}\sum_{k=0}^{n}\omega_{n-k}G^k,
\end{aligned}
\tag{3.116}
$$

where $\omega_k=(-1)^k\binom{\alpha}{k}$, θ_k is defined as in (3.95)–(3.97), and B_n and C_n are defined by (3.106) and (3.114), respectively.

Since B_n and C_n in (3.116) satisfy $B_n=O(n^{-1})$ and $C_n=O(n^{-1})$. So one can similarly prove that the improved algorithm (3.116) is unconditionally stable and convergent of order two.

Other related works on the linear stability of the model problem (3.58) can be found in [53, 54, 55, 58, 59], where the linear stability with the stability region of the explicit Adams multistep methods, the fractional Adams–Moulton methods, and the predictor-corrector algorithms were investigated. The implicit Adams product quadrature rules and their stability properties were studied in [103].

Chapter 4

Finite Difference Methods for Fractional Partial Differential Equations

In this chapter, several kinds of finite difference methods are derived for fractional evolutional equations, including time-fractional equations in one space dimension, space-fractional equations in one space dimension, time-space fractional equations in one space dimension, and fractional partial differential equations in two space dimensions. Numerical examples are presented which are in line with the theoretical analysis.

4.1 Introduction

This chapter is divided into four sections. In the first section, we investigate the finite difference methods for the time-fractional equation in one spatial dimension, for example see [11, 35]. In the second section, we construct the finite difference methods for the space-fractional equations in one spatial dimension, e.g. [145]. In the following section, we derive the finite difference methods for time-space fractional equations in one space dimension, say, [10, 101]. In the last section of this chapter, we establish the finite difference methods for the two-dimensional case, for example, see [4]. Some other topics, such as the *homotopy perturbation method* for solving fractional differential equations [93], *inverse problems for fractional differential equations* [153], etc., are not going to be presented in this book.

4.2 One-Dimensional Time-Fractional Equations

Denote by $I = (a,b)$. Let Δt be the time step size and n_T be a positive integer with $\Delta t = T/n_T$ and $t_n = n\tau$ for $n = 0,1,\cdots,n_T$. Denote by $t_{n+\frac{1}{2}} = (t_n + t_{n+1})/2$ for $n = 0,1,\cdots,n_T - 1$. One can define the space step size $\Delta x = (b-a)/N$, N is a positive integer. The space grid point x_i is given by $x_i = a + i\Delta x, i = 0,1,\cdots,N$. Let $x_{i+\frac{1}{2}} =$

$(x_i + x_{i+1})/2$. For the function $U(x,t)\in C(I\times[0,T])$, denote by $U^n = U^n(\cdot) = U(\cdot,t_n)$ and $U_i^n = U(x_i,t_n)$.

Next, we introduce the following notations that will be used in the description of the numerical schemes.

$$\delta_x U^n_{i+\frac{1}{2}} = \frac{U^n_{i+1} - U^n_i}{\Delta x}, \quad \delta_x^2 U_i^n = \frac{U^n_{i+1} - 2U^n_i + U^n_{i-1}}{\Delta x^2}. \tag{4.1}$$

$$\delta_t U_i^{n+\frac{1}{2}} = \frac{U_i^{n+1} - U_i^n}{\Delta t}. \tag{4.2}$$

Next, we list some formulas for the discretization of the Riemann–Liouville derivatives and Caputo derivatives.

- The γth-order ($\gamma > 0$) Riemann–Liouville derivative of $U(t), t\in(0,T]$ at $t = t_n$ can be discretized by the Grünwald–Letnikov formula

$$^{GL}\delta_t^{(\gamma)} U^n = \frac{1}{\Delta t^\gamma}\sum_{k=0}^{n}\omega_{n-k}^{(\gamma)}U^k, \qquad \omega_k^{(\gamma)} = (-1)^k\binom{\gamma}{k}. \tag{4.3}$$

- The γth-order ($0 < \gamma < 1$) Riemann–Liouville derivative of $U(t), t\in(0,T]$ at $t = t_n$ can be discretized by the L1 method as

$$^{L1}_{RL}\delta_t^{(\gamma)} U^n = \frac{1}{\Delta t^\gamma}\left(\sum_{k=0}^{n-1} b_{n-k-1}^{(\gamma)}(U^{k+1} - U^k) + \frac{n^{-\gamma}}{\Gamma(1-\gamma)}U^0\right), \; b_k^{(\gamma)} = \frac{(k+1)^{1-\gamma} - k^{1-\gamma}}{\Gamma(2-\gamma)}. \tag{4.4}$$

  The γth-order ($0 < \gamma < 1$) Caputo derivative of $U(t), t\in(0,T]$ at $t = t_n$ can be discretized by the L1 method as

$$^{L1}_{C}\delta_t^{(\gamma)} U^n = \frac{1}{\Delta t^\gamma}\left(\sum_{k=0}^{n-1} b_{n-k-1}^{(\gamma)}(U^{k+1} - U^k)\right), \quad b_k^{(\gamma)} = \frac{(k+1)^{1-\gamma} - k^{1-\gamma}}{\Gamma(2-\gamma)}. \tag{4.5}$$

- The γth-order ($\gamma > 0$) Riemann–Liouville derivative of $U(t), t\in(0,T]$ at $t = t_n$ can be discretized by the fractional backward difference formula (BDF) as

$$^{B}_{p}\delta_t^{(\gamma)} U^n = \frac{1}{\Delta t^\gamma}\sum_{k=0}^{n}\omega_{n-k}^{(-\gamma)}U^k, \tag{4.6}$$

  where $\omega_k^{(-\gamma)}$ are coefficients of the Taylor expansions of the generating functions $w_p^{(-\gamma)}(z)$ defined by

$$w_p^{(-\gamma)}(z) = \left(w_p(z)\right)^\gamma, \quad w_p(z) = \sum_{j=1}^{p}\frac{1}{j}(1-z)^j, \quad p = 1,2,\cdots,6. \tag{4.7}$$

- The αth-order ($\alpha > 0$) left Riemann–Liouville derivative of $U(x), x\in(a,b)$ at $x = x_i$ can be discretized by the Grünwald–Letnikov formula as follows

$$
{}_L\delta_x^{(\alpha)} U_i = \frac{1}{\Delta x^\alpha}\sum_{j=0}^{i}\omega_j^{(\alpha)} U_{i-j}, \qquad \omega_j^{(\alpha)} = (-1)^j\binom{\alpha}{j}. \tag{4.8}
$$

- The αth-order ($\alpha > 0$) right Riemann–Liouville derivative of $U(x), x\in(a,b)$ at $x = x_i$ can be discretized by the Grünwald–Letnikov formula as

$$
{}_R\delta_x^{(\alpha)} U_i = \frac{1}{\Delta x^\alpha}\sum_{j=0}^{N-i}\omega_j^{(\alpha)} U_{i+j}, \qquad \omega_j^{(\alpha)} = (-1)^j\binom{\alpha}{j}. \tag{4.9}
$$

4.2.1 Riemann–Liouville Type Subdiffusion Equations

Consider the following type of *time-fractional diffusion equation*

$$
\begin{cases}
\partial_t U = {}_{RL}\mathrm{D}_{0,t}^{1-\gamma}\left(K_\gamma \partial_x^2 U\right) + f(x,t), & (x,t)\in(a,b)\times(0,T],\\
U(x,0) = \phi_0(x), \quad x\in(a,b), \\
U(a,t) = U_a(t),\ U(b,t) = U_b(t), \quad t\in(0,T],
\end{cases} \tag{4.10}
$$

where $K_\gamma > 0$ and $0 < \gamma < 1$.

Clearly, if $\gamma \to 1$, the above equation is reduced to the classical diffusion equation. It is known that there are many numerical techniques to solve such an equation, such as the forward and backward Euler methods, the Crank–Nicolson method, etc. For the subdiffusion equation (4.10), there also exist several analogs such as the forward and backward Euler methods, and the Crank–Nicolson methods.

Since the spatial direction is the classical second-order differential operator, almost all the classical numerical methods (such as the finite difference method, the finite element method, the spectral method, the discontinuous Galerkin method, etc.) can be used to discretize the space derivative of (4.10). Here we mainly focus on the time discretization of (4.10). In the following, the second-order spatial derivative in (4.10) is discretized by the second-order central difference method for brevity.

4.2.1.1 Explicit Euler Type Methods

The explicit method is particularly of interest because of its simplicity, easy implementation, and low cost in real computation. Like the explicit Euler method for the heat equation ($\gamma = 1$ in (4.10)), we can present the corresponding explicit method for the fractional subdiffusion equation (4.10), which can be seen as an extension of the forward Euler method.

Letting $(x,t) = (x_i,t_n)$ in (4.10) leads to

$$
\partial_t U(x_i,t_n) = K_\gamma\left({}_{RL}D_{0,t}^{1-\gamma}\partial_x^2 U\right)(x_i,t_n) + f(x_i,t_n). \tag{4.11}
$$

The integer-order time derivative and fractional derivative in (4.11) are discretized by the forward Euler method and the Grünwald–Letnikov formula, i.e.,

$$\partial_t U(x_i,t_n) = \frac{U(x_i,t_{n+1}) - U(x_i,t_n)}{\Delta t} + O(\Delta t) = \delta_t U_i^{n+\frac{1}{2}} + O(\Delta t), \quad (4.12)$$

$$\left({}_{RL}\mathrm{D}_{0,t}^{1-\gamma}\partial_x^2 U\right)(x_i,t_n) = {}^{GL}\delta_t^{(1-\gamma)}(\partial_x^2 U^n(x_i)) + O(\Delta t), \quad (4.13)$$

where ${}^{GL}\delta_t^{(1-\gamma)}$ is defined by (4.3), and the space is discretized by the central difference scheme, i.e.,

$$\partial_x^2 U(x_i,t_n) = \partial_x^2 U^n(x_i) = \delta_x^2 U_i^n + O(\Delta x^2). \quad (4.14)$$

Hence, one can obtain

$$\delta_t U_i^{n+\frac{1}{2}} = K_\gamma {}^{GL}\delta_t^{(1-\gamma)}\delta_x^2 U_i^n + f_i^n + O(\Delta t + \Delta x^2). \quad (4.15)$$

Replacing U_i^n by u_i^n and neglecting the truncation error in the above equation, one can get the following *explicit Euler method* for (4.10) as: Find $u_i^{n+1}(i = 1,2,\cdots,N-1, n = 0,1,\cdots,n_T-1)$, such that

$$\begin{cases} \delta_t u_i^{n+\frac{1}{2}} = K_\gamma {}^{GL}\delta_t^{(1-\gamma)}(\delta_x^2 u_i^n) + f_i^n, \quad i = 1,2,\cdots,N-1, \quad n = 0,1,\cdots,n_T-1, \\ u_i^0 = \phi_0(x_i), \quad i = 0,1,2,\cdots,N, \\ u_0^n = U_a(t_n),\ u_N^n = U_b(t_n), \end{cases} \quad (4.16)$$

where ${}^{GL}\delta_t^{(1-\gamma)}$is defined by (4.3).

Of course, the fractional derivative can be approximated by other methods such as the fractional BDF methods (see [104], ${}^{GL}\delta_t^{(1-\gamma)}$ in (4.16) is replaced by ${}_p^B\delta_t^{(1-\gamma)}$ defined by (4.6)) or the L1 method (see Eq. (2.63), ${}^{GL}\delta_t^{(1-\gamma)}$ in (4.16) is replaced by ${}_{RL}^{L1}\delta_t^{(1-\gamma)}$ defined in (4.4)), which lead to the different schemes that have similar forms as (4.16); we do not list these methods here.

If $\gamma \to 1$, method (4.16) is reduced to the classical forward Euler method.

Let $\mu = \frac{K_\gamma \Delta t^\gamma}{\Delta x^2}$. Then method (4.16) can be written as

$$u_i^{n+1} = u_i^n + \mu \sum_{k=0}^{n} \omega_{n-k}^{(1-\gamma)}(u_{i+1}^k - 2u_i^k + u_{i-1}^k) + \Delta t f_i^n. \quad (4.17)$$

Therefore, the unknowns u_i^{n+1} can be solved if $u_i^k\,(k = 0,1,\cdots,n)$ and f_i^n are given.

Next, we analyze the stability of the explicit method (4.16). There are several methods for the stability analysis, such as the energy method [56], the Fourier method (also called the fractional von Neumann analysis) [18, 24, 160, 161], and the matrix method [98]. The Fourier method is relatively simple, which is suitable for the linear equations with constant coefficients. Therefore, we first use the *Fourier method* for the stability analysis for scheme (4.16).

The ***fractional von Neumann analysis*** for the stability analysis of scheme (4.16) is illustrated below.

Let $f_i^n = 0$ and $u_i^k = \rho_k e^{j\sigma i \Delta x} (j^2 = -1)$. Inserting u_i^k into (4.17) yields

$$\rho_{n+1} = \rho_n - 4\mu \sin^2\left(\frac{\sigma \Delta x}{2}\right) \sum_{k=0}^{n} \omega_{n-k}^{(1-\gamma)} \rho_k. \tag{4.18}$$

According to the von Neumann method [24, 161], we can first assume that $\rho_{n+1} = \xi(\sigma)\rho_n$ and $\xi(\sigma)$ is independent of time. Then (4.18) implies a closed equation for the amplification factor ξ as:

$$\xi = 1 - 4\mu \sin^2\left(\frac{\sigma \Delta x}{2}\right) \sum_{k=0}^{n} \omega_k^{(1-\gamma)} \xi^{-k}. \tag{4.19}$$

If $|\xi| > 1$ for some σ, ρ_n grows to infinity and the method is unstable. Considering the extreme value $\xi = -1$, we obtain from (4.19) the following stability bound on μ:

$$\mu \sin^2\left(\frac{\sigma \Delta x}{2}\right) \leq \frac{1}{2 \sum_{k=0}^{n} \omega_k^{(1-\gamma)} (-1)^{-k}} \equiv S_{\gamma,n}. \tag{4.20}$$

The bound defined by (4.20) depends on the number n of iterations. Nevertheless, this dependence is weak: $S_{\gamma,n}$ approaches $S_\gamma = \lim_{n\to\infty} S_{\gamma,n}$ in the form of oscillations with small decaying amplitudes [161]. Since $\sum_{k=0}^{\infty} \omega_k^{(1-\gamma)} z^{-k} = (1-z^{-1})^{1-\gamma} = w_1^{(1-\gamma)}(z^{-1})$ (see the first equation in (2.43)). Therefore, we find that the explicit method (4.16) is stable as long as

$$\mu \sin^2\left(\frac{\sigma \Delta x}{2}\right) \leq S_\gamma = \frac{1}{2 w_1^{(1-\gamma)}(-1)}. \tag{4.21}$$

Since $\sin^2\left(\frac{\sigma \Delta x}{2}\right) \leq 1$, we can give a more conservative but simple bound: the explicit method (4.16) is stable when

$$\mu = \frac{K_\gamma \Delta t^\gamma}{\Delta x^2} \leq S_\gamma = \frac{1}{2 w_1^{(1-\gamma)}(-1)} = \frac{1}{2^{2-\gamma}}. \tag{4.22}$$

Obviously, the stability bound in (4.22) is reduced to that of the forward Euler method if $\gamma \to 1$.

If the fractional derivative is discretized by the fractional backward difference formula (see (4.6)), i.e., ${}^{GL}\delta_t^{(1-\gamma)}$ in (4.16) is replaced by ${}^{B}_{p}\delta_t^{(1-\gamma)}$, one can obtain a series of explicit methods. For example, for $p = 2$ with $w_2^{(1-\gamma)}(z) = (3/2 - 2z + z^2/2)^{1-\gamma}$, one can obtain that the explicit method (4.16) is stable when

$$\mu = \frac{K_\gamma \Delta t^\gamma}{\Delta x^2} \leq S_\gamma = \frac{1}{2 w_2^{(1-\gamma)}(-1)} = \frac{1}{4^{3/2-\gamma}}. \tag{4.23}$$

Next, we consider the convergence. Let $e_i^n = U(x_i, t_n) - u_i^n$. Then one can derive the error equation from (4.15) and (4.16) as

$$e_i^{n+1} = e^n + \mu \sum_{k=0}^{n-1} \omega_{n-k}^{(1-\gamma)} (e_{i+1}^k - 2e_i^k + e_{i-1}^k) + \Delta t R_i^n. \tag{4.24}$$

Let $e_i^n = \eta^n e^{j\sigma i\Delta x}$, $R_i^n = r^n e^{j\sigma i\Delta x}$, and $\mu^* = 4\mu \sin^2\left(\frac{\sigma\Delta x}{2}\right)$. Then one has

$$\eta_{n+1} = \eta_n - \mu^* \sum_{k=0}^{n} \omega_{n-k}^{(1-\gamma)} \eta_k + \Delta t r^n. \tag{4.25}$$

From (4.15), we find that the local truncation error of the method (4.16) is $O(\Delta t(\Delta t + \Delta x^2))$. It is a little difficult to prove the global truncation error. In the following sections, some techniques will be introduced to prove the convergence of the numerical schemes for the subdiffusion equation (4.10).

4.2.1.2 Implicit Euler Type Methods

The idea for constructing the backward Euler method can be extended to establish the corresponding method for the subdiffusion equation (4.10).

In (4.11), if the integer time derivative, the Riemann–Liouville derivative, and the space derivative are approximated by the backward Euler formula, the Grünwald–Letnikov formula, and the central difference method, respectively, i.e.,

$$\partial_t U(x_i,t_n) = \frac{U(x_i,t_n) - U(x_i,t_{n-1})}{\Delta t} + O(\Delta t) = \delta_t U_i^{n-\frac{1}{2}} + O(\Delta t), \tag{4.26}$$

$$\left({}_{RL}\mathrm{D}_{0,t}^{1-\gamma} \partial_x^2 U\right)(x_i,t_n) = {}^{GL}\delta_t^{(1-\gamma)}(\partial_x^2 U^n(x_i)) + O(\Delta t), \tag{4.27}$$

$$\partial_x^2 U(x_i,t_n) = \delta_x^2 U_i^n + O(\Delta x^2), \tag{4.28}$$

where ${}^{GL}\delta_t^{(1-\gamma)}$ is defined by (4.3), then one can obtain

$$\delta_t U_i^{n-\frac{1}{2}} = K_\gamma {}^{GL}\delta_t^{(1-\gamma)}(\delta_x^2 U_i^n) + f_i^n + O(\Delta t + \Delta x^2). \tag{4.29}$$

Removing the truncation error $O(\Delta t + \Delta x^2)$ in (4.29), and replacing U_i^k with u_i^k, we can obtain the *backward Euler method* for (4.10) as: Find $u_i^n (i = 1,2,\cdots,N-1, n = 1,\cdots,n_T)$, such that

$$\begin{cases} \delta_t u_i^{n-\frac{1}{2}} = K_\gamma {}^{GL}\delta_t^{(1-\gamma)} \delta_x^2 u_i^n + f_i^n, \quad i = 1,2,\cdots,N-1, \\ u_i^0 = \phi_0(x_i), \quad i = 0,1,2,\cdots,N, \\ u_0^n = U_a(t_n),\ u_N^n = U_b(t_n), \end{cases} \tag{4.30}$$

where ${}^{GL}\delta_t^{(1-\gamma)}$is given by (4.3).

Next, we give a simple implementation of the method (4.30). We first rewrite the scheme (4.30) as

$$u_i^n = u_i^{n-1} + \mu \sum_{k=0}^{n} \omega_{n-k}^{(1-\gamma)} (u_{i+1}^k - 2u_i^k + u_{i-1}^k) + \Delta t f_i^n, \tag{4.31}$$

where $\mu = K_\gamma \Delta t^\gamma / \Delta x^2$. Denote $E_N \in R^{N\times N}$ as the identity matrix, and

$$B = \begin{bmatrix} 1 & 0 \\ 0 & 0 \\ \vdots & \vdots \\ 0 & 0 \\ 0 & 1 \end{bmatrix}_{(N-1)\times 2}, \tag{4.32}$$

$$S_{N-1} = \begin{bmatrix} -2 & 1 & 0 & \cdots & 0 & 0 \\ 1 & -2 & 1 & \cdots & 0 & 0 \\ 0 & 1 & -2 & \cdots & 0 & 0 \\ \vdots & \vdots & \vdots & \ddots & \vdots & \vdots \\ 0 & 0 & 0 & \cdots & -2 & 1 \\ 0 & 0 & 0 & \cdots & 1 & -2 \end{bmatrix}_{(N-1)\times(N-1)}. \tag{4.33}$$

Let $\underline{\mathbf{u}}^n = (u_1^n, \cdots, u_{N-1}^n)^T$, $\mathbf{u}_b^n = (u_0^n, u_N^n)^T = (U_a(t_n), U_b(t_n))^T$, and $\underline{\mathbf{f}}^n = (f_1^n, \cdots, f_{N-1}^n)^T$. Then the matrix representation of (4.31) can be written as

$$(E_{N-1} - \mu S_{N-1})\underline{\mathbf{u}}^n = \underline{\mathbf{u}}^{n-1} + \mu \sum_{k=0}^{n-1} \omega_{n-k}^{(1-\gamma)} (S_{N-1}\underline{\mathbf{u}}^k) + \mu \sum_{k=0}^{n} \omega_{n-k}^{(1-\gamma)} (B\mathbf{u}_b^k) + \underline{\mathbf{f}}^n.$$

We consider the stability of the finite difference scheme (4.30). The Fourier method and energy method are two powerful tools for the stability and convergence analysis of the numerical methods for fractional differential equations [18, 160, 161, 183]. We mainly focus on the stability analysis, and the convergence analysis is somewhat equivalent to the stability analysis for the linear problems.

- **Fourier method**

We first use the *Fourier method* [18, 160, 161] for the stability analysis of the method (4.30).

Let us rewrite the scheme (4.30) as the following form

$$u_i^n = u_i^{n-1} + \mu \sum_{k=0}^{n} \omega_{n-k}^{(1-\gamma)} (u_{i+1}^k - 2u_i^k + u_{i-1}^k) + \Delta t f_i^n, \tag{4.34}$$

where $\mu = K_\gamma \Delta t^\gamma / \Delta x^2$. Supposing that u_i^n has perturbation $\tilde{u}_i^n$, we can obtain the perturbation equation as follows

$$\tilde{u}_i^n = \tilde{u}_i^{n-1} + \mu \sum_{k=0}^{n} \omega_{n-k}^{(1-\gamma)} (\tilde{u}_{i+1}^k - 2\tilde{u}_i^k + \tilde{u}_{i-1}^k). \tag{4.35}$$

Letting $\tilde{u}_i^n = \rho_n e^{j\sigma i \Delta x} (j^2 = -1)$ and inserting $\tilde{u}_i^n$ into (4.35), one gets

$$\left(1 + 4\mu \sin^2\left(\frac{\sigma \Delta x}{2}\right)\right)\rho_n = \rho_{n-1} - 4\mu \sin^2\left(\frac{\sigma \Delta x}{2}\right) \sum_{k=1}^{n} \omega_k^{(1-\gamma)} \rho_{n-k}. \tag{4.36}$$

Next, we introduce a lemma which is useful to prove that $|\rho^n| \le |\rho^0|$ from (4.36).

Lemma 4.2.1 ([18]) *Let* $\omega_k^{(1-\gamma)} = (-1)^k\binom{1-\gamma}{k}, 0 < \gamma < 1$. *Then one has*

$$\begin{aligned} &\omega_0^{(1-\gamma)} = 1, \quad \omega_k^{(1-\gamma)} < 0,\ k > 0; \\ &\sum_{k=0}^{\infty} \omega_k^{(1-\gamma)} = 0, \quad -\sum_{k=1}^{n} \omega_k^{(1-\gamma)} < 1,\ n \in \mathbb{N}. \end{aligned} \tag{4.37}$$

Based on the above lemma, we give the following stability theorem.

Theorem 22 *The finite difference method* (4.30) *is unconditionally stable.*

Proof. We use the mathematical induction to complete the proof. Let $\mu^* = 4\mu\sin^2\left(\frac{\sigma\Delta x}{2}\right)$. Then we have from (4.36)

$$\rho_n = \frac{1}{1+\mu^*}\rho_{n-1} - \frac{\mu^*}{1+\mu^*}\sum_{k=1}^{n} \omega_k^{(1-\gamma)}\rho_{n-k}. \tag{4.38}$$

For $n = 1$, it follows from (4.38) and Lemma 4.2.1 that

$$|\rho_1| = \frac{|1-\mu^*\omega_1^{(1-\gamma)}|}{1+\mu^*}|\rho_0| = \frac{1+\mu^*(1-\gamma)}{1+\mu^*}|\rho_0| \le |\rho_0|. \tag{4.39}$$

Suppose that $|\rho_k| \le |\rho_0|$ $(0 \le k \le n-1)$. For $k = n$, we get from (4.38) and Lemma 4.2.1

$$\begin{aligned} |\rho_n| &\le \frac{1}{1+\mu^*}|\rho_{n-1}| + \frac{\mu^*}{1+\mu^*}\sum_{k=1}^{n} |\omega_{n-k}^{(1-\gamma)}||\rho_{n-k}| \\ &\le \frac{1}{1+\mu^*}|\rho_0| + \frac{\mu^*}{1+\mu^*}\sum_{k=1}^{n} |\omega_{n-k}^{(1-\gamma)}||\rho_0| \\ &= \frac{1}{1+\mu^*}|\rho_0| + \frac{\mu^*}{1+\mu^*}\left(-\sum_{k=1}^{n} \omega_{n-k}^{(1-\gamma)}\right)|\rho_0| \\ &\le \frac{1}{1+\mu^*}|\rho_0| + \frac{\mu^*}{1+\mu^*}|\rho_0| = |\rho_0|. \end{aligned} \tag{4.40}$$

Therefore, $|\rho_n| \le |\rho_0|$. The proof is completed. □

- **Energy method**

Now we introduce the *energy method* to prove the stability of the scheme (4.30). Let $\mathbf{u} = (u_0, u_1, \cdots, u_N)^T$ and $\mathbf{u}^n = (u_0^n, u_1^n, \cdots, u_N^n)^T$. Denote by the discrete inner product $(\cdot,\cdot)_N$ and the norm $\|\cdot\|_N$ as

$$(\mathbf{u},\mathbf{v})_N = \Delta x \sum_{i=0}^{N-1} u_i v_i, \quad \mathbf{u},\mathbf{v} \in R^{(N+1)\times 1}$$

and

$$\|\mathbf{u}\|_N = \sqrt{(\mathbf{u},\mathbf{u})_N}. \tag{4.41}$$

For brevity, we also introduce the following notations

$$\begin{aligned}(\delta_x\mathbf{u},\delta_x\mathbf{v})_N =&\Delta x\sum_{i=0}^{N-1}\delta_x u_{i+\frac{1}{2}}\delta_x v_{i+\frac{1}{2}},\\(\delta_x^2\mathbf{u},\mathbf{v})_N =&\Delta x\sum_{i=1}^{N-1}(\delta_x^2 u_i)v_i,\\|\mathbf{u}|_{1,N} =& \sqrt{(\delta_x\mathbf{u},\delta_x\mathbf{u})_N}.\end{aligned} \tag{4.42}$$

Lemma 4.2.2 *Let* $\mathbf{u} = (u_0,u_1,\cdots,u_N)^T$ *and* $\mathbf{v} = (v_0,v_1,\cdots,v_N)^T$. *If* $u_0 = u_N = v_0 = v_N = 0$, *then*

$$(\delta_x^2\mathbf{u},\mathbf{v})_N = (\mathbf{u},\delta_x^2\mathbf{v})_N = -(\delta_x\mathbf{u},\delta_x\mathbf{v})_N.$$

Proof. It is easy to calculate

$$\begin{aligned}(\delta_x^2\mathbf{u},\mathbf{v})_N =&\Delta x\sum_{i=1}^{N-1}v_i\delta_x^2u_i = \sum_{i=1}^{N-1}v_i(\delta_x u_{i+\frac{1}{2}} - \delta_x u_{i-\frac{1}{2}})\\=&\sum_{i=1}^{N-1}v_i\delta_x u_{i+\frac{1}{2}} - \sum_{i=0}^{N-2}v_{i+1}\delta_x u_{i+\frac{1}{2}}\\=&\sum_{i=0}^{N-1}v_i\delta_x u_{i+\frac{1}{2}} - \sum_{i=0}^{N-1}v_{i+1}\delta_x u_{i+\frac{1}{2}}\\=&-\sum_{i=0}^{N-1}(v_{i+1}-v_i)\delta_x u_{i+\frac{1}{2}} = -(\delta_x\mathbf{u},\delta_x\mathbf{v})_N.\end{aligned}$$

One can similarly get $(\mathbf{u},\delta_x^2\mathbf{v})_N = -(\delta_x\mathbf{u},\delta_x\mathbf{v})_N$, which ends the proof. □

Lemma 4.2.3 ([183]) *Let* $\mathbf{u} = (u_0,u_1,\cdots,u_N)^T$. *If* $u_0 = u_N = 0$, *then there exists a positive constant C such that*

$$\|\mathbf{u}\|_N \leq C|\mathbf{u}|_{1,N}.$$

Next, we use the energy method to prove that the scheme (4.30) is unconditionally stable.

Theorem 23 *Let* $\mathbf{u}^n = (u_0^n,u_1^n,\cdots,u_N^n)^T$ *be the solutions to the finite difference scheme* (4.30), $u_0^n = u_N^n = 0$, *and* $\mathbf{f}^n = (f_0,f_1^n,\cdots,f_N^n)^T$. *Then there exists a positive constant C independent of n,* Δt *and* Δx, *such that*

$$\|\mathbf{u}^n\|_N^2 \leq \|\mathbf{u}^0\|_N^2 + \Delta t^\gamma K_\gamma|\mathbf{u}^0|_{1,N}^2 + C\max_{0\leq k\leq n_T}\|\mathbf{f}^k\|_N^2.$$

Proof. We rewrite the scheme (4.30) in the form below

$$u_i^n = u_i^{n-1} + \Delta t^\gamma K_\gamma \sum_{k=0}^{n} \omega_{n-k}^{(1-\gamma)} \delta_x^2 u_i^k + \Delta t f_i^n. \tag{4.43}$$

From (4.43), one can immediately get

$$\begin{aligned}(\mathbf{u}^n,\mathbf{u}^n)_N &= (\mathbf{u}^{n-1},\mathbf{u}^n)_N + \Delta t^\gamma K_\gamma \sum_{k=0}^{n} \omega_{n-k}^{(1-\gamma)} (\delta_x^2 \mathbf{u}^k,\mathbf{u}^n)_N + \Delta t(\mathbf{f}^n,\mathbf{u}^n)_N \\ &= (\mathbf{u}^{n-1},\mathbf{u}^n)_N - \Delta t^\gamma K_\gamma \sum_{k=0}^{n} \omega_{n-k}^{(1-\gamma)} (\delta_x \mathbf{u}^k,\delta_x\mathbf{u}^n)_N + \Delta t(\mathbf{f}^n,\mathbf{u}^n)_N,\end{aligned} \tag{4.44}$$

where Lemma 4.2.2 is used. Denote by

$$b_n = \sum_{k=0}^{n} \omega_k^{(1-\gamma)} = \frac{\Gamma(n+\gamma)}{\Gamma(\gamma)\Gamma(n+1)} = \frac{(n+1)^{\gamma-1}}{\Gamma(\gamma)} + O((n+1)^{-2+\gamma}).$$

Then one has $b_n - b_{n-1} = \omega_n^{(1-\gamma)}$ and b_n satisfies $C_0 b_n \Delta t^\gamma \le \Delta t \le C_1 b_n \Delta t^\gamma$, C_0, C_1 are positive constants independent of n [183].

Using the Cauchy–Schwarz inequality, one obtains

$$\begin{aligned}&\|\mathbf{u}^n\|_N^2 + \Delta t^\gamma K_\gamma |\mathbf{u}^n|_{1,N} \\ &\le \frac{1}{2}(\|\mathbf{u}^{n-1}\|_N^2 + \|\mathbf{u}^n\|_N^2) + \frac{\Delta t^\gamma K_\gamma}{2} \sum_{k=0}^{n-1} (b_{n-k-1} - b_{n-k})(|\mathbf{u}^k|_{1,N}^2 + |\mathbf{u}^n|_{1,N}^2) \\ &+ \Delta t(\epsilon\|\mathbf{u}^n\|_N^2 + \frac{1}{4\epsilon}\|\mathbf{f}^n\|_N^2),\end{aligned}$$

where ϵ is a suitable positive constant. Denote

$$E^n = \|\mathbf{u}^n\|_N^2 + \Delta t^\gamma K_\gamma \sum_{k=0}^{n} b_{n-k} |\mathbf{u}^k|_{1,N}^2.$$

Then one gets

$$\begin{aligned}E^n + \Delta t^\gamma K_\gamma b_n |\mathbf{u}^n|_{1,N} &\le E^{n-1} + \Delta t\left(\frac{1}{2\epsilon}\|\mathbf{f}^n\|_N^2 + 2\epsilon\|\mathbf{u}^n\|_N^2\right) \\ &\le E^{n-1} + \Delta t\left(\frac{1}{2\epsilon}\|\mathbf{f}^n\|_N^2 + 2C_2\epsilon\|\mathbf{u}^n\|_N^2\right),\end{aligned} \tag{4.45}$$

where Lemma 4.2.3 is utilized. Choosing suitable $\epsilon = \frac{K_\gamma}{2C_1C_2}$ satisfies

$$2C_2\epsilon\Delta t \le 2C_2\epsilon C_1 b_n \Delta t^\gamma \le K_\gamma b_n \Delta t^\gamma.$$

Therefore, one obtains

$$\begin{aligned} E^n &\le E^{n-1} + C\Delta t \|\mathbf{f}^n\|_N^2 \le E^0 + C\Delta t \sum_{k=1}^{n} \|\mathbf{u}^k\|_N^2 \\ &= \|\mathbf{u}^0\|_N^2 + \Delta t^\gamma K_\gamma |\mathbf{u}^0|_{1,N}^2 + C\Delta t \sum_{k=1}^{n} \|\mathbf{f}^k\|_N^2. \end{aligned} \tag{4.46}$$

By the definition of E^n, we have

$$\|\mathbf{u}^n\|_N^2 \le \|\mathbf{u}^0\|_N^2 + \Delta t^\gamma K_\gamma |\mathbf{u}^0|_{1,N}^2 + C \max_{0\le k\le n_T} \|\mathbf{f}^k|_N^2. \tag{4.47}$$

The proof is completed. □

Remark 4.2.1 *If $\omega_k^{(1-\gamma)}$ in* (4.44) *satisfies $\omega_0^{(1-\gamma)} > 0$ and $\sum_{k=1}^n |\omega_k^{(1-\gamma)}| \le \omega_0^{(1-\gamma)}$, then the inequality* (4.47) *holds.*

Next, we consider the convergence analysis for (4.30). From Theorem 23, we can obtain the error bounds for the method (4.30).

Theorem 24 *Let $U(x,t)$ and u_i^n $(i = 0, 1, 2, \cdots, N, n = 1, 2, \cdots, n_T)$ be solutions to equations* (4.10) *and* (4.30)*, respectively. Denote by $e_i^n = u_i^n - U(x_i, t_n)$ and $\mathbf{e}^n = (e_0^n, e_1^n, \cdots, e_N^n)^T$. Then there exists a positive constant C independent of $n, \Delta t$ and Δx, such that*

$$\|\mathbf{e}^n\|_N \le C(\Delta t + \Delta x^2).$$

Proof. One can get the error equation as follows

$$e_i^n = e_i^{n-1} + \Delta t^\gamma K_\gamma \sum_{k=0}^{n} \omega_{n-k}^{(1-\gamma)} \delta_x^2 e_i^k + \Delta t R_i^n, \tag{4.48}$$

in which R_i^n is the truncation error satisfying $R_i^n = O(\Delta t + \Delta x^2)$ from (4.29).

By Theorem 23, we only need to estimate

$$\|\mathbf{e}^0\|_N^2 + \Delta t^\gamma K_\gamma |\mathbf{e}^0|_{1,N}^2 + C \max_{0\le k\le n_T} \|\mathbf{R}^k\|_N^2$$

to get the error bound, where $\mathbf{R}^n = (R_0^n, R_1^n, \cdots, R_N^n)^T$ with $R_i^n = O(\Delta t + \Delta x^2)$. Obviously, $\|\mathbf{e}^0\|_N = |\mathbf{e}^0|_{1,N} = 0$, and $\|\mathbf{R}^k\|_N \le C(\Delta t + \Delta x^2)$. Hence $\|\mathbf{e}^n\|_N \le C(\Delta t + \Delta x^2)$, which ends the proof. □

Obviously, the time fractional derivative in (4.10) can be discretized by different methods, which yield different backward Euler type methods. For example, the time fractional derivative in (4.10) can be discretized by the L1 method (4.4), the fractional backward difference method (4.6).

In (4.30), if the time-fractional derivative in (4.10) is approximated by the L1 method (4.4), one has the following implicit scheme: Find u_i^n ($i = 1,2,\cdots,N-1, n = 1,2,\cdots,n_T$), such that

$$\begin{cases} \delta_t u_i^{n-\frac{1}{2}} = K_\gamma {}_{RL}^{L1}\delta_t^{(1-\gamma)} \delta_x^2 u_i^n + f_i^n, \quad i = 1,2,\cdots,N-1, \\ u_i^0 = \phi_0(x_i), \quad i = 0,1,2,\cdots,N, \\ u_0^n = U_a(t_n),\ u_N^n = U_b(t_n), \end{cases} \tag{4.49}$$

where ${}_{RL}^{L1}\delta_t^{(1-\gamma)}$ is given by (4.4).

Using the Fourier method or the energy method, we can similarly prove that the finite difference method (4.49) is unconditionally stable and convergent of order $O(\Delta t + \Delta x^2)$; readers can refer to [71, 86] for more information.

If $\gamma = 1$, then the two methods (4.30) and (4.49) are reduced to the backward Euler method. The two methods (4.30) and (4.49) have only first-order accuracy in time for $\gamma \in (0,1)$.

Cui [19] proposed a compact finite difference scheme to solve (4.10), in which the time discretization is the same as the method (4.30), while the space was discretized by the fourth-order compact finite difference scheme. In [17], the explicit and implicit finite difference methods were presented to solve the fractional reaction-subdiffusion equation. The implicit method is similar to (4.30), which is unconditionally stable and convergent of order $O(\Delta t + \Delta x^2)$. The explicit method is also similar to (4.30), except that the integer-order time derivative was discretized by the forward Euler method, i.e., $\delta_t u_i^{n-\frac{1}{2}}$ in (4.30) is replaced by $\delta_t u_i^{n+\frac{1}{2}}$.

4.2.1.3 Crank–Nicolson Type Methods

We know that the Crank–Nicolson (CN) method for the classical equation (See (4.10) with $\gamma = 1$) has second-order accuracy in time. The CN method for the classical diffusion equation can be constructed by the following direct methods:

- **Method I**: Letting $t = t_{n+\frac{1}{2}}$ in (4.10) with $\gamma = 1$ yields

  $$\partial_t U(t_{n+\frac{1}{2}}) = \mu \partial_x^2 U(t_{n+\frac{1}{2}}) + f(t_{n+\frac{1}{2}}).$$

  Note that $\partial_t U(t_{n+\frac{1}{2}}) = \delta_t U^{n+\frac{1}{2}} + O(\Delta t^2)$ and $U(t_{n+\frac{1}{2}}) = U^{n+\frac{1}{2}} + O(\Delta t^2)$, one has

  $$\delta_t U^{n+\frac{1}{2}} = \mu \partial_x^2 U^{n+\frac{1}{2}} + f(t_{n+\frac{1}{2}}) + O(\Delta t^2).$$

  Letting $x = x_i$ and using $\partial_x^2 U^n = \delta_x^2 U_i^n + O(\Delta x^2)$, we have

  $$\delta_t U_i^{n+\frac{1}{2}} = \mu \delta_x^2 U_i^{n+\frac{1}{2}} + f(x_i, t_{n+\frac{1}{2}}) + O(\Delta t^2 + \Delta x^2).$$

  Neglecting the truncation error and letting $u_i^n = U_i^n$ yields the classical CN method below:

  $$\delta_t u_i^{n+\frac{1}{2}} = \mu \delta_x^2 u_i^{n+\frac{1}{2}} + f(x_i, t_{n+\frac{1}{2}}). \tag{4.50}$$

- **Method II**: Letting $x = x_i, t = t_k, k = n, n+1$ in (4.10) with $\gamma = 1$ gives

$$\partial_t U(x_i, t_n) = \mu \partial_x^2 U(x_i, t_n) + f(x_i, t_n), \tag{4.51}$$

$$\partial_t U(x_i, t_{n+1}) = \mu \partial_x^2 U(x_i, t_{n+1}) + f(x_i, t_{n+1}). \tag{4.52}$$

Adding the two equations leads to

$$\partial_t U(x_i, t_n) + \partial_t U(x_i, t_{n+1}) = \mu(\partial_x^2 U(x_i, t_{n+1}) + \partial_x^2 U(x_i, t_n)) + f(x_i, t_n) + f(x_i, t_{n+1}). \tag{4.53}$$

Noting that $\partial_t U(x_i, t_n) + \partial_t U(x_i, t_{n+1}) = 2\delta_t U^{n+\frac{1}{2}} + O(\Delta t^2)$ and $\partial_x^2 U(x_i, t_n) = 2\delta_x^2 U^{n+\frac{1}{2}} + O(\Delta x^2)$, one has

$$\delta_t U_i^{n+\frac{1}{2}} = \mu \delta_x^2 U_i^{n+\frac{1}{2}} + f_i^{n+\frac{1}{2}} + O(\Delta t^2 + \Delta x^2).$$

Dropping the truncation error and letting $u_i^n = U_i^n$ yields the following CN method:

$$\delta_t u_i^{n+\frac{1}{2}} = \mu \delta_x^2 u_i^{n+\frac{1}{2}} + f_i^{n+\frac{1}{2}}. \tag{4.54}$$

For the classical diffusion equation, the two methods (4.50) and (4.54) are unconditionally stable and convergent of order 2 for the suitably smooth solutions. For the fractional subdiffusion equation (4.10), we can use similar techniques to derive the *Crank–Nicolson type methods*, which yield different properties of the desired scheme.

Similar to (4.50), we first let $(x, t) = (x_i, t_{n-\frac{1}{2}})$ in (4.10), which gives

$$\partial_t U(x_i, t_{n-\frac{1}{2}}) = K_\gamma \left({}_{RL}\mathrm{D}_{0,t}^{1-\gamma} \partial_x^2 U \right)(x_i, t_{n-\frac{1}{2}}) + f(x_i, t_{n-\frac{1}{2}}). \tag{4.55}$$

Clearly, the first-order time derivative in (4.55) is discretized by the central difference method, and the space derivative can be discretized by the cental difference scheme. Now we should adopt a technique to discretize the fractional operator ${}_{RL}\mathrm{D}_{0,t}^{1-\gamma}$ at $t = t_{n-\frac{1}{2}}$ such that higher accuracy can be obtained. Fortunately, we indeed have a method, named the modified L1 method (2.70), with $(1+\gamma)$th-order accuracy to approximate ${}_{RL}\mathrm{D}_{0,t}^{1-\gamma}$ at $t = t_{n-\frac{1}{2}}$. Hence, we have

$$\delta_t U_i^{n-\frac{1}{2}} = K_\gamma \delta_t^{(1-\gamma)} \delta_x^2 U_i^{n-\frac{1}{2}} + f(x_i, t_{n-\frac{1}{2}}) + (\Delta t^{1+\gamma} + \Delta x^2), \tag{4.56}$$

where $\delta_t^{(1-\gamma)}$ is defined by

$$\delta_t^{(1-\gamma)} U_i^{n-\frac{1}{2}} = \frac{1}{\Delta t^{1-\gamma}} \left[b_0 U_i^{n-\frac{1}{2}} - \sum_{k=1}^{n-1} (b_{n-1-k} - b_{n-k}) U_i^{k-\frac{1}{2}} - (b_{n-1} - B_{n-1}) U_i^{\frac{1}{2}} - A_{n-1} U_i^0 \right], \tag{4.57}$$

in which $A_n = B_n - \frac{\gamma(n+1/2)^{\gamma-1}}{\Gamma(1+\gamma)}$, b_n and B_n are defined by

$$b_n = \frac{1}{\Gamma(1+\gamma)} \left[(n+1)^\gamma - n^\gamma \right], \quad B_n = \frac{2}{\Gamma(1+\gamma)} \left[(n+1/2)^\gamma - n^\gamma \right]. \tag{4.58}$$

Therefore, the first CN type method is given by:

$$\begin{cases} \delta_t u_i^{n-\frac{1}{2}} = K_\gamma \delta_t^{(1-\gamma)} \delta_x^2 u_i^{n-\frac{1}{2}} + f(x_i, t_{n-\frac{1}{2}}), \quad i = 1, 2, \cdots, N-1, \quad n = 1, 2, \cdots, n_T, \\ u_i^0 = \phi_0(x_i), \quad i = 0, 1, 2, \cdots, N, \\ u_0^n = U_a(t_n), \ u_N^n = U_b(t_n), \end{cases} \tag{4.59}$$

where $\delta_t^{(1-\gamma)}$ is defined by (4.57).

The CN type method (4.59) is reduced to the classical CN method (4.50) if $\gamma \to 1$. Of course, one can also use

$$\frac{1}{2}\left[\left({}_{RL}\mathrm{D}_{0,t}^{1-\gamma} \partial_x^2 U\right)(x_i, t_n) + \left({}_{RL}\mathrm{D}_{0,t}^{1-\gamma} \partial_x^2 U\right)(x_i, t_{n-1})\right]$$

to replace $\left({}_{RL}\mathrm{D}_{0,t}^{1-\gamma} \partial_x^2 U\right)(x_i, t_{n-\frac{1}{2}})$ in (4.55) as in the classical CN method (4.50). Then the appropriate discretization for the time fractional derivative operator ${}_{RL}\mathrm{D}_{0,t}^{1-\gamma}$ at $t = t_n, t_{n-1}$ is applied. So we can derive the following CN type method

$$\begin{cases} \delta_t u_i^{n-\frac{1}{2}} = \dfrac{K_\gamma}{2}\left[\delta_t^{(1-\gamma)} \delta_x^2 u_i^{n-1} + \delta_t^{(1-\gamma)} \delta_x^2 u_i^n\right] + f(x_i, t_{n+\frac{1}{2}}), \\ \qquad\qquad i = 1, 2, \cdots, N-1, \quad n = 1, 2, \cdots, n_T, \\ u_i^0 = \phi_0(x_i), \quad i = 0, 1, 2, \cdots, N, \\ u_0^n = U_a(t_n), \ u_N^n = U_b(t_n), \end{cases} \tag{4.60}$$

where $\delta_t^{(1-\gamma)}$ is the approximate operator of the time fractional derivative operator ${}_{RL}\mathrm{D}_{0,t}^{1-\gamma}$.

It is known that $\frac{1}{2}\left(\left[{}_{RL}\mathrm{D}_{0,t}^{1-\gamma} u(t)\right]_{t=t_{n-1}} + \left[{}_{RL}\mathrm{D}_{0,t}^{1-\gamma} u(t)\right]_{t=t_n}\right)$ is not a good approximation to $\left[{}_{RL}\mathrm{D}_{0,t}^{1-\gamma} u(t)\right]_{t=t_{n-\frac{1}{2}}}$. For example, $u(t) = t^\nu, \nu \geq 0$, so we can derive

$$\begin{aligned} &\frac{1}{2}\left(\left[{}_{RL}\mathrm{D}_{0,t}^{1-\gamma} u(t)\right]_{t=t_{n-1}} + \left[{}_{RL}\mathrm{D}_{0,t}^{1-\gamma} u(t)\right]_{t=t_n}\right) - \left[{}_{RL}\mathrm{D}_{0,t}^{1-\gamma} u(t)\right]_{t=t_{n-\frac{1}{2}}} \\ =&\frac{\Gamma(\nu+1)}{2\Gamma(\nu+\gamma)}\left(t_{n-1}^{\nu+\gamma-1} + t_n^{\nu+\gamma-1} - 2t_{n-\frac{1}{2}}^{\nu+\gamma-1}\right) = O(\Delta t^2 t_n^{\nu+\gamma-3}). \end{aligned} \tag{4.61}$$

Clearly, if $(\nu+\gamma)$ is not a positive integer or the noninteger number $(\nu+\gamma)$ is small, then the error introduced in (4.61) is of order $O(\Delta t^{\nu+\gamma-1})$ for small n. For example, ν is a nonnegative integer and $\gamma \neq 1$, which implies that the smooth enough $u(t)$ can not guarantee second-order approximation. Hence, even if the high-order method is used to discretize the time fractional operator ${}_{RL}\mathrm{D}_{0,t}^{1-\gamma}$, we can not obtain the satisfactory numerical solutions when n is small.

Next, we adopt the second technique used in (4.54) to construct the corresponding CN method for (4.10), which can avoid using (4.61) in the construction of the numerical algorithm.

Letting $(x,t) = (x_i, t_n)$ and $(x,t) = (x_i, t_{n-1})$ in (4.10) leads to

$$\partial_t U(x_i,t_n) = K_\gamma \left({}_{RL}\mathrm{D}_{0,t}^{1-\gamma}\partial_x^2 U\right)(x_i,t_n) + f(x_i,t_n), \tag{4.62}$$

$$\partial_t U(x_i,t_{n-1}) = K_\gamma \left({}_{RL}\mathrm{D}_{0,t}^{1-\gamma}\partial_x^2 U\right)(x_i,t_{n-1}) + f(x_i,t_{n-1}). \tag{4.63}$$

Adding (4.62) to (4.63), one has

$$\begin{aligned}&\partial_t U(x_i,t_n) + \partial_t U(x_i,t_{n-1})\\ =&K_\gamma\left[\left({}_{RL}\mathrm{D}_{0,t}^{1-\gamma}\partial_x^2 U\right)(x_i,t_n) + \left({}_{RL}\mathrm{D}_{0,t}^{1-\gamma}\partial_x^2 U\right)(x_i,t_{n-1})\right] + f(x_i,t_n) + f(x_i,t_{n-1}).\end{aligned} \tag{4.64}$$

If the high-order method is used to discretize ${}_{RL}D_{0,t}^{1-\gamma}\partial_x^2 U(x,t)$ at $t = t_n$, then we can get more accurate numerical algorithms than the Euler type methods introduced before. One choice is to use L1 method to discretize ${}_{RL}D_{0,t}^{1-\gamma}\partial_x^2 U(x,t)$ at $t = t_{n-1}$ and $t = t_n$, which gives

$$\begin{aligned}\delta_t U_i^{n-\frac{1}{2}} =&\frac{K_\gamma}{2}\left[{}_{RL}^{L1}\delta_t^{(1-\gamma)}\delta_x^2 U_i^n + {}_{RL}^{L1}\delta_t^{(1-\gamma)}\delta_x^2 U_i^{n-1}\right]\\ &+\frac{1}{2}(f_i^n + f_i^{n-1}) + O(\Delta t^{1+\gamma} + \Delta x^2), \qquad n > 1,\end{aligned} \tag{4.65}$$

where ${}_{RL}^{L1}\delta_t^{(1-\gamma)}$ is defined by (4.4). When $U(x,t)$ is smooth in time, $f(x_i,t_0)$ may be unbounded. Therefore, for $n = 1$, one can use the following relation

$$\delta_t U_i^{\frac{1}{2}} = K_\gamma {}_{RL}^{L1}\delta_t^{(1-\gamma)}\delta_x^2 U_i^1 + f_i^1 + O(\Delta t + \Delta x^2). \tag{4.66}$$

Removing the truncation error in (4.65) and (4.66), and replacing U_i^n with u_i^n, we can get the following CN type method

$$\begin{cases}\delta_t u_i^{\frac{1}{2}} = K_\gamma \delta_t^{(1-\gamma)}\delta_x^2 u_i^1 + f_i^1, \quad i = 1,2,\cdots,N-1,\\ \delta_t u_i^{n-\frac{1}{2}} = \dfrac{K_\gamma}{2}\left({}_{RL}^{L1}\delta_t^{(1-\gamma)}\delta_x^2 u_i^n + {}_{RL}^{L1}\delta_t^{(1-\gamma)}\delta_x^2 u_i^{n-1}\right) + \dfrac{1}{2}(f_i^n + f_i^{n-1}),\\ \qquad i = 1,2,\cdots,N-1, \quad n = 2,3,\cdots,n_T,\\ u_i^0 = \phi_0(x_i), \quad i = 0,1,2,\cdots,N,\\ u_0^n = U_a(t_n),\ u_N^n = U_b(t_n),\end{cases} \tag{4.67}$$

where ${}_{RL}^{L1}\delta_t^{(1-\gamma)}$ is defined by (4.4).

If $\gamma \to 1$, the CN type method (4.60) is reduced to the classical CN method (4.50). For the CN type method (4.67), the classical CN method (4.54) can not be recovered when $\gamma \to 1$ and $n = 1$.

Next, we consider the stability and convergence of the three CN methods (4.59), (4.60), and (4.67).

Lemma 4.2.4 ([166]) *Let b_n, B_n be defined by (4.58), $A_n = B_n - \frac{\gamma(n+1/2)^{\gamma-1}}{\Gamma(1+\gamma)\Delta t^{1-\gamma}}$, $0<\gamma\leq 1$. Then we have*

$$\begin{aligned}&0\leq b_n \leq b_{n-1}, \quad B_n \leq b_{n-1}, \qquad n \geq 1;\\ &A_n \leq B_n, \qquad\quad\ \ b_n \leq B_n, \qquad n \geq 0,\end{aligned} \tag{4.68}$$

and

$$\sum_{j=1}^{n} A_j \leq \frac{2^{\gamma}\Delta t^{\gamma-1}}{\Gamma(1-\gamma)}. \tag{4.69}$$

Theorem 25 *Let* $\mathbf{u}^n = (u_0^n, u_1^n, \cdots, u_N^n)^T$ *be the solution to the finite difference scheme* (4.59), $u_0^n = u_N^n = 0$, *and* $\mathbf{f}^{n-\frac{1}{2}} = (0, f_1^{n-\frac{1}{2}}, \cdots, f_{N-1}^{n-\frac{1}{2}}, 0)^T$. *Then*

$$||\mathbf{u}^{n+1}||_N^2 \leq 2||\mathbf{u}^0||_N^2 + C_1\Delta t^{\gamma}|\mathbf{u}^0|_{1,N}^2 + C_2\Delta t\sum_{j=0}^{n}||\mathbf{f}^{k+\frac{1}{2}}||_N^2,$$

where C_1 *is a positive constant independent of* n, h, τ *and* T, *and* C_2 *is a positive constant independent of* n, h *and* τ.

Proof. Let $\delta_t\mathbf{u}^{n+\frac{1}{2}} = (\delta_t u_0^{n+\frac{1}{2}}, \delta_t u_1^{n+\frac{1}{2}}, \cdots, \delta_t u_N^{n+\frac{1}{2}})^T$ and $\mathbf{u}^{n+\frac{1}{2}} = (u_0^{n+\frac{1}{2}}, \cdots, u_N^{n+\frac{1}{2}})^T$. Then from (4.59) and Lemma 4.2.2 we have

$$\begin{aligned}(\delta_t\mathbf{u}^{n+\frac{1}{2}}, \mathbf{u}^{n+\frac{1}{2}})_N = \mu\Big[&-b_0(\delta_x\mathbf{u}^{n+\frac{1}{2}}, \delta_x\mathbf{u}^{n+\frac{1}{2}})_N + \sum_{j=2}^{n-1}(b_{n-j}-b_{n+1-j})(\delta_x\mathbf{u}^{j-\frac{1}{2}}, \delta_x\mathbf{u}^{n+\frac{1}{2}})_N \\ &+ (b_{n-1}-B_n)(\delta_x\mathbf{u}^{\frac{1}{2}}, \delta_x\mathbf{u}^{n+\frac{1}{2}})_N + A_n(\delta_x\mathbf{u}^0, \delta_x\mathbf{u}^{n+\frac{1}{2}})_N\Big] + (\mathbf{f}^{n+\frac{1}{2}}, \mathbf{u}^{n+\frac{1}{2}})_N.\end{aligned} \tag{4.70}$$

Using Cauchy–Schwarz inequality and Lemma 4.2.4 brings about

$$\begin{aligned}&(\delta_t\mathbf{u}^{n+\frac{1}{2}}, \mathbf{u}^{n+\frac{1}{2}})_N \\ \leq&\frac{\mu}{2}\Big[-2b_0|\mathbf{u}^{n+\frac{1}{2}}|_{1,N}^2 + \sum_{j=2}^{n-1}(b_{n-j}-b_{n+1-j})(|\mathbf{u}^{j-\frac{1}{2}}|_{1,N}^2 + |\mathbf{u}^{n+\frac{1}{2}}|_{1,N}^2) \\ &+ (b_{n-1}-B_n)(|\mathbf{u}^{\frac{1}{2}}|_{1,N}^2 + |\mathbf{u}^{n+\frac{1}{2}}|_{1,N}^2) + A_n(|\mathbf{u}^0|_{1,N}^2 + |\mathbf{u}^{n-\frac{1}{2}}|_{1,N}^2)\Big] + (\mathbf{f}^{n+\frac{1}{2}}, \mathbf{u}^{n+\frac{1}{2}})_N \\ =&\frac{\mu}{2}\Big[(-b_0-B_n+A_n)|\mathbf{u}^{n+\frac{1}{2}}|_{1,N}^2 + \sum_{j=2}^{n-1}(b_{n-j}-b_{n+1-j})|\mathbf{u}^{j-\frac{1}{2}}|_{1,N}^2 \\ &+ (b_{n-1}-B_n)|\mathbf{u}^{\frac{1}{2}}|_{1,N}^2 + A_n|\mathbf{u}^0|_{1,N}^2\Big] + (\mathbf{f}^{n+\frac{1}{2}}, \mathbf{u}^{n+\frac{1}{2}})_N \\ \leq&\frac{\mu}{2}\Big[-b_0|\mathbf{u}^{n+\frac{1}{2}}|_{1,N}^2 + \sum_{j=1}^{n-1}(b_{n-j}-b_{n+1-j})|\mathbf{u}^{j-\frac{1}{2}}|_{1,N}^2 + A_n|\mathbf{u}^0|_{1,N}^2\Big] + (\mathbf{f}^{n+\frac{1}{2}}, \mathbf{u}^{n+\frac{1}{2}})_N.\end{aligned} \tag{4.71}$$

Rewriting (4.71) leads to

$$\begin{aligned}&||\mathbf{u}^{n+1}||_N^2 + \mu\Delta t\sum_{j=1}^{n+1}b_{n+1-j}|\mathbf{u}^{j-\frac{1}{2}}|_{1,N}^2 \\ \leq&||\mathbf{u}^n||_N^2 + \mu\Delta t\sum_{j=1}^{n}b_{n-j}|\mathbf{u}^{j-\frac{1}{2}}|_{1,N}^2 + \mu\Delta tA_n|\mathbf{u}^0|_{1,N}^2 + 2\Delta t(\mathbf{f}^{n+\frac{1}{2}}, \mathbf{u}^{n+\frac{1}{2}})_N.\end{aligned} \tag{4.72}$$

Denote by

$$E^{n+1} = \|\mathbf{u}^{n+1}\|_N^2 + \mu\Delta t \sum_{j=1}^{n+1} b_{n+1-j} |\mathbf{u}^{j-\frac{1}{2}}|_{1,N}^2.$$

Then, one can obtain from (4.72) that

$$\begin{aligned}
E^{n+1} &\leq E^n + \mu\Delta t |\mathbf{u}^0|_{1,N}^2 A_n + 2\Delta t (\mathbf{f}^{n+\frac{1}{2}}, \mathbf{u}^{n+\frac{1}{2}})_N \\
&\leq E^{n-1} + \mu\Delta t |\mathbf{u}^0|_{1,N}^2 (A_n + A_{n-1}) + 2\Delta t \left[(\mathbf{f}^{n+\frac{1}{2}}, \mathbf{u}^{n+\frac{1}{2}})_N + (\mathbf{f}^{n-\frac{1}{2}}, \mathbf{u}^{n-\frac{1}{2}})_N\right] \\
&\leq E^1 + \mu\Delta t |\mathbf{u}^0|_{1,N}^2 \sum_{j=1}^{n} A_n + 2\Delta t \sum_{k=1}^{n} (\mathbf{f}^{n+\frac{1}{2}}, \mathbf{u}^{k+\frac{1}{2}})_N \\
&\leq E^1 + \mu\Delta t |\mathbf{u}^0|_{1,N}^2 \sum_{j=1}^{n} A_n + \epsilon\mu\Delta t \sum_{j=1}^{n+1} b_{n+1-j} |\mathbf{u}^{j-\frac{1}{2}}|_{1,N}^2 + \sum_{j=1}^{n+1} \frac{\Delta t}{\epsilon\mu b_{n+1-j}} \|\mathbf{f}^{j-\frac{1}{2}}\|_N^2.
\end{aligned}$$

Here we have used the Cauchy–Schwarz inequality and $\epsilon\|\mathbf{u}^{j-\frac{1}{2}}\|_N^2 \leq |\mathbf{u}^{j-\frac{1}{2}}|_{1,N}^2$ from Lemma 4.2.3. ϵ is a suitable positive constant independent of j, h, τ and u_h^j. Therefore, we have

$$\|\mathbf{u}^{n+\frac{1}{2}}\|_N^2 \leq E^1 + \mu\Delta t^{\gamma} \|\mathbf{u}^0\|^2 \sum_{j=1}^{n} A_n + C\Delta t \sum_{j=1}^{n+1} \|\mathbf{f}^{j-\frac{1}{2}}\|_N^2, \tag{4.73}$$

where $1/b_n \leq C_\gamma n^{1-\gamma} \Delta t^{1-\gamma} \leq C_\gamma T^{1-\gamma}$, C_γ only depends on γ.

E^1 can be estimated in the following way. Letting $n = 0$ in (4.70) and using the Cauchy–Schwarz inequality gives

$$\begin{aligned}
&\|\mathbf{u}^1\|_N^2 + \mu\Delta t B_0 |\mathbf{u}^{\frac{1}{2}}|_{1,N}^2 \\
=&\|\mathbf{u}^0\|_N^2 + \mu\Delta t A_0 (\delta_x \mathbf{u}^{\frac{1}{2}}, \delta_x \mathbf{u}^0)_N + 2\Delta t (\mathbf{f}^{\frac{1}{2}}, \mathbf{u}^{\frac{1}{2}})_N \\
\leq&\|\mathbf{u}^0\|_N^2 + \mu\Delta t A_0 \left(\epsilon_1 |\mathbf{u}^{\frac{1}{2}}|_{1,N}^2 + \frac{1}{4\epsilon_1} |\mathbf{u}^0|_{1,N}^2\right) + 2\Delta t \left(\frac{1}{4\epsilon_2} \|\mathbf{f}^{\frac{1}{2}}\|_N^2 + \epsilon_2 \|\mathbf{u}^{\frac{1}{2}}\|_N^2\right),
\end{aligned} \tag{4.74}$$

where $\epsilon_1, \epsilon_2 > 0$ are suitable constants such that

$$\epsilon_1 \mu A_0 |\mathbf{u}^{\frac{1}{2}}|_{1,N}^2 + 2\epsilon_2 \|\mathbf{u}^{\frac{1}{2}}\|_N^2 \leq \frac{1}{2} \mu B_0 |\mathbf{u}^{\frac{1}{2}}|_{1,N}^2.$$

Therefore, we obtain

$$\begin{aligned}
E^1 &= \|\mathbf{u}^1\|_N^2 + \mu\Delta t B_0 |\mathbf{u}^{\frac{1}{2}}|_{1,N}^2 \leq 2\|\mathbf{u}^1\|_N^2 + \mu\Delta t B_0 |\mathbf{u}^{\frac{1}{2}}|_{1,N}^2 \\
&\leq 2\|\mathbf{u}^0\|_N^2 + \frac{\mu\Delta t A_0}{2\epsilon_1} |\mathbf{u}^0|_{1,N}^2 + \frac{\Delta t}{\epsilon_2} \|\mathbf{f}^{\frac{1}{2}}\|_N^2.
\end{aligned} \tag{4.75}$$

Combining (4.73) and (4.75) yields

$$\|\mathbf{u}^{n+1}\|_N^2 \leq 2\|\mathbf{u}^0\|_N^2 + C_1 \Delta t^{\gamma} |\mathbf{u}^0|_{1,N}^2 + C_2 \Delta t \sum_{j=0}^{n} \|\mathbf{f}^{k+\frac{1}{2}}\|_N^2, \tag{4.76}$$

in which C_1 is independent of n, h, τ and T, and C_2 is independent of n, h and τ. The proof is completed. □

Theorem 25 states that the CN type method (4.59) is unconditionally stable.

Let $e_i^n = U(x_i, t_n) - u_i^n$. Then we can derive the error equation of (4.59) as

$$\delta_t e_i^{n-\frac{1}{2}} = K_\gamma \delta_t^{(1-\gamma)} \delta_x^2 e_i^{n-\frac{1}{2}} + R_i^n, \tag{4.77}$$

where $R_i^n = O(\Delta t^{1+\gamma} + \Delta x^2)$ and $e_0^n = e_N^n = 0$ and $e_i^0 = 0, i = 0, 1, \cdots, N$. Denote $\mathbf{e}^n = (e_0^n, e_1^n, \cdots, e_N^n)^T$ and $\mathbf{R}^n = (R_0^n, R_1^n, \cdots, R_N^n)^T$. Then from (4.77) and Theorem 25, we can easily obtain

$$||\mathbf{e}^{n+1}||_N^2 \leq 2||\mathbf{e}^0||_N^2 + C_1 \Delta t^\gamma |\mathbf{e}^0|_{1,N}^2 + C_2 \Delta t \sum_{k=0}^{n} ||\mathbf{R}^k||_N^2 \leq C(\Delta t^{1+\gamma} + \Delta x^2). \tag{4.78}$$

In a similar manner, we can show that the CN type method (4.67) is unconditionally stable and convergent of order $O(\Delta t^{\min\{2-\gamma/2, 1+\gamma\}} + \Delta x^2)$. This result was derived in [173]. For the CN type method (4.60), one can prove that it is unconditionally stable if the time fractional operator is discretized by the Grünwald–Letnikov formula (see (4.3)). But the convergence rate is not very satisfactory for the smooth enough solutions; see (4.61). We can see that the CN type method (4.59) is more natural than the other two CN methods (4.60) and (4.67).

The CN type method (4.60) can be seen as a special case of the following *weighted average finite difference method*

$$\begin{cases} \delta_t u_i^{n-\frac{1}{2}} = K_\gamma \left[(1-\theta) \delta_t^{(1-\gamma)} \delta_x^2 u_i^n + \theta \delta_t^{(1-\gamma)} \delta_x^2 u_i^{n-1} \right] + f(x_i, t_{n-\frac{1}{2}}), \\ \qquad i = 1, 2, \cdots, N-1, n = 1, 2, \cdots, n_T \\ u_i^0 = \phi_0(x_i), \quad i = 0, 1, 2, \cdots, N, \\ u_0^n = U_a(t_n),\ u_N^n = U_b(t_n), \end{cases} \tag{4.79}$$

where $0 \leq \theta \leq 1$, and the operator $\delta_t^{(1-\gamma)}$ can be defined by any approximation operator to the time fractional derivative operator ${}_{RL}D_{0,t}^{1-\gamma}$.

If $\theta = \frac{1}{2}$, the method (4.79) is reduced to (4.60). The stability was studied by using the *fractional von Neumann analysis* in [160] when $\delta_t^{(1-\gamma)} = {}_p^B\delta_t^{(1-\gamma)}$ is defined by (4.6). The method (4.79) is unconditionally stable for $0 \leq \theta \leq \frac{1}{2}$, and stable for $\frac{1}{2} < \theta \leq 1$ under the condition $\frac{K_\gamma \Delta t^\gamma}{\Delta x^2} \leq \frac{1}{2(2\theta-1) w^{(\gamma-1)}(-1)}$, where $w^{(\gamma-1)}(z)$ is defined by (4.7). If $\theta = 0$ (or $\theta = 1$), the explicit (or implicit) Euler type method (4.16) (or (4.30)) is recovered.

4.2.1.4 Integration Methods

Next, we introduce an indirect method (or the integration method) to discretize the subdiffusion equation (4.10).

Letting $x = x_i$ and integrating both sides of (4.10) on the interval $[t_{n-1}, t_n]$ in time direction, one can obtain

$$U(x_i,t_n) = U(x_i,t_{n-1}) + K_\gamma\left\{\left[\mathrm{D}_{0,t}^{-(1-\gamma)}\partial_x^2 U(x_i,t)\right]_{t=t_n} - \left[\mathrm{D}_{0,t}^{-(1-\gamma)}\partial_x^2 U(x_i,t)\right]_{t=t_{n-1}}\right\} + F_i^n, \tag{4.80}$$

where $F_i^n = \int_{t_{n-1}}^{t_n} f(x_i,s)\,\mathrm{d}s$.

The fractional integral $\left[\mathrm{D}_{0,t}^{-(1-\gamma)}\partial_x^2 U(x_i,t)\right]_{t=t_n}$ can be discretized by different methods which lead to different numerical schemes. We can naturally think of using the left fractional rectangular formula (2.6), the right fractional rectangular formula (2.8), or the fractional trapezoidal formula (2.12) to approximate the time fractional integral operator in (4.80).

If the *left fractional rectangular formula* (2.6) is used to discretize $\mathrm{D}_{0,t}^{-(1-\gamma)}\partial_x^2 U(x_i,t)$ at $t = t_n, t_{n-1}$ in (4.80), and the space is discretized by the central difference, then one has the following explicit method

$$U_i^n = U_i^{n-1} + \Delta t^\gamma K_\gamma\left[\sum_{k=1}^{n} b_{n-k}^{(1-\gamma)}\delta_x^2 U_i^{k-1} - \sum_{k=1}^{n-1} b_{n-k-1}^{(1-\gamma)}\delta_x^2 U_i^{k-1}\right] + F_i^n + O(\Delta t(\Delta t + \Delta x^2)), \tag{4.81}$$

where $b_k^{(1-\gamma)} = \frac{1}{\Gamma(1+\gamma)}[(k+1)^\gamma - k^\gamma]$.

Replacing U_i^n with u_i^n and removing the truncation error, one obtains the following explicit method

$$\begin{cases} u_i^n = u_i^{n-1} + \Delta t^\gamma K_\gamma\left[\sum_{k=1}^{n} b_{n-k}^{(1-\gamma)}\delta_x^2 u_i^{k-1} - \sum_{k=1}^{n-1} b_{n-k-1}^{(1-\gamma)}\delta_x^2 u_i^{k-1}\right] + F_i^n, \quad i = 1,2,\cdots,N-1, \\ u_i^0 = \phi_0(x_i), \quad i = 0,1,2,\cdots,N, \\ u_0^n = U_a(t_n), \quad u_N^n = U_b(t_n), \end{cases} \tag{4.82}$$

where $b_k^{(1-\gamma)} = \frac{1}{\Gamma(1+\gamma)}[(k+1)^\gamma - k^\gamma]$.

Similarly, if the *right fractional rectangular formula* (2.8) is used in (4.80), we can derive the following implicit method

$$\begin{cases} u_i^n = u_i^{n-1} + \Delta t^\gamma K_\gamma\left[\sum_{k=1}^{n} b_{n-k}^{(1-\gamma)}\delta_x^2 u_i^{k} - \sum_{k=1}^{n-1} b_{n-k-1}^{(1-\gamma)}\delta_x^2 u_i^{k}\right] + F_i^n, \quad i = 1,2,\cdots,N-1, \\ u_i^0 = \phi_0(x_i), \quad i = 0,1,2,\cdots,N, \\ u_0^n = U_a(t_n), \quad u_N^n = U_b(t_n), \end{cases} \tag{4.83}$$

where $b_k^{(1-\gamma)} = \frac{1}{\Gamma(1+\gamma)}[(k+1)^\gamma - k^\gamma]$.

If the *fractional trapezoidal formula* (2.12) is used in (4.80), we can obtain the following algorithm

$$\begin{cases} u_i^n = u_i^{n-1} + \Delta t^{\gamma} K_{\gamma} \left(\sum_{k=0}^{n} a_{k,n} \delta_x^2 u_i^k - \sum_{k=0}^{n-1} a_{k,n-1} \delta_x^2 u_i^k \right) + F_i^n, \quad i = 1, 2, \cdots, N-1, \\ u_i^0 = \phi_0(x_i), \quad i = 0, 1, 2, \cdots, N, \\ u_0^n = U_a(t_n), \; u_N^n = U_b(t_n), \end{cases} \tag{4.84}$$

where $a_{n,k}$ is defined by

$$a_{k,n} = \frac{1}{\Gamma(\gamma+2)} \begin{cases} (n-1)^{\gamma+1} - (n-1-\gamma)n^{\gamma}, & k = 0, \\ (n-k+1)^{\gamma+1} + (n-1-k)^{\gamma+1} - 2(n-k)^{\gamma+1}, & 1 \le k \le n-1, \\ 1, & k = n. \end{cases}$$

If $\gamma \to 1$, these three methods (4.82), (4.83), and (4.84) are reduced to the forward Euler method, the backward Euler method, and the CN method, respectively. If $U(x,t)$ is smooth enough in time, the truncation errors of the three methods (4.82), (4.83), and (4.84) are $O(\Delta t + \Delta x^2)$, $O(\Delta t + \Delta x^2)$, and $O(\Delta t^2 + \Delta x^2)$, respectively, which can be derived from (2.20).

One can prove by the Fourier method that the explicit method (4.82) is stable if $\frac{K_{\gamma}\Delta t^{\gamma}}{\Delta x^2} \le \frac{1}{4}$, and the implicit method (4.83) is unconditionally stable. One can also refer to [183], where the unconditional stability and the convergence of the implicit method (4.83) were proved by the energy method. The convergence of method (4.84) can be found in [158].

The integration technique can be used to solve other fractional equations, such as the *fractional Fokker–Planck equation* [154]

$$\partial_t U = {}_{RL}\mathrm{D}_{0,t}^{1-\gamma} \big(\partial_x (d(x) \partial_x U) + f(x,t) \big),$$

the *fractional cable equation* [62, 99]

$$\partial_t U = {}_{RL}\mathrm{D}_{0,t}^{1-\gamma_1} \big(K_1 \partial_x^2 U \big) - K_2 {}_{RL}\mathrm{D}_{0,t}^{1-\gamma_2} U + f(x,t), \; 0 < \gamma_1, \gamma_2 < 1,$$

and the *Stokes' first problem for a heated generalized second grade fluid with fractional derivative* [16]

$$\partial_t U = {}_{RL}\mathrm{D}_{0,t}^{1-\gamma_1} \big(K_1 \partial_x^2 U \big) + K_2 {}_{RL}\mathrm{D}_{0,t}^{1-\gamma_2} U + f(x,t), \; 0 < \gamma_1, \gamma_2 < 1.$$

4.2.1.5 Numerical Examples

We present numerical examples to illustrate the effectiveness of the numerical methods in this subsection.

Example 4 *Consider the time-fractional subdiffusion equation*

$$\begin{cases} \partial_t u = {}_{RL}\mathrm{D}_{0,t}^{1-\gamma} u + f(x,t), \quad (x,t) \in (0,\pi) \times (0,1], \\ u(x,0) = 0, \quad x \in [0,\pi], \\ u(0,t) = 0, \quad u(\pi,t) = 0, \quad t \in [0,1], \end{cases} \tag{4.85}$$

where $f = \left(2.5t^{1.5} + \frac{\Gamma(3.5)}{\Gamma(3.5-\alpha)}t^{2.5-\alpha}\right)\sin(x)$. *The above equation* (4.85) *has the analytical solution* $u(x,t) = t^{2.5}\sin(x)$.

Next, we check the stability and convergence of the explicit type Euler methods (4.16) and (4.82), the implicit methods (4.30), (4.59), (4.83), and (4.84). The maximum L^2 error is defined as follows

$$\max_{0\le n\le n_T} \|\mathbf{u}-\mathbf{U}^n\|_N = \max_{0\le n\le n_T} \sqrt{\Delta x \sum_{j=0}^{N-1} (U(x_j,t_n) - u_j^n)^2}.$$

We first test the stability of the explicit Euler type methods (4.16) and (4.82), the maximum L^2 errors are shown in Tables 4.1 and 4.2, respectively. From (4.22), one finds that the method (4.16) is stable when $\frac{\Delta t^\gamma}{\Delta x^2} \le 2^{\gamma-2}$ for solving (4.85), which is illustrated in Table 4.1, where "NaN" means that the stability condition $\frac{\Delta t^\gamma}{\Delta x^2} \le 2^{\gamma-2}$ is not satisfied. Table 4.2 shows the similar results as Table 4.1.

Obviously, the explicit methods (4.16) and (4.82) need strict restriction on $\frac{\Delta t^\gamma}{\Delta x^2}$ in the stability. When the space step size Δx is reduced, the much smaller time step size Δt is needed in order to keep the methods (4.16) and (4.82) stable, which requires large storage in the real computation.

Next, we test the stability and convergence of implicit methods such as (4.30), (4.59), (4.83), and (4.84), which show better stability than the explicit methods (4.16) and (4.82). Tables 4.3–4.6 display the maximum L^2 errors of the implicit methods (4.30), (4.59), (4.83), and (4.84) for Example 4, which show good performances, and the observed experimental convergence orders are in line with the theoretical analysis.

TABLE 4.1: The maximum L^2 error of the explicit Euler type method (4.16), $\Delta t = 1/40000$.

N	$\gamma = 0.2$	$\frac{\Delta t^\gamma}{\Delta x^2}$	$2^{\gamma-2}$	$\gamma = 0.5$	$\frac{\Delta t^\gamma}{\Delta x^2}$	$2^{\gamma-2}$	$\gamma = 0.8$	$\gamma = 0.95$
10	NaN	1.22		3.7919e−3	0.05		2.9621e−3	2.5676e−3
20	NaN	4.87		9.8971e−4	0.20		7.8085e−4	6.8137e−4
30	NaN	10.95	0.29	NaN	0.46	0.35	3.7647e−4	3.3157e−4
40	NaN	19.47		NaN	0.81		2.3491e−4	2.0911e−4
50	NaN	30.42		NaN	1.27		1.6938e−4	1.5242e−4

TABLE 4.2: The maximum L^2 error of the explicit Euler type method (4.82), $\Delta t = 1/40000$.

N	$\gamma = 0.2$	$\gamma = 0.5$	$\gamma = 0.8$	$\gamma = 0.95$
10	NaN	3.7998e−3	2.9649e−3	2.5682e−3
20	NaN	9.9768e−4	7.8359e−4	6.8199e−4
30	NaN	NaN	3.7921e−4	3.3219e−4
40	NaN	NaN	2.3765e−4	2.0973e−4
50	NaN	NaN	1.7212e−4	1.5304e−4

TABLE 4.3: The maximum L^2 error of the implicit Euler type method (4.30), $N = 1000$.

$1/\Delta t$	$\gamma = 0.2$	order	$\gamma = 0.5$	order	$\gamma = 0.8$	order
8	1.8195e−1		1.5777e−1		1.4051e−1	
16	8.9917e−2	1.0169	7.8387e−2	1.0092	7.0388e−2	0.9973
32	4.4687e−2	1.0087	3.9064e−2	1.0048	3.5224e−2	0.9988
64	2.2275e−2	1.0045	1.9499e−2	1.0025	1.7619e−2	0.9994
128	1.1120e−2	1.0022	9.7410e−3	1.0012	8.8112e−3	0.9997

TABLE 4.4: The maximum L^2 error of the Crank–Nicolson type method (4.82), $N = 1000$.

$1/\Delta t$	$\gamma = 0.2$	order	$\gamma = 0.5$	order	$\gamma = 0.8$	order
8	4.2600e−2		9.6759e−3		2.6338e−3	
16	2.0452e−2	1.0586	4.4424e−3	1.1231	4.8959e−4	2.4275
32	9.3682e−3	1.1264	1.8185e−3	1.2886	9.4388e−5	2.3749
64	4.1927e−3	1.1599	7.0398e−4	1.3692	1.8835e−5	2.3252
128	1.8536e−3	1.1775	2.6416e−4	1.4141	8.4538e−6	1.1558

TABLE 4.5: The maximum L^2 error of the implicit Euler type method (4.83), $N = 1000$.

$1/\Delta t$	$\gamma = 0.2$	order	$\gamma = 0.5$	order	$\gamma = 0.8$	order
8	2.2569e−1		2.4567e−1		2.4628e−1	
16	1.1822e−1	0.9328	1.2830e−1	0.9372	1.2735e−1	0.9515
32	6.1127e−2	0.9516	6.5855e−2	0.9622	6.4805e−2	0.9746
64	3.1352e−2	0.9633	3.3465e−2	0.9767	3.2703e−2	0.9867
128	1.5994e−2	0.9710	1.6904e−2	0.9852	1.6431e−2	0.9930

TABLE 4.6: The maximum L^2 error of the implicit method (4.84), $N = 1000$.

$1/\Delta t$	$\gamma = 0.2$	order	$\gamma = 0.5$	order	$\gamma = 0.8$	order
8	4.0242e−3		3.1760e−3		3.0590e−3	
16	9.8681e−4	2.0279	7.8837e−4	2.0103	7.7210e−4	1.9862
32	2.4196e−4	2.0280	1.9645e−4	2.0047	1.9479e−4	1.9868
64	5.9575e−5	2.0220	4.9238e−5	1.9963	4.9222e−5	1.9846
128	1.4937e−5	1.9958	1.2563e−5	1.9706	1.2582e−5	1.9680

4.2.2 Caputo Type Subdiffusion Equations

If $U(x,t)$ is suitably smooth in time, and one applies the fractional integral $\mathrm{D}_{0,t}^{-(1-\gamma)}$ to both sides of (4.10), one can obtain the following *Caputo type time-fractional diffusion equation* [56, 113]

$$\begin{cases} {}_C\mathrm{D}_{0,t}^{\gamma}U = K_\gamma \partial_x^2 U + g(x,t), & (x,t) \in (a,b)\times(0,T], \\ U(x,0) = \phi_0(x), & x \in (a,b), \\ U(a,t) = U_a(t),\ U(b,t) = U_b(t), & t \in (0,T], \end{cases} \tag{4.86}$$

where $g(x,t) = D_{0,t}^{\gamma-1} f(x,t)$.

If $0 < \gamma < 1$ and the solution $U(x,t)$ is suitably smooth in time, then ${}_C\mathrm{D}_{0,t}^{\gamma}U(x,t) = {}_{RL}\mathrm{D}_{0,t}^{\gamma}(U(x,t) - U(x,0))$. Hence, a natural way to discretize the Caputo derivative in (4.86) is to use the Grünwald–Letnikov approximation, or the L1 method, or the fractional linear multistep method, etc., and the space is discretized by the classical methods such as the central difference method or the compact difference method [47, 56, 62, 176].

4.2.2.1 Explicit Euler Type Methods

Here, we only introduce two explicit methods that are reduced to the classical Euler methods when $\gamma \to 1$.

Let $(x,t) = (x_i, t_n)$ in (4.86). Then one has

$$ {}_C\mathrm{D}_{0,t}^{\gamma}U(x_i,t_n) = K_\gamma \partial_x^2 U(x_i,t_n) + g_i^n = K_\gamma \partial_x^2 U(x_i,t_{n-1}) + g_i^n + O(\Delta t). \tag{4.87}$$

The Caputo derivative in (4.87) can be discretized by the known methods, i.e., the Grünwald–Letnikov formula, the L1 method, or the FLMM, etc.; the space direction is discretized by the central difference method. One has

$$\delta_t^{(\gamma)} U_i^n = K_\gamma \delta_x^2 U_i^{n-1} + g_i^n + O(\Delta t + \Delta x^2), \tag{4.88}$$

where $\delta_t^{(\gamma)}$ is the approximate operator in time that is to be defined.

Next, we provide two ways for the discretization of the Caputo derivative.

The Caputo derivative is discretized by the Grünwald–Letnikov formula and the space direction is discretized by the central difference method, one can get the finite difference method for (4.86) as: Find u_i^n $(i = 1,2,\cdots,N-1, n = 1,2,\cdots,n_T)$, such that

$$\begin{cases} \delta_t^{(\gamma)}(u_i^n - u_i^0) = K_\gamma \delta_x^2 u_i^{n-1} + g_i^n, & i = 1,2,\cdots,N-1, \\ u_i^0 = \phi_0(x_i), \quad i = 0,1,2,\cdots,N, \\ u_0^n = U_a(t_n),\ u_N^n = U_b(t_n), \end{cases} \tag{4.89}$$

where $\delta_t^{(\gamma)}(u_i^n - u_i^0)$ is defined by

$$\delta_t^{(\gamma)}(u_i^n - u_i^0) = \frac{1}{\Delta t^\gamma}\sum_{k=0}^{n} \omega_{n-k}^{(\gamma)}(u_i^k - u_i^0), \quad \omega_k^{(\gamma)} = (-1)^k\binom{\gamma}{k}.$$

The Caputo derivative is discretized by the L1 method with the space direction approximated by the central difference scheme; one can derive the method for (4.86) as: Find u_i^n $(i = 1,2,\cdots,N-1, n = 1,2,\cdots,n_T)$, such that

$$\begin{cases} \delta_t^{(\gamma)} u_i^n = K_\gamma \delta_x^2 u_i^{n-1} + g_i^n, & i = 1,2,\cdots,N-1, \\ u_i^0 = \phi_0(x_i), & i = 0,1,2,\cdots,N, \\ u_0^n = U_a(t_n),\ u_N^n = U_b(t_n), \end{cases} \tag{4.90}$$

where $\delta_t^{(\gamma)}$ is defined by

$$\delta_t^{(\gamma)} u_i^n = \frac{1}{\Delta t^\gamma} \sum_{k=0}^{n-1} b_{n-k-1}^{(\gamma)} (u_i^{k+1} - u_i^k) = \frac{1}{\Delta t^\gamma} \left[b_0 u_i^n - \sum_{k=0}^{n-1} (b_{n-k-1}^{(\gamma)} - b_{n-k}^{(\gamma)}) u_i^k - b_n u_i^0 \right],$$

$$b_k^{(\gamma)} = \frac{1}{\Gamma(2-\gamma)} \left[(k+1)^{1-\gamma} - k^{1-\gamma} \right].$$

Now, we discuss the stability of the two methods (4.89) and (4.90).

Theorem 26 *Suppose that* u_i^n $(i = 1,2,\cdots,N-1, n = 0,1,2,\cdots,n_T)$ *is the solution to* (4.89). *Let* $\mu = K_\gamma \Delta t^\gamma / \Delta x^2$. *If* $\mu \le \gamma/2$, *then the method* (4.89) *is stable.*

Proof. Suppose that u_i^n $(i = 1,2,\cdots,N-1)$ and g_i^n $(i = 1,2,\cdots,N-1)$ have perturbations $\tilde{u}_i^n$ $(i = 1,2,\cdots,N-1)$ and $\tilde{g}_i^n$ $(i = 1,2,\cdots,N-1)$, respectively. Denote by $\tilde{\mathbf{u}}^n = (\tilde{u}_1^n, \tilde{u}_2^n, \cdots, \tilde{u}_{N-1}^n)^T$, $\tilde{\mathbf{g}}^n = (\tilde{g}_1^n, \tilde{g}_2^n, \cdots, \tilde{g}_{N-1}^n)^T$, and

$$A = \begin{bmatrix} \omega_1^{(\gamma)} + 2\mu & -\mu & 0 & \cdots & 0 & 0 \\ -\mu & \omega_1^{(\gamma)} + 2\mu & -\mu & \cdots & 0 & 0 \\ 0 & -\mu & \omega_1^{(\gamma)} + 2\mu & \cdots & 0 & 0 \\ \vdots & \vdots & \vdots & \ddots & \vdots & \vdots \\ 0 & 0 & 0 & \cdots & \omega_1^{(\gamma)} + 2\mu & -\mu \\ 0 & 0 & 0 & \cdots & -\mu & \omega_1^{(\gamma)} + 2\mu \end{bmatrix}_{(N-1)\times(N-1)},$$

$$B = \begin{bmatrix} 1-2\mu & \mu & 0 & \cdots & 0 & 0 \\ \mu & 1-2\mu & \mu & \cdots & 0 & 0 \\ 0 & \mu & 1-2\mu & \cdots & 0 & 0 \\ \vdots & \vdots & \vdots & \ddots & \vdots & \vdots \\ 0 & 0 & 0 & \cdots & 1-2\mu & \mu \\ 0 & 0 & 0 & \cdots & \mu & 1-2\mu \end{bmatrix}_{(N-1)\times(N-1)}.$$

Expand the equation (4.89) in the following form

$$\sum_{k=0}^{n} \omega_{n-k}^{(\gamma)} (\tilde{u}_i^k - \tilde{u}_i^0) = \mu \tilde{u}_i^{n-1} (\tilde{u}_{i+1}^{n-1} - 2\tilde{u}_i^{n-1} + \tilde{u}_{i-1}^{n-1}) + \Delta t^\gamma \tilde{g}_i^n. \tag{4.91}$$

Then the matrix representation of the perturbation equation (4.91) can be expressed as

$$\begin{cases} \tilde{\mathbf{u}}^1 = B\tilde{\mathbf{u}}^0 + \Delta t^\gamma \tilde{\mathbf{g}}^1, \quad n = 1, \\ \tilde{\mathbf{u}}^n = -A\tilde{\mathbf{u}}^{n-1} - \sum_{k=1}^{n-2} \omega_{n-k}^{(\gamma)} \tilde{\mathbf{u}}^k + \sum_{k=1}^{n} \omega_{n-k}^{(\gamma)} \tilde{\mathbf{u}}^0 + \Delta t^\gamma \tilde{\mathbf{g}}^n, \quad n > 1. \end{cases} \tag{4.92}$$

Since $2\mu \le \gamma = -\omega_1^{(\gamma)}$, it is easy to obtain $||A|| \le -\omega_1^{(\gamma)}$ and $||B|| \le 1$ by the Greschgorin's theorem. Here $||A||$ denotes the *spectral norm* (or *2-norm*) of the matrix A, which is equal to the absolute largest eigenvalue of A when A is symmetric.

For convenience, we also denote it by

$$b_{n-1} = \sum_{k=0}^{n-1} \omega_k^{(\gamma)} = \frac{\Gamma(n-\gamma)}{\Gamma(1-\gamma)\Gamma(n)} = \frac{n^{-\gamma}}{\Gamma(1-\gamma)} + O(n^{-1-\gamma}).$$

Then one can easily prove that $\Delta t^\gamma \le C b_{n-1}$, C is a positive constant only dependent on γ and T. Next, we prove that

$$||\tilde{\mathbf{u}}^n|| \le ||\tilde{\mathbf{u}}^0|| + C \max_{1 \le n \le n_T} ||\tilde{\mathbf{g}}^n|| = E, \tag{4.93}$$

where $||\cdot||$ is the discrete L^2 norm for the vector, which is defined by

$$||\mathbf{u}|| = \left(\sum_{i=1}^{N-1} u_i^2 \right)^{1/2}, \quad \mathbf{u} = (u_1, u_2, \cdots, u_{N-1})^T \in R^{N-1}.$$

We use the mathematical induction method to prove (4.93). For $n = 1$, one has from (4.92)

$$||\tilde{\mathbf{u}}^1|| = ||B\tilde{\mathbf{u}}^0 + \Delta t^\gamma \tilde{\mathbf{g}}^1|| \le ||B|| ||\tilde{\mathbf{u}}^0|| + C b_0 ||\tilde{\mathbf{g}}^1|| \le ||\tilde{\mathbf{u}}^0|| + C ||\tilde{\mathbf{g}}^1|| \le E.$$

Suppose that $||\tilde{\mathbf{u}}^n|| \le E, n = 1, 2, \cdots, m-1$. For $n = m$, one has from (4.92)

$$\begin{aligned} ||\tilde{\mathbf{u}}^m|| &\le ||A\tilde{\mathbf{u}}^{m-1}|| - \sum_{k=1}^{m-2} \omega_{m-k}^{(\gamma)} ||\tilde{\mathbf{u}}^k|| + \sum_{k=1}^{m} \omega_{m-k}^{(\gamma)} ||\tilde{\mathbf{u}}^0|| + \Delta t^\gamma ||\tilde{\mathbf{g}}^m|| \\ &\le ||A|| E - \sum_{k=1}^{m-2} \omega_{m-k}^{(\gamma)} E + b_{m-1} ||\tilde{\mathbf{u}}^0|| + \Delta t^\gamma ||\tilde{\mathbf{g}}^m|| \\ &\le -\omega_1^{(\gamma)} E - \sum_{k=1}^{m-2} \omega_{m-k}^{(\gamma)} E + b_{m-1} E \\ &= b_0 E - \sum_{k=0}^{m-1} \omega_1^{(\gamma)} E + b_{m-1} E = E. \end{aligned} \tag{4.94}$$

Hence, $||\tilde{\mathbf{u}}^m|| \le E$. Therefore, (4.93) holds for all $n > 0$, which ends the proof. □

From Theorem 26, we can obtain that the explicit method (4.89) is convergent with order $O(\Delta t+\Delta x^2)$ if $\frac{K_\gamma \Delta t^\gamma}{\Delta x^2} \le -\frac{1}{2}\omega_1^{(\gamma)} = \frac{\gamma}{2}$. The convergence of the explicit method (4.89) was also proved by Gorenflo and Abdel–Rehim [60] in the Fourier–Laplace domain.

One can similarly prove that the explicit method (4.90) is conditionally stable and convergent with order $O(\Delta t+\Delta x^2)$ if $\frac{K_\gamma \Delta t^\gamma}{\Delta x^2} \le \frac{b_0^{(\gamma)}-b_1^{(\gamma)}}{2} = \frac{1-2^{-\gamma}}{\Gamma(2-\gamma)}$.

Obviously, the two explicit methods (4.89) and (4.90) are reduced to the explicit Euler method if $\gamma = 1$. When γ is small, the time step Δt is sufficiently small in order to meet the stability condition; this may need more iterations to get the numerical solutions. Generally speaking, implicit methods have a larger stability region, which has a weaker restriction on the step size in time; and the implicit methods may have higher accuracy than the explicit ones.

4.2.2.2 Implicit Euler Type Methods

Next, we introduce the typical implicit methods. Let $(x,t) = (x_i,t_n)$ in (4.86). Then one has

$$ {}_C\mathrm{D}_{0,t}^{\gamma} U(x_i,t_n) = K_\gamma \partial_x^2 U(x_i,t_n) + g_i^n. \tag{4.95}$$

The time derivative in (4.95) is discretized by the Grünwald–Letnikov formula, and the space derivative is discretized by the central difference method. One can obtain the following finite difference method for (4.86), which is given by: Find u_i^n $(i = 1,2,\cdots,N-1, n = 1,2,\cdots,n_T)$, such that

$$\begin{cases} \delta_t^{(\gamma)}(u_i^n - u_i^0) = K_\gamma \delta_x^2 u_i^n + g_i^n, \quad i = 1,2,\cdots,N-1, \\ u_i^0 = \phi_0(x_i), \quad i = 0,1,2,\cdots,N, \\ u_0^n = U_a(t_n),\ u_N^n = U_b(t_n), \end{cases} \tag{4.96}$$

where $\delta_t^{(\gamma)}(u_i^n - u_i^0)$ is defined by

$$\delta_t^{(\gamma)}(u_i^n - u_i^0) = \frac{1}{\Delta t^\gamma}\sum_{k=0}^{n} \omega_{n-k}^{(\gamma)}(u_i^n - u_i^0), \quad \omega_k^{(\gamma)} = (-1)^k \binom{\gamma}{k}.$$

The L1 method can be used to discretize the Caputo derivative in (4.86) or (4.95), and the space derivative is discretized by the central difference. The corresponding method is given by: Find u_i^n $(i = 1,2,\cdots,N-1, n = 1,2,\cdots,n_T)$, such that

$$\begin{cases} \delta_t^{(\gamma)} u_i^n = K_\gamma \delta_x^2 u_i^n + g_i^n, \quad i = 1,2,\cdots,N-1, \\ u_i^0 = \phi_0(x_i), \quad i = 0,1,2,\cdots,N, \\ u_0^n = U_a(t_n),\ u_N^n = U_b(t_n), \end{cases} \tag{4.97}$$

where $\delta_t^{(\gamma)}$ is defined by

$$\delta_t^{(\gamma)} u_i^n = \frac{1}{\Delta t^\gamma}\sum_{k=0}^{n-1} b_{n-k}^{(\gamma)}(u_i^{k+1} - u_i^k),$$

$$b_k^{(\gamma)} = \frac{1}{\Gamma(2-\gamma)}\left[(k+1)^{1-\gamma} - k^{1-\gamma}\right].$$

It is easy to prove that the two difference methods (4.96) and (4.97) are unconditionally stable using the Fourier method or the energy method, and are convergent of order $O(\Delta t+\Delta x^2)$ and $O(\Delta t^{2-\gamma}+\Delta x^2)$, respectively.

Next, we just give the stability and convergence analysis for (4.97); the stability and convergence for (4.96) is very similar. The Fourier method and the energy method are also illustrated in the proof.

Theorem 27 *The finite difference method* (4.97) *is unconditionally stable.*

Proof. We first use the Fourier method. Suppose that $g_i^n=0$ and $u_i^n=\rho_n e^{j\sigma i\Delta x}(j^2=-1)$. Inserting u_i^n into (4.97) yields

$$\left(b_0^{(\gamma)}+4\mu^*\right)\rho_n=\sum_{k=1}^{n-1}(b_{n-k-1}^{(\gamma)}-b_{n-k}^{(\gamma)})\rho_k+b_n^{(\gamma)}\rho_0, \tag{4.98}$$

where $\mu^*=\frac{\Delta t^\gamma}{\Delta x^2}K_\gamma\sin^2\left(\frac{\sigma\Delta x}{2}\right)$. Next, we use the mathematical induction method to prove that $|\rho_n|\le|\rho_0|$. It is easy to verify that $0\le b_{k+1}^{(\gamma)}\le b_k^{(\gamma)}, k=0,1,\cdots$.

If $n=1$, one can get

$$|\rho_1|=\frac{b_1^{(\gamma)}}{b_0^{(\gamma)}+4\mu^*}|\rho_0|\le|\rho_0|.$$

Hence, $|\rho_1|\le|\rho_0|$. Suppose that $|\rho_k|\le|\rho_0|, n=1,2,\cdots,m-1$. For $n=m$, one has

$$\begin{aligned}|\rho_m|&\le\frac{1}{b_0^{(\gamma)}+4\mu\sin^2\frac{\sigma\Delta x}{2}}\sum_{k=1}^{m-1}(b_{m-k-1}^{(\gamma)}-b_{m-k}^{(\gamma)})|\rho_k|+\frac{b_{n-1}^{(\gamma)}}{b_0^{(\gamma)}+4\mu^*}|\rho_0|\\&\le\frac{1}{b_0^{(\gamma)}+4\mu^*}\left(\sum_{k=1}^{m-1}(b_{m-k-1}^{(\gamma)}-b_{m-k}^{(\gamma)})+b_{m-1}^{(\gamma)}\right)|\rho_0|\\&=\frac{b_0^{(\gamma)}}{b_0^{(\gamma)}+4\mu^*}|\rho_0|\le|\rho_0|.\end{aligned} \tag{4.99}$$

Therefore, $|\rho_m|\le|\rho_0|$, so that $|\rho_n|\le|\rho_0|$ for all $0\le n\le n_T$. □

Next, we use the energy method to prove Theorem 27. We first introduce a lemma.

Lemma 4.2.5 *Let* $\mathbf{u}^k=(u_0^k,u_1^k,\cdots,u_N^k)$ *and* $\mathbf{g}^k=(g_0^k,g_1^k,\cdots,g_N^k)$. *The series* $\{b_k\}$ *satisfies* $b_0>0$, $\sum_{k=1}^{\infty}|b_k|\le b_0$, $b_k=O(k^{-\gamma})$, $B_n=O(n^{-\gamma})$, *and* $\Delta t^\gamma\le Cb_n$, *C is independent of n and* Δt. *If*

$$b_0(\mathbf{u}^n,\mathbf{u}^n)_N\le\sum_{k=1}^{n-1}b_{n-k}(\mathbf{u}^k,\mathbf{u}^n)_N+B_n(\mathbf{u}^0,\mathbf{u}^n)_N+\Delta t^\gamma(\mathbf{g}^n,\mathbf{u}^n)_N, \tag{4.100}$$

then

$$||\mathbf{u}^n||_N^2\le C_\gamma||\mathbf{u}^0||_N^2+C_1\max_{0\le k\le n_T}||\mathbf{g}^k||_N^2, \tag{4.101}$$

where C_γ *is only dependent on* γ *and* C_1 *is independent of* $n,\Delta t$.

Proof. Denote by $\mu = \Delta t^{\gamma}/b_0^{(\gamma)}$ and $c_k = b_k/b_0 = O(k^{-\gamma})$, so $c_0 = 1$, $|c_k| \le 1$ and $\sum_{k=1}^{\infty} |c_k| \le 1$. From (4.100) and the Cauchy inequality, one has

$$\begin{aligned}||\mathbf{u}^n||_N^2 \le & \frac{1}{2}\sum_{k=1}^{n-1} |c_{n-k}|(||\mathbf{u}^k||_N^2 + ||\mathbf{u}^n||_N^2) \\ & + \frac{|c_n|}{4}||\mathbf{u}^n||_N^2 + \frac{B_n^2}{b_0|c_n|}||\mathbf{u}^0||_N^2 + \frac{|c_n|}{4}||\mathbf{u}^n||_N^2 + \frac{\mu^2}{|c_n|}||\mathbf{g}^n||_N^2.\end{aligned} \tag{4.102}$$

One can immediately get from (4.102)

$$\begin{aligned}||\mathbf{u}^n||_N^2 \le & \sum_{k=1}^{n-1} |c_{n-k}|\,||\mathbf{u}^k||_N^2 + \frac{2B_n^2}{b_0|c_n|}||\mathbf{u}^0||_N^2 + \frac{2\mu^2}{|c_n|}||\mathbf{g}^n||_N^2 \\ \le & \sum_{k=1}^{n-1} |c_{n-k}|\,||\mathbf{u}^k||_N^2 + |c_n|\left(C_0||\mathbf{u}^0||_N^2 + C_1||\mathbf{g}^n||_N^2\right),\end{aligned} \tag{4.103}$$

where we have used $\frac{2B_n^2}{b_0^2|c_n|} \le C_\gamma|c_n|$ and $\frac{\mu^2}{c_n^2} \le C$, here C_γ is only dependent on γ, and C_1 is independent of Δt and n.

Now, we use the mathematical induction method to prove that

$$||\mathbf{u}^1||_N^2 \le C_\gamma||\mathbf{u}^0||_N^2 + C_1 \max_{0\le k\le n_T} ||\mathbf{g}^k||_N^2 = E. \tag{4.104}$$

For $n = 1$, one has from (4.103)

$$||\mathbf{u}^1||_N^2 \le |c_1|\left(C_0||\mathbf{u}^0||_N^2 + C_1||\mathbf{g}^0||_N^2\right) \le E. \tag{4.105}$$

Hence, (4.104) holds for $n = 1$. Suppose that (4.104) holds for $n = 1, 2, \cdots, m-1$. For $n = m$, one has from (4.103)

$$||\mathbf{u}^n||_N^2 \le \sum_{k=1}^{n-1} |c_{n-k}|\,||\mathbf{u}^k||_N^2 + |c_n|E \le E\sum_{k=1}^{n} |c_k| \le c_0E = E. \tag{4.106}$$

Therefore, $||\mathbf{u}^n||_N^2 \le E$ holds for all n. The proof is completed. □

Remark 4.2.2 *If B_n in (4.100) satisfies $|B_n| \le |b_n|$, then $C_\gamma = 2$. If $b_{n-k} = b_{n-k-1}^{(\gamma)} - b_{n-k}^{(\gamma)}$ and $B_n = b_{n-1}^{(\gamma)}$, $0 \le b_{k+1}^{(\gamma)} \le b_k^{(\gamma)}, k = 0, 1, \cdots$, then $C_\gamma = 2$.*

Proof. Here we use the energy method to prove it. Expanding the equation (4.97) yields

$$b_0^{(\gamma)}u_i^n - \sum_{k=1}^{n-1}(b_{n-k-1}^{(\gamma)} - b_{n-k}^{(\gamma)})u_i^k - b_{n-1}^{(\gamma)}u_i^0 = K_\gamma\Delta t^\gamma\delta_x^2u_i^n + \Delta t^\gamma g_i^n. \tag{4.107}$$

Assume that $u_0^n = u_N^n = 0$. Like Eq. (4.44) in Theorem 23, one can easily get

$$\begin{aligned} b_0^{(\gamma)}(\mathbf{u}^n,\mathbf{u}^n)_N \leq & b_0^{(\gamma)}(\mathbf{u}^n,\mathbf{u}^n)_N + K_\gamma \Delta t^\gamma K_\gamma(\delta_x \mathbf{u}^n, \delta_x \mathbf{u}^n)_N \\ = & \sum_{k=1}^{n-1}(b_{n-k-1}^{(\gamma)} - b_{n-k}^{(\gamma)})(\mathbf{u}^k,\mathbf{u}^n)_N + b_{n-1}^{(\gamma)}(\mathbf{u}^0,\mathbf{u}^n)_N + \Delta t^\gamma(\mathbf{g}^n,\mathbf{u}^n)_N, \end{aligned} \tag{4.108}$$

where $\mathbf{u}^n = (u_0^n, u_1^n, \cdots, u_N^n)^T$ and $\mathbf{g}^n = (g_0^n, g_1^n, \cdots, g_N^n)^T$. Applying Lemma 4.2.5 and Remark 4.2.2 yields

$$||\mathbf{u}^n||_N^2 \leq 2||\mathbf{u}^0||_N^2 + C \max_{0\leq k\leq n_T} ||\mathbf{g}^k||_N^2. \tag{4.109}$$

The proof is finished. □

Theorem 28 *Let $U(x_i,t_n)$ and u_i^n $(i = 0,1,\cdots,N, n = 1,2,\cdots,n_T)$ are the solutions to the equations* (4.86) *and* (4.97)*, respectively. Denote by $e_i^n = U(x_i,t_n) - u_i^n$ and $\mathbf{e}^n = (e_0^n, e_1^n, \cdots, e_N^n)^T$. Then there exists a positive constant C independent of $n, \Delta t$ and Δx, such that*

$$||\mathbf{e}^n||_N \leq C(\Delta t^{2-\gamma} + \Delta x^2).$$

Proof. One can get the error equation as follows

$$b_0^{(\gamma)} e_i^n - \sum_{k=1}^{n-1}(b_{n-k-1}^{(\gamma)} - b_{n-k}^{(\gamma)})e_i^k - b_{n-1}^{(\gamma)} e_i^0 = K_\gamma \Delta t^\gamma \delta_x^2 e_i^n + \Delta t^\gamma R_i^n,$$

where $|R_i^n| \leq C(\Delta t^{2-\gamma} + \Delta x^2)$. By (4.109), one has

$$||\mathbf{e}^n||_N^2 \leq C_\gamma ||\mathbf{e}^0||_N^2 + C \max_{0\leq k\leq n_T} ||\mathbf{R}^k||_N^2 \leq C(\Delta t^{2-\gamma} + \Delta x^2).$$

The proof is completed. □

4.2.2.3 FLMM Finite Difference Methods

In this subsection, we introduce the fractional linear multistep methods for the time discretization of the equation (4.10).

Consider the following fractional ordinary differential equation (FODE)

$${}_C\mathrm{D}_{0,t}^\gamma y(t) = \lambda y(t) + g(t), \quad y(0) = y_0, \quad 0 < \gamma < 1. \tag{4.110}$$

From (3.105), (3.108), and (3.109), we have the discretization for (4.110)

$$\frac{1}{\Delta t^\gamma}\sum_{k=0}^{n}\omega_{n-k}(y(t_k) - y_0) = \lambda \sum_{k=0}^{n}\theta_{n-k}^{(m)} y(t_k) + \lambda B_n^{(m)} y_0 + \frac{1}{\Delta t^\gamma}\sum_{k=0}^{n}\omega_{n-k} G_k + R_n, \tag{4.111}$$

where $\omega_n = (-1)^n\binom{\gamma}{n}$, $G_n = \left[\mathrm{D}_{0,t}^{\gamma}g(t)\right]_{t=t_n}$, $R^n = O(n^{-1}\Delta t)$, and $\theta_n^{(m)}, m = 1,2$ are defined by

$$\theta_n^{(1)} = \frac{1}{2^\gamma}(-1)^n\omega_n, \tag{4.112}$$

$$\theta_0^{(2)} = 1 - \frac{\gamma}{2}, \quad \theta_1^{(2)} = \frac{\gamma}{2}, \quad \theta_n^{(2)} = 0,\ n > 1, \tag{4.113}$$

and $B_n^{(m)}$, $m = 1,2$ is given by

$$B_n^{(m)} = \frac{1}{\Gamma(1+\gamma)}\sum_{k=0}^{n}\omega_{n-k}k^\gamma - \sum_{k=0}^{n}\theta_k^{(m)} = O(n^{-1}). \tag{4.114}$$

Letting $x = x_i, t = t_n$ in (4.86) and using (4.111), we have

$$\frac{1}{\Delta t^\gamma}\sum_{k=0}^{n}\omega_{n-k}(U_i^k - U_i^0) = \lambda\sum_{k=0}^{n}\theta_{n-k}^{(m)}\delta_x^2 U_i^k + \lambda B_n^{(m)}\delta_x^2 U_i^0 + \frac{1}{\Delta t^\gamma}\sum_{k=0}^{n}\omega_{n-k}G_i^k + R_i^n, \tag{4.115}$$

where $R_i^n = O(n^{-1}\Delta t + \Delta x^2)$.

From (4.115), we can obtain two types of fractional linear multistep finite difference methods for (4.86) as follows.

- FLMM-FDM I: Find u_i^n $(i = 1,2,\cdots,N-1, n = 1,2,\cdots,n_T)$, such that

$$\begin{cases} \dfrac{1}{\Delta t^\gamma}\left[\displaystyle\sum_{k=0}^{n}\omega_k(u_i^{n-k} - u_i^0)\right] = \dfrac{K_\gamma}{2^\gamma}\displaystyle\sum_{k=0}^{n}(-1)^k\omega_k\delta_x^2 u_i^{n-k} + K_\gamma B_n^{(1)}\delta_x^2 u_i^0 \\ \qquad\qquad\qquad\qquad + \dfrac{1}{\Delta t^\gamma}\displaystyle\sum_{k=0}^{n}\omega_{n-k}G_i^{n-k}, \\ u_i^0 = \phi_0(x_i), \quad i = 0,1,2,\cdots,N, \\ u_0^n = U_a(t_n),\ u_N^n = U_b(t_n), \end{cases} \tag{4.116}$$

  where $\omega_k = (-1)^k\binom{\gamma}{k}$, $G_i^n = \left[\mathrm{D}_{0,t}^{-\gamma}g(x_i,t)\right]_{t=t_n}$, and $B_n^{(1)}$ is defined by (4.114) with $m = 1$.

- FLMM-FDM II: Find u_i^n $(i = 1,2,\cdots,N-1, n = 1,2,\cdots,n_T)$, such that

$$\begin{cases} \dfrac{1}{\Delta t^\gamma}\left[\displaystyle\sum_{k=0}^{n}\omega_k(u_i^{n-k} - u_i^0)\right] = K_\gamma\left[(1-\dfrac{\gamma}{2})\delta_x^2 u_i^n + \dfrac{\gamma}{2}\delta_x^2 u_i^{n-1}\right] + K_\gamma B_n^{(2)}\delta_x^2 u_i^0 \\ \qquad\qquad\qquad\qquad + \dfrac{1}{\Delta t^\gamma}\left[(1-\dfrac{\gamma}{2})G_i^n + \dfrac{\gamma}{2}G_i^{n-1}\right], \\ u_i^0 = \phi_0(x_i), \quad i = 0,1,2,\cdots,N, \\ u_0^n = U_a(t_n),\ u_N^n = U_b(t_n), \end{cases} \tag{4.117}$$

  where $\omega_k = (-1)^k\binom{\gamma}{k}$, $G_i^n = \left[\mathrm{D}_{0,t}^{-\gamma}g(x_i,t)\right]_{t=t_n}$, and $B_n^{(2)}$ is defined by (4.114) with $m = 2$.

From (4.115), we know that the truncation errors of the two methods (4.116) and (4.117) are $O(n^{-1}\Delta t+\Delta x^2)$, which show first-order accuracy when n is small. In real computation, the two methods (4.116) and (4.117) show second-order accuracy when the exact solutions of (4.86) are smooth enough. In the following, we will show that the two methods are unconditionally stable and convergent of order $O(\Delta t+\Delta x^2)$.

For the classical case of diffusion equation (4.86) with $\gamma = 1$, there exists the Crank–Nicolson method for such an equation, which is unconditionally stable and convergent of order $O(\Delta t^2+\Delta x^2)$. Next, we construct the corresponding algorithms for (4.86) with unconditional stability and convergence of order $O(\Delta t^2+\Delta x^2)$.

We can use the time discretization (3.113) for (4.86), which reads

$$\begin{aligned}\frac{1}{\Delta t^{\gamma}}\sum_{k=0}^{n}\omega_{n-k}(U_i^k-U_i^0) =&K_\gamma\sum_{k=0}^{n}\theta_{n-k}^{(m)}\delta_x^2U_i^k+K_\gamma B_n^{(m)}\delta_x^2U_i^0+K_\gamma C_n^{(m)}\delta_x^2(U_i^1-U_i^0)\\&+\frac{1}{\Delta t^{\gamma}}\sum_{k=0}^{n}\omega_{n-k}G_i^k+R_i^n,\end{aligned} \tag{4.118}$$

where $R_i^n = O(\Delta t^2+\Delta x^2)$, ω_n and $\theta_n^{(m)}$ are defined in (4.115), and $C_n^{(m)} (m = 1,2)$ is defined by

$$C_n^{(m)} = \frac{\Gamma(2)}{\Gamma(\alpha+2)}\sum_{k=0}^{n}\omega_{n-k}k^{\gamma+1}-\sum_{k=1}^{n}\theta_{n-k}^{(m)}k = O(n^{-1}). \tag{4.119}$$

We can see from (4.115) and (4.118) that the term $K_\gamma C_n^{(m)}\delta_x^2(U_i^1-U_i^0)$ is added to the right-hand side of (4.115), which ensures that the method (4.118) has second-order accuracy for all time level n, which is inline with the classical Crank–Nicolson method, while in the time discretization (4.115), the method has first-order accuracy when time level n is small.

From (4.118), we can derive two improved FLMM finite difference methods for (4.86) as follows.

- Improved FLMM-FDM I: Find u_i^n $(i = 1,2,\cdots,N-1, n = 1,2,\cdots,n_T)$, such that

$$\begin{cases}\dfrac{1}{\Delta t^{\gamma}}\left[\displaystyle\sum_{k=0}^{n}\omega_k(u_i^{n-k}-u_i^0)\right]=\dfrac{K_\gamma}{2^\gamma}\displaystyle\sum_{k=0}^{n}(-1)^k\omega_k\delta_x^2u_i^{n-k}+K_\gamma B_n^{(1)}\delta_x^2u_i^0\\ \qquad\qquad\qquad +K_\gamma C_n^{(1)}\delta_x^2(u_i^1-u_i^0)+\dfrac{1}{\Delta t^{\gamma}}\displaystyle\sum_{k=0}^{n}\omega_{n-k}G_i^{n-k},\\ u_i^0=\phi_0(x_i),\quad i=0,1,2,\cdots,N,\\ u_0^n=U_a(t_n),\ u_N^n=U_b(t_n),\end{cases} \tag{4.120}$$

where $\omega_k = (-1)^k\binom{\gamma}{k}$, $G_i^n = \left[\mathrm{D}_{0,t}^{\gamma}g(x_i,t)\right]_{t=t_n}$, $B_n^{(1)}$ is defined by (4.114) with $m = 1$, and $C_n^{(1)}$ is defined by (4.119) with $m = 1$.

- Improved FLMM-FDM II: Find u_i^n $(i = 1,2,\cdots,N-1, n = 1,2,\cdots,n_T)$, such that

$$\begin{cases} \frac{1}{\Delta t^{\gamma}}\left[\sum_{k=0}^{n}\omega_k(u_i^{n-k}-u_i^0)\right] = K_\gamma\left[(1-\frac{\gamma}{2})\delta_x^2 u_i^n + \frac{\gamma}{2}\delta_x^2 u_i^{n-1}\right] + K_\gamma B_n^{(2)}\delta_x^2 u_i^0 \\ \qquad\qquad + K_\gamma C_n^{(2)}\delta_x^2(u_i^1-u_i^0) + \frac{1}{\Delta t^{\gamma}}\left[(1-\frac{\gamma}{2})G_i^n + \frac{\gamma}{2}G_i^{n-1}\right], \\ u_i^0 = \phi_0(x_i), \quad i = 0,1,2,\cdots,N, \\ u_0^n = U_a(t_n),\ u_N^n = U_b(t_n), \end{cases} \tag{4.121}$$

where $\omega_k = (-1)^k\binom{\gamma}{k}$, $G_i^n = \left[\mathrm{D}_{0,t}^{\gamma} g(x_i,t)\right]_{t=t_n}$, $B_n^{(2)}$ is defined by (4.114) with $m = 2$, and $C_n^{(2)}$ is defined by (4.119) with $m = 2$.

Remark 4.2.3 *If $\gamma = 1$, the four schemes* (4.116), (4.117), (4.120), *and* (4.121) *are reduced to the typical Crank–Nicolson finite difference method for the classical PDE* (4.86) *with $\gamma = 1$.*

Next, we will find that the four FLMM FDMs (4.116), (4.117), (4.120), and (4.121) are all unconditionally stable. The methods (4.116) and (4.117) are convergent of order $O(\Delta t + \Delta x^2)$, and the improved methods (4.120) and (4.121) are convergent of order $O(\Delta t^2 + \Delta x^2)$. We list the stability and convergence results below.

Theorem 29 *Suppose that u_i^n $(i = 1,2,\cdots,N-1, n = 1,2,\cdots,n_T)$ is the solution of* (4.116), (4.117), (4.120) *or* (4.121), $\mathbf{u}^n = (u_0^n, u_1^n, \cdots, u_N^n)^T$. *Then*

$$||\mathbf{u}^n||_N^2 + K_\gamma \Delta t^{\gamma}(1/2)|\mathbf{u}^n|_{1,N} \le C_1(||\mathbf{u}^0||_N^2 + \Delta t^{\gamma}|\mathbf{u}^0|_{1,N}) + C_2 \max_{0\le k\le n_T} ||\mathbf{g}^n||_N^2, \tag{4.122}$$

where C_1 is independent of $n, \Delta t, \Delta x$ and T, and C_2 is independent of $n, \Delta t, \Delta x$.

Theorem 30 *Suppose that u_i^n $(i = 1,2,\cdots,N-1, n = 1,2,\cdots,n_T)$ is the solution of* (4.116) *or* (4.117), *$U(x,t)$ is the exact solution of* (4.86), *$U \in C^2(0,T;H(I)), I = (a,b)$. Then there exists a positive constant C independent of n, Δx, and Δt, such that*

$$||\mathbf{u}^n - \mathbf{U}^n||_N \le C(\Delta t + \Delta x^2)$$

and

$$\sqrt{\Delta t \sum_{k=0}^{n} ||\mathbf{u}^k - \mathbf{U}^k||_N^2} \le C(\Delta t^{1.5} + \Delta x^2),$$

where $\mathbf{u}^n = (u_0^n, u_1^n, \cdots, u_N^n)^T$ and $\mathbf{U}^n = (U_0^n, U_1^n, \cdots, U_N^n)^T$.

Theorem 31 *Suppose that u_i^n $(i = 1,2,\cdots,N-1, n = 1,2,\cdots,n_T)$ is the solution of* (4.120) *or* (4.121), *$U(x,t)$ is the exact solution of* (4.86), *$U \in C^2(0,T;H(I)), I = (a,b)$. Then there exists a positive constant C independent of n, Δx, and Δt, such that*

$$||\mathbf{u}^n - \mathbf{U}^n||_N \le C(\Delta t^2 + \Delta x^2),$$

where $\mathbf{u}^n = (u_0^n, u_1^n, \cdots, u_N^n)^T$ and $\mathbf{U}^n = (U_0^n, U_1^n, \cdots, U_N^n)^T$.

TABLE 4.7: The maximum L^2 error of the method (4.97), $N = 5000$.

$1/\Delta t$	$\gamma = 0.2$	order	$\gamma = 0.5$	order	$\gamma = 0.8$	order
16	1.0205e−4		8.2444e−4		4.3359e−3	
32	3.1317e−5	1.7043	2.9980e−4	1.4594	1.9158e−3	1.1784
64	9.5398e−6	1.7149	1.0806e−4	1.4721	8.4078e−4	1.1882
128	2.9211e−6	1.7075	3.8756e−5	1.4794	3.6766e−4	1.1934
256	9.2947e−7	1.6520	1.3880e−5	1.4814	1.6048e−4	1.1960

TABLE 4.8: The maximum L^2 error of the scheme (4.116), $N = 5000$.

$1/\Delta t$	$\gamma = 0.2$	order	$\gamma = 0.5$	order	$\gamma = 0.8$	order
16	1.2264e−3		1.1742e−3		3.5197e−4	
32	6.9022e−4	0.8292	6.7594e−4	0.7967	1.4258e−4	1.3036
64	3.5544e−4	0.9575	3.1157e−4	1.1174	6.1553e−5	1.2119
128	1.7656e−4	1.0094	1.3176e−4	1.2416	2.0892e−5	1.5589
256	8.6224e−5	1.0340	5.3113e−5	1.3107	6.5110e−6	1.6820

Theorems 30 and 31 can be directly deduced from Theorem 29. For the detailed proofs, readers can refer to [168, 169]. The FLMM difference methods used in (4.116)–(4.117) and (4.120)–(4.120) have also been extended to the time-fractional diffusion wave equation in [163]. The numerical methods based the second-order fractional backward difference method to solve the time-fractional diffusion and diffusion/wave equations can be found in some literatures, see e.g. [36, 63, 67, 156].

4.2.2.4 Numerical Examples

This subsection provides the numerical examples to verify the finite difference schemes in Section 4.2.2.

Example 5 *Consider the following subdiffusion equation*

$$\begin{cases} {}_C\mathrm{D}_{0,t}^{\gamma} U = \partial_x^2 U + f(x,t), \quad (x,t) \in (0,1)\times(0,1], \quad 0<\gamma\le 1, \\ U(x,0) = 2\sin(\pi x), \quad x\in[0,1], \\ U(0,t) = U(1,t) = 0, \quad t\in[0,1]. \end{cases} \tag{4.123}$$

Choose the suitable right-hand side function $f(x,t)$ such that the above equation (4.123) *has the exact solution $U(x,t) = (t^{2+\gamma} + t + 2)\sin(\pi x)$.*

We use the methods (4.97) and (4.116)–(4.121) to solve this problem, the numerical results are shown in Tables 4.7–4.13. It is shown that about $(2-\gamma)$th-order experimental accuracy is observed in Table 4.7, which is in line with the theoretical analysis. In Tables 4.8 and 4.9, a little better numerical results than the theoretical analyses are shown. Tables 4.10 and 4.11 also show a little better numerical results. In Tables 4.12 and 4.13, we present the L^2 errors at $t = 1$, which show second-order accuracy. It is better than the theoretical result presented in Theorem 30.

TABLE 4.9: The maximum L^2 error of the scheme (4.117), $N = 5000$.

$1/\Delta t$	$\gamma = 0.2$	order	$\gamma = 0.5$	order	$\gamma = 0.8$	order
16	7.9954e−5		5.3590e−4		7.2152e−4	
32	7.2395e−5	0.1433	6.6261e−5	3.0157	1.1475e−4	2.6525
64	5.7008e−5	0.3447	1.1206e−5	2.5639	2.1759e−5	2.3988
128	3.2334e−5	0.8181	5.3558e−6	1.0651	4.9059e−6	2.1490
256	1.6649e−5	0.9576	2.3970e−6	1.1599	1.2455e−6	1.9778

TABLE 4.10: The maximum L^2 error of the scheme (4.120), $N = 5000$.

$1/\Delta t$	$\gamma = 0.2$	order	$\gamma = 0.5$	order	$\gamma = 0.8$	order
16	3.8041e−4		6.2221e−4		6.2077e−4	
32	8.1801e−5	2.2174	1.0391e−4	2.5820	9.0134e−5	2.7839
64	1.7541e−5	2.2214	2.2495e−5	2.2077	2.2715e−5	1.9884
128	3.7303e−6	2.2334	5.7013e−6	1.9803	5.7557e−6	1.9806
256	7.6683e−7	2.2823	1.4901e−6	1.9359	1.5035e−6	1.9366

TABLE 4.11: The maximum L^2 error of the scheme (4.121), $N = 5000$.

$1/\Delta t$	$\gamma = 0.2$	order	$\gamma = 0.5$	order	$\gamma = 0.8$	order
16	3.8041e−4		6.2221e−4		6.2077e−4	
32	8.1801e−5	2.2174	1.0391e−4	2.5820	7.9829e−5	2.9591
64	1.7541e−5	2.2214	1.7223e−5	2.5930	1.0414e−5	2.9384
128	3.7303e−6	2.2334	2.8273e−6	2.6068	1.8163e−6	2.5194
256	7.6682e−7	2.2823	4.5046e−7	2.6500	5.1733e−7	1.8119

TABLE 4.12: The L^2 error at $t = 1$ for the scheme (4.116), $N = 5000$.

$1/\Delta t$	$\gamma = 0.2$	order	$\gamma = 0.5$	order	$\gamma = 0.8$	order
16	1.8973e−5		2.6723e−4		3.5197e−4	
32	5.2333e−6	1.8582	7.9187e−5	1.7547	9.0025e−5	1.9671
64	1.7679e−6	1.5657	2.1478e−5	1.8824	2.2755e−5	1.9842
128	6.7348e−7	1.3923	5.6189e−6	1.9345	5.7706e−6	1.9794
256	2.8699e−7	1.2306	1.4885e−6	1.9165	1.5076e−6	1.9364

TABLE 4.13: The L^2 error at $t = 1$ for the scheme (4.117), $N = 5000$.

$1/\Delta t$	$\gamma = 0.2$	order	$\gamma = 0.5$	order	$\gamma = 0.8$	order
16	2.9439e−5		6.3211e−5		1.0855e−4	
32	8.1144e−6	1.8592	1.6493e−5	1.9383	2.7600e−5	1.9756
64	2.1149e−6	1.9399	4.2104e−6	1.9698	6.9926e−6	1.9807
128	5.8476e−7	1.8547	1.1142e−6	1.9180	1.8138e−6	1.9468
256	2.0685e−7	1.4992	3.4069e−7	1.7094	5.1661e−7	1.8118

4.3 One-Dimensional Space-Fractional Differential Equations

In this section, we consider finite difference methods for the one-dimensional space-fractional partial differential equations. These equations include the space fractional diffusion equation [136, 138, 145, 151], the fractional advection-dispersion equation [24, 100, 111], fractional advection-diffusion equation [137, 141], the space fractional Fokker–Planck equation [97], the fractional partial differential equations with Riesz space fractional derivatives [14, 40, 157], and so on.

4.3.1 One-Sided Space-Fractional Diffusion Equation

We consider the following *space-fractional diffusion equation* with Dirichlet boundary conditions [112]

$$\begin{cases} \partial_t U = d(x)_{RL}\mathrm{D}_{a,x}^{\alpha} U + g(x,t), \quad (x,t) \in (a,b)\times(0,T], \\ U(x,0) = \phi_0(x), \quad x \in (a,b), \\ U(a,t) = U_a(t),\ U(b,t) = U_b(t), \quad t \in (0,T], \end{cases} \tag{4.124}$$

where $1 < \alpha \leq 2$ and $d(x) > 0$.

Since the time derivative is the classical one, all the classical numerical methods for time discretization can be used. Therefore, we mainly focus on the space discretization for (4.124).

Since the Grünwald–Letnikov derivative of a given function is convergent to the Riemann–Liouville derivative when the function is smooth, a natural way to discretize the space-fractional Riemann–Liouville derivative is to use the definition of the Grünwald–Letnikov formula (2.51)

$$\left({}_{RL}\mathrm{D}_{a,x}^{\alpha} U\right)(x_i,t) = \frac{1}{\Delta x^{\alpha}} \sum_{j=0}^{i} \omega_j^{(\alpha)} U(t, x_{i-j}) + O(\Delta x). \tag{4.125}$$

The first-order time derivative in (4.124) can be discretized by the classical methods such as the explicit Euler method, the implicit Euler method and the Crank–Nicolson method, etc. Unfortunately, the explicit Euler method, the implicit Euler method, and the Crank–Nicolson method based on the standard Grünwald–Letnikov formula for (4.124) are often unstable [111].

Proposition 4.3.1 ([111]) *The explicit Euler method solution to Eq.* (4.124)*, based on the Grünwald–Letnikov approximation* (4.125) *to the fractional derivative, is unstable.*

Proof. Let u_i^n be the approximate solutions to (4.124). Then the explicit Euler method on the Grünwald–Letnikov approximation (4.125) for (4.124) is given by

$$\frac{u_i^{n+1} - u_i^n}{\Delta t} = \frac{d_i}{\Delta x^{\alpha}} \sum_{j=0}^{i} \omega_j^{(\alpha)} u_{i-j}^n + g_i^n, \quad i = 1,2,\cdots,N-1, \tag{4.126}$$

where $\omega_j^{(\alpha)} = (-1)^j \binom{\alpha}{j}$. u_i^{n+1} in (4.126) can be explicitly expressed as

$$u_i^{n+1} = u_i^n + \frac{\Delta t}{\Delta x^\alpha} d_i \sum_{j=0}^{i} \omega_j^{(\alpha)} u_{i-j}^n + \Delta t g_i^n. \tag{4.127}$$

Assume that u_i^0 is the only term that has an error, so the perturbed value is $\underline{u}_i^0 = u_i^0 + \varepsilon_i^0$. This perturbation produces a perturbed value for u_i^1 given by $\underline{u}_i^1 = u_i^1 + \varepsilon_i^1$. So (4.127) yields

$$\underline{u}_i^1 = \mu_i \underline{u}_i^0 + \frac{\Delta t}{\Delta x^\alpha} d_i \sum_{j=1}^{i} \omega_j^{(\alpha)} u_{i-j}^0 + \Delta t g_i^0 = \mu_i \varepsilon_i^0 + u_i^1, \tag{4.128}$$

where the factor

$$\mu_i = 1 + \frac{\Delta t}{\Delta x^\alpha} d_i.$$

Therefore we have $\varepsilon_i^1 = \mu_i \varepsilon_i^0$. That is, the error is amplified by the factor μ_i when the finite difference equation is advanced by one timestep. After n timesteps, one may write $\varepsilon_i^n = \mu_i^n \varepsilon_i^0$. We refer to μ_i as the amplification factor (or magnification factor). In order for the explicit Euler method to be stable, it is necessary that $\mu_i \le 1$ for all Δx sufficiently small. Obviously, $|\mu_i| > 1$. Hence, although it is true that the errors may not grow for larger values of Δx, the method is not stable as Δx is refined, and therefore the numerical solution does not converge to the exact solution of the differential equation. □

Proposition 4.3.2 ([111]) *The implicit Euler method solution to Eq.* (4.124)*, based on the Grünwald–Letnikov approximation* (4.125) *to the fractional derivative, is unstable.*

Proof. Let u_i^n be the approximate solutions to (4.124). Then the implicit Euler method for (4.124) is given by

$$\frac{u_i^{n+1} - u_i^n}{\Delta t} = \frac{d_i}{\Delta x^\alpha} \sum_{j=0}^{i} \omega_j^{(\alpha)} u_{i-j}^{n+1} + g_i^{n+1}, \quad i = 1, 2, \cdots, N-1, \tag{4.129}$$

where $\omega_j^{(\alpha)} = (-1)^j \binom{\alpha}{j}$. Similar to Proposition 4.3.1, we can get $\varepsilon_i^n = \mu_i^n \varepsilon_i^0$, where the amplification factor $\mu_i = \frac{1}{1 - \frac{\Delta t}{\Delta x^\alpha} d_i} > 1$ for all Δx. So the implicit Euler method is unstable in this case, and hence its numerical solution does not converge to the exact solution of the differential equation. □

One can similarly prove that the Crank–Nicolson method solution to Eq. (4.124), based on the Grünwald–Letnikov approximation (4.125) to the fractional derivative, is unstable [111]. To remedy this situation, the shifted Grünwald formula (2.52) can be used to overcome this drawback.

The shifted Grünwald formula with p shifts is defined as

$$\left({}_{RL}\mathrm{D}_{a,x}^{\alpha}U\right)(x_i,t) = \frac{1}{\Delta x^{\alpha}}\sum_{j=0}^{i+p}\omega_j^{(\alpha)}U(t,x_{i-j}+p\Delta x)+O(\Delta x). \tag{4.130}$$

In [111], the error bound for the shifted Grünwald formula (4.130) was proved with the following form

$$C(p-\alpha/2)\Delta x + O(\Delta x^2),$$

where C is independent of Δx. Therefore, the best performance comes from minimizing $|p-\alpha/2|$. For $1<\alpha\leq 2$, the optimal choice is $p=1$. If $\alpha=2$, then this coincides with the centered second difference estimator of the second derivative.

From Eq. (4.8), we know that the right-shifted Grünwald formula with p shifts for the αth-order left Riemann–Liouville derivative of $U(x), x\in[a,b]$ at $x=x_i$ can be expressed by

$${}_L\delta_x^{(\alpha)}U_{i+p} = \frac{1}{\Delta x^{\alpha}}\sum_{j=0}^{i+p}\omega_j^{(\alpha)}U_{i+p-j}, \quad U_j = U(x_j).$$

Next, we introduce the Euler method and the Crank–Nicolson method based on the shifted Grünwald formula for (4.124).

(1) The **explicit Euler method** solution to Eq. (4.124), based on the shifted (1 shift) Grünwald–Letnikov approximation (4.130) to the fractional derivative, is given by: Find u_i^n $(i=1,2,\cdots,N-1, n=0,1,2,\cdots,n_T-1)$, such that

$$\begin{cases} \delta_t u_i^{n+\frac{1}{2}} = d_{iL}\delta_x^{(\alpha)}u_{i+1}^n + g_i^n, \quad i=1,2,\cdots,N-1,\\ u_i^0 = \phi_0(x_i), \quad i=0,1,2,\cdots,N,\\ u_0^n = U_a(t_n),\ u_N^n = U_b(t_n), \end{cases} \tag{4.131}$$

where ${}_L\delta_x^{(\alpha)}u_{i+1}^n = \frac{1}{\Delta x^{\alpha}}\sum_{j=0}^{i+1}\omega_j^{(\alpha)}u_{i+1-j}^n$, $\omega_j^{(\alpha)} = (-1)^j\binom{\alpha}{j}$.

(2) The **implicit Euler method** solution to Eq. (4.124), based on the shifted (1 shift) Grünwald–Letnikov approximation (4.130) to the fractional derivative, is given by: Find u_i^n $(i=1,2,\cdots,N-1, n=0,1,2,\cdots,n_T-1)$, such that

$$\begin{cases} \delta_t u_i^{n+\frac{1}{2}} = d_{iL}\delta_x^{(\alpha)}u_{i+1}^{n+1} + g_i^{n+1}, \quad i=1,2,\cdots,N-1,\\ u_i^0 = \phi_0(x_i), \quad i=0,1,2,\cdots,N,\\ u_0^n = U_a(t_n),\ u_N^n = U_b(t_n), \end{cases} \tag{4.132}$$

where ${}_L\delta_x^{(\alpha)}u_{i+1}^n = \frac{1}{\Delta x^{\alpha}}\sum_{j=0}^{i+1}\omega_j^{(\alpha)}u_{i+1-j}^n$, $\omega_j^{(\alpha)} = (-1)^j\binom{\alpha}{j}$.

(3) The **Crank–Nicolson method** solution to Eq. (4.124), based on the shifted (1

shift) Grünwald–Letnikov approximation (4.130) to the fractional derivative, is given by: Find u_i^n $(i = 1,2,\cdots,N-1, n = 0,1,2,\cdots,n_T-1)$, such that

$$\begin{cases} \delta_t u_i^{n+\frac{1}{2}} = d_{iL}\delta_x^{(\alpha)} u_{i+1}^{n+\frac{1}{2}} + g(x_i, t_{n+\frac{1}{2}}), \quad i = 1,2,\cdots,N-1, \\ u_i^0 = \phi_0(x_i), \quad i = 0,1,2,\cdots,N, \\ u_0^n = U_a(t_n),\ u_N^n = U_b(t_n), \end{cases} \tag{4.133}$$

where ${}_L\delta_x^{(\alpha)} u_{i+1}^n = \frac{1}{\Delta x^\alpha}\sum_{j=0}^{i+1}\omega_j^{(\alpha)} u_{i+1-j}^n$, $\omega_j^{(\alpha)} = (-1)^j\binom{\alpha}{j}$.

In order to simply give the matrix representations of (4.131)–(4.133), we introduce some notations. We adopt the symbol '.∗' used in MATLAB to express

$$(A. * B)_{i,j} = a_{i,j} b_{i,j}, \tag{4.134}$$

where $(A)_{i,j} = a_{i,j}$ and $(B)_{i,j} = b_{i,j}$ are the matrices with the same sizes.

Denote by the matrix $S_{N-1}^{(\alpha)}$ as

$$S_{N-1}^{(\alpha)} = \begin{pmatrix} \omega_1^{(\alpha)} & \omega_0^{(\alpha)} & 0 & \cdots & 0 \\ \omega_2^{(\alpha)} & \omega_1^{(\alpha)} & \omega_0^{(\alpha)} & \cdots & 0 \\ \vdots & \vdots & \vdots & \ddots & \vdots \\ \omega_{N-2}^{(\alpha)} & \omega_{N-3}^{(\alpha)} & \omega_{N-4}^{(\alpha)} & \cdots & \omega_0^{(\alpha)} \\ \omega_{N-1}^{(\alpha)} & \omega_{N-2}^{(\alpha)} & \omega_{N-3}^{(\alpha)} & \cdots & \omega_1^{(\alpha)} \end{pmatrix}_{(N-1)\times(N-1)}. \tag{4.135}$$

Define the two vectors $B_L^{(\alpha)}$ and $B_R^{(\alpha)}$ as

$$B_L^{(\alpha)} = \begin{pmatrix} \omega_2^{(\alpha)} \\ \omega_3^{(\alpha)} \\ \vdots \\ \omega_{N-1}^{(\alpha)} \\ \omega_N^{(\alpha)} \end{pmatrix}_{(N-1)\times 1}, \quad B_R^{(\alpha)} = \begin{pmatrix} 0 \\ 0 \\ \vdots \\ 0 \\ \omega_0^{(\alpha)} \end{pmatrix}_{(N-1)\times 1}. \tag{4.136}$$

Next, we give the matrix representation of the three methods (4.131)–(4.133).

(1) Matrix representation of the explicit Euler method (4.131):

$$\underline{\mathbf{u}}^{n+1} = (E + \mu S)\underline{\mathbf{u}}^n + \Delta t \underline{\mathbf{g}}^n + \mu(B_L u_0^n + B_R u_N^n), \tag{4.137}$$

where $\mu = \frac{\Delta t}{\Delta x^\alpha}$, $\underline{\mathbf{u}}^n = (u_1^n,\cdots,u_{N-1}^n)^T$, $\underline{\mathbf{g}}^n = (g_1^n,\cdots,g_{N-1}^n)^T$, E is an $(N-1)\times(N-1)$ identity matrix,

$$S = D_{N-1} S_{N-1}^{(\alpha)}, \quad D_{N-1} = diag(d_1, d_2, \cdots, d_{N-1})$$

and

$$B_L = B_0^{(\alpha)}. * d, \quad B_R = B_N^{(\alpha)}. * d, \quad d = (d_1,\cdots,d_{N-1})^T. \tag{4.138}$$

(2) Matrix representation of the implicit Euler method (4.132):

$$(E-\mu S)\underline{\mathbf{u}}^{n+1}=\underline{\mathbf{u}}^{n}+\Delta t\underline{\mathbf{g}}^{n+1}+\mu(B_L u_0^{n+1}+B_R u_N^{n+1}), \tag{4.139}$$

where μ, E, S, B_L, B_R are defined as in (4.137).

(3) Matrix representation of the Crank–Nicolson method (4.133):

$$(E-\frac{\mu}{2}S)\underline{\mathbf{u}}^{n+1}=(E+\frac{\mu}{2}S)\underline{\mathbf{u}}^{n}+\frac{\Delta t}{2}(\underline{\mathbf{g}}^{n}+\underline{\mathbf{g}}^{n+1})+\mu(B_L u_0^{n+\frac{1}{2}}+B_R u_N^{n+\frac{1}{2}}), \tag{4.140}$$

where μ, E, S, B_L, B_R are defined as in (4.137).

Next, we investigate the stability of the explicit Euler method (4.131), the implicit Euler method (4.132), and the Crank–Nicolson method (4.133). From (4.135) and (4.138), and the *Gerschgorin theorem*, we can obtain that the eigenvalues λ of the matrix S satisfy

$$|\lambda-d_i\omega_1^{(\alpha)}|\le d_i\omega_0^{(\alpha)}+d_i\sum_{j=2}^{i}\omega_j^{(\alpha)}\le -d_i\omega_1^{(\alpha)},$$

which implies

$$-2\alpha d_{\max}=2d_{\max}\omega_1^{(\alpha)}\le 2d_i\omega_1^{(\alpha)}\le\lambda<0,$$

where $d_{\max}=\max\limits_{0\le i\le N} d_i$.

Therefore, the eigenvalue of the matrix $(E+\mu S)$ lies in $[1-2\mu\alpha d_{\max},1]$, and the eigenvalue of the matrix $(E-\mu S)$ lies in $[1,1+2\mu\alpha d_{\max}]$.

From (4.137), we can obtain that the explicit Euler method (4.131) is stable if

$$1-2\mu\alpha d_{\max}\ge -1 \iff \frac{\Delta t}{\Delta x^{\alpha}}=\mu\le\frac{1}{\alpha d_{\max}}.$$

From (4.139), we obtain

$$\underline{\mathbf{u}}^{n+1}=(E-\mu S)^{-1}\underline{\mathbf{u}}^{n}+(E-\mu S)^{-1}(\Delta t\underline{\mathbf{g}}^{n+1}+\mu(B_L u_0^{n+1}+B_R u_N^{n+1})). \tag{4.141}$$

Since the eigenvalues of the matrix $(E-\mu S)$ are all equal to or greater than 1, the eigenvalues of $(E-\mu S)^{-1}$ are not greater than 1. Hence, the implicit Euler method (4.132) is unconditionally stable.

For the Crank–Nicolson method (4.133), we have from (4.140)

$$\underline{\mathbf{u}}^{n+1}=(E-\frac{\mu}{2}S)^{-1}(E+\frac{\mu}{2}S)\underline{\mathbf{u}}^{n}+\underline{\mathbf{b}}^{n}, \tag{4.142}$$

where $\underline{\mathbf{b}}^{n}=\Delta t(E-\frac{\mu}{2}S)^{-1}\left(\frac{1}{2}(\underline{\mathbf{g}}^{n}+\underline{\mathbf{g}}^{n+1})+\mu(B_L u_0^{n+\frac{1}{2}}+B_R u_N^{n+\frac{1}{2}})\right)$. It is known that if λ is the eigenvalue of S, then $\frac{2-\mu\lambda}{2+\mu\lambda}$ is the eigenvalue of $(E-\frac{\mu}{2}S)^{-1}(E+\frac{\mu}{2}S)$. Since λ has negative real part, so we have $\frac{|2-\mu\lambda|}{|2+\mu\lambda|}<1$. Hence the Crank–Nicolson method

(4.133) is unconditionally stable. For more detailed information, one can refer to [111, 112, 145].

From (4.130), one knows that the truncation errors of the explicit Euler method, the implicit Euler method, and the Crank–Nicolson method are $O(\Delta t+\Delta x)$, $O(\Delta t+\Delta x)$, and $O(\Delta t^2+\Delta x)$, respectively, the proof of which is almost the same as that of the classical methods, i.e., $\alpha=2$ in (4.131)–(4.132).

The explicit Euler method (4.131), the implicit Euler method (4.132) and the Crank–Nicolson method (4.133) can be seen as the special cases of the following weighted difference methods [41]:

$$\begin{cases} \delta_t u_i^{n+\frac{1}{2}} = d_i\left[(1-\theta)_L\delta_x^{(\alpha)} u_{i+1}^{n+1} + \theta_L\delta_x^{(\alpha)} u_{i+1}^{n}\right] + (1-\theta)g_i^{n+1} + \theta g_i^n, \quad i=1,2,\cdots,N-1, \\ u_i^0 = \phi_0(x_i), \quad i=0,1,2,\cdots,N, \\ u_0^n = U_a(t_n),\ u_N^n = U_b(t_n), \end{cases} \tag{4.143}$$

where $0\le\theta\le 1$, and where ${}_L\delta_x^{(\alpha)} u_{i+1}^n = \frac{1}{\Delta x^\alpha}\sum_{j=0}^{i+1}\omega_j^{(\alpha)} u_{i+1-j}^n$, $\omega_j^{(\alpha)} = (-1)^j\binom{\alpha}{j}$.

The weighted difference methods (4.143) can be derived in the following way. Letting $(x,t)=(x_i,t_{n+\frac{1}{2}})$ in (4.124) yields

$$\partial_t U(x_i,t_{n+\frac{1}{2}}) = d(x)({}_{RL}\mathrm{D}_{a,x}^{\alpha}U)(x_i,t_{n+\frac{1}{2}}) + g(x_i,t_{n+\frac{1}{2}}). \tag{4.144}$$

The right-hand side of (4.144) is evaluated by the weighted average values of $U(x,t)$ at the time levels n and $(n+1)$, which leads to

$$\begin{aligned} \partial_t U(x_i,t_{n+\frac{1}{2}}) =& d(x_i)\left[(1-\theta)({}_{RL}\mathrm{D}_{a,x}^{\alpha}U)(x_i,t_{n+1}) + \theta({}_{RL}\mathrm{D}_{a,x}^{\alpha}U)(x_i,t_n)\right] \\ &+ (1-\theta)g_i^{n+1} + \theta g_i^n + \tilde{R}_i^{n+\frac{1}{2}}. \end{aligned} \tag{4.145}$$

The time derivative is discretized by the central difference, and the space fractional derivative is discretized by the right-shifted (1 shift) Grünwald–Letnikov approximation (4.130), so one has

$$\begin{aligned} \delta_t U_i^{n+\frac{1}{2}} =& d_i\left[(1-\theta)_L\delta_x^{(\alpha)} U_{i+1}^{n+1} + \theta_L\delta_x^{(\alpha)} U_{i+1}^{n}\right] \\ &+ (1-\theta)g(x_i,t_{n+1}) + \theta g(x_i,t_n) + R_i^{n+\frac{1}{2}}. \end{aligned} \tag{4.146}$$

Dropping the truncation error $R_i^{n+\frac{1}{2}}$ and replacing U_i^n by u_i^n in (4.146) yields the weighted finite difference method (4.143).

Obviously, the weighted finite difference method (4.143) is reduced to the explicit Euler method (4.131) if $\theta=1$, the implicit Euler method (4.132) if $\theta=0$, and the Crank–Nicolson method (4.133) if $\theta=1/2$. One can similarly prove that the weighted finite difference methods (4.143) are unconditionally stable when $0\le\theta\le 1/2$, conditionally stable when $1/2<\theta\le 1$ and $\frac{\Delta t}{\Delta x^\alpha}\le\frac{1}{\alpha d_{\max}(2\theta-1)}$.

In Chapter 2, we introduced the $L2$ and $L2C$ methods for the discretization of the αth-order Riemann–Liouville derivative, see (2.73) and (2.78), which can yield a series of finite difference methods as (4.131)–(4.133) and (4.143). The explicit and implicit methods similar to (4.131) and (4.132) based on the L1 and L2C methods for the space discretization can be found in [105]. Readers can also refer to [136, 157] for the related information.

It is known that the methods (4.131)–(4.133) and (4.143) have second-order accuracy if $\alpha = 2$. Next, we introduce the second-order approximation to the left and right Riemann–Liouville derivative operators.

Lemma 4.3.1 ([111]) *Suppose that $f(x) \in L_1(\mathbb{R})$, ${}_{RL}D^{\alpha+2}_{-\infty,x}f(x)$ and its Fourier transform belong to $L_1(\mathbb{R})$, and let*

$$ {}_{L}\delta^{(\alpha)}_{\Delta x,p}f(x) = \frac{1}{\Delta x^{\alpha}}\sum_{k=0}^{\infty}\omega_k^{(\alpha)}f(x-(k-p)\Delta x), \quad \omega_k^{(\alpha)} = \frac{1}{\Gamma(-\alpha)}\frac{\Gamma(k-\alpha)}{\Gamma(k+1)}, $$

where p is a nonnegative integer. Then

$$ {}_{L}\delta^{(\alpha)}_{\Delta x,p}f(x) = {}_{RL}\mathrm{D}^{\alpha}_{-\infty,x}f(x) + C\left(p - \frac{\alpha}{2}\right)\Delta x + O(\Delta x^2), \tag{4.147} $$

where C is a constant independent of p.

From Lemma 4.3.1, we can get that

$$ \frac{\alpha}{2}{}_{L}\delta^{(\alpha)}_{\Delta x,1}f(x) + (1-\frac{\alpha}{2}){}_{L}\delta^{(\alpha)}_{\Delta x,0}f(x) = {}_{RL}\mathrm{D}^{\alpha}_{-\infty,x}f(x) + O(\Delta x^2). \tag{4.148} $$

Hence, $\frac{\alpha}{2}{}_{L}\delta^{(\alpha)}_{\Delta x,1}f(x) + (1-\frac{\alpha}{2}){}_{L}\delta^{(\alpha)}_{\Delta x,0}f(x)$ has second-order accuracy for approximating ${}_{RL}\mathrm{D}^{\alpha}_{-\infty,x}f(x)$; see also (2.58) and (2.59).

A more general second-order discretization of the left Riemann–Liouville operator was developed in [146], which can be given as

$$ \frac{\alpha-2q}{2(p-q)}{}_{L}\delta^{(\alpha)}_{\Delta x,p}f(x) + \frac{2p-\alpha}{2(p-q)}{}_{L}\delta^{(\alpha)}_{\Delta x,q}f(x) = {}_{RL}\mathrm{D}^{\alpha}_{-\infty,x}f(x) + O(\Delta x^2), \tag{4.149} $$

where p and q are integers. Eq. (4.149) can be also derived from Lemma 4.3.1 by eliminating Δx from (4.147) through setting $p = p, q$. Obviously, Eq. (4.149) is reduced to (4.148) when $p = 1$ and $q = 0$.

Let

$$ {}_{R}\delta^{(\alpha)}_{\Delta x,p}f(x) = \frac{1}{\Delta x^{\alpha}}\sum_{k=0}^{\infty}\omega_k^{(\alpha)}f(x+(k-p)\Delta x), \quad \omega_k^{(\alpha)} = \frac{1}{\Gamma(-\alpha)}\frac{\Gamma(k-\alpha)}{\Gamma(k+1)}. $$

Then we can similarly get the following second-order discretization for the right Riemann–Liouville operator [146]

$$ \frac{\alpha-2q}{2(p-q)}{}_{R}\delta^{(\alpha)}_{\Delta x,p}f(x) + \frac{2p-\alpha}{2(p-q)}{}_{R}\delta^{(\alpha)}_{\Delta x,q}f(x) = {}_{RL}\mathrm{D}^{\alpha}_{x,\infty}f(x) + O(\Delta x^2), \tag{4.150} $$

where p and q are integers.

Let $f(x)$ be well defined on the interval $[a,b]$. If $f(a)=0$, then the left Riemann–Liouville operator ${}_{RL}\mathrm{D}_{a,x}^{\alpha}$ at $x=x_i$ can be discretized by the following formula

$$\left[{}_{RL}\mathrm{D}_{a,x}^{\alpha}f(x)\right]_{x=x_i}=\frac{\alpha-2q}{2(p-q)}{}_{L}\delta_x^{(\alpha)}f_{i+p}+\frac{2p-\alpha}{2(p-q)}{}_{L}\delta_x^{(\alpha)}f_{i+q}+O(\Delta x^2), \qquad (4.151)$$

where $f_j=f(x_j)$ and the operator ${}_{L}\delta_x^{(\alpha)}$ is defined by (4.8).

If $f(b)=0$, then the right Riemann–Liouville operator ${}_{RL}\mathrm{D}_{x,b}^{\alpha}$ at $x=x_i$ can be similarly discretized as

$$\left[{}_{RL}\mathrm{D}_{x,b}^{\alpha}f(x)\right]_{x=x_i}=\frac{\alpha-2q}{2(p-q)}{}_{R}\delta_x^{(\alpha)}f_{i-p}+\frac{2p-\alpha}{2(p-q)}{}_{R}\delta_x^{(\alpha)}f_{i-q}+O(\Delta x^2), \qquad (4.152)$$

where $f_j=f(x_j)$ and the operator ${}_{R}\delta_x^{(\alpha)}$ is defined by (4.9).

We are interested in the two cases of (p,q), in which (4.151) and (4.152) are reduced to the central difference when $\alpha=2$.

- Case I: $(p,q)=(1,0)$, the left and right Riemann–Liouville derivatives ${}_{RL}\mathrm{D}_{a,x}^{\alpha}f(x)$ and ${}_{RL}\mathrm{D}_{x,b}^{\alpha}f(x)$ at $x=x_i$ can be discretized by the following weighted shifted Grünwald formulas

$${}_{L}\delta_x^{(\alpha,1)}f_i=\frac{\alpha}{2}{}_{L}\delta_x^{(\alpha)}f_{i+1}+\frac{2-\alpha}{2}{}_{L}\delta_x^{(\alpha)}f_i=\sum_{j=0}^{i+1}g_j^{(\alpha,1)}f_{i+1-j}, \qquad (4.153)$$

and

$${}_{R}\delta_x^{(\alpha,1)}f_i=\frac{\alpha}{2}{}_{R}\delta_x^{(\alpha)}f_{i-1}+\frac{2-\alpha}{2}{}_{R}\delta_x^{(\alpha)}f_i=\sum_{j=0}^{N-i+1}g_j^{(\alpha,1)}f_{i-1+j}, \qquad (4.154)$$

respectively, where

$$g_0^{(\alpha,1)}=\frac{\alpha}{2}\omega_0^{(\alpha)},\quad g_k^{(\alpha,1)}=\frac{\alpha}{2}\omega_k^{(\alpha)}+\frac{2-\alpha}{2}\omega_{k-1}^{(\alpha)},\quad k\geq 1. \qquad (4.155)$$

- Case II: $(p,q)=(1,-1)$, the left and right Riemann–Liouville derivatives ${}_{RL}\mathrm{D}_{a,x}^{\alpha}f(x)$ and ${}_{RL}\mathrm{D}_{x,b}^{\alpha}f(x)$ at $x=x_i$ can be discretized by the following weighted shifted Grünwald formula

$${}_{L}\delta_x^{(\alpha,2)}f_i=\frac{2+\alpha}{4}{}_{L}\delta_x^{(\alpha)}f_{i+1}+\frac{2-\alpha}{4}{}_{L}\delta_x^{(\alpha)}f_{i-1}=\sum_{j=0}^{i+1}g_j^{(\alpha,2)}f_{i+1-j}, \qquad (4.156)$$

and

$${}_{R}\delta_x^{(\alpha,2)}f_i=\frac{2+\alpha}{4}{}_{R}\delta_x^{(\alpha)}f_{i-1}+\frac{2-\alpha}{4}{}_{R}\delta_x^{(\alpha)}f_{i+1}=\sum_{j=0}^{N-i+1}g_j^{(\alpha,2)}f_{i-1+j}, \qquad (4.157)$$

respectively, where

$$g_0^{(\alpha,2)}=\frac{2+\alpha}{4}\omega_0^{(\alpha)},\quad g_1^{(\alpha,2)}=\frac{2+\alpha}{4}\omega_1^{(\alpha)},\quad g_k^{(\alpha,2)}=\frac{2+\alpha}{4}\omega_k^{(\alpha)}+\frac{2-\alpha}{4}\omega_{k-2}^{(\alpha)},\quad k\geq 1. \qquad (4.158)$$

Lemma 4.3.2 ([146]) *The coefficients defined by* (4.155) *and* (4.158) *satisfy the following properties for* $1<\alpha\leq 2$:

(1) Case I: $(p,q)=(1,0)$

$$\begin{cases} g_0^{(\alpha,1)}=\frac{\alpha}{2}, \quad g_1^{(\alpha,1)}=\frac{2-\alpha-\alpha^2}{2}<0, \quad g_2^{(\alpha,1)}=\frac{\alpha(\alpha^2+\alpha-4)}{4}, \\ 1\geq g_0^{(\alpha,1)}\geq g_3^{(\alpha,1)}\geq g_4^{(\alpha,1)}\geq\cdots\geq 0, \\ \sum_{k=0}^{\infty}g_k^{(\alpha,1)}=0, \quad \sum_{k=m}^{\infty}g_k^{(\alpha,1)}<0, \quad m\geq 2. \end{cases} \tag{4.159}$$

(2) Case II: $(p,q)=(1,-1)$

$$\begin{cases} g_0^{(\alpha,2)}=\frac{2+\alpha}{4}, \quad g_1^{(\alpha,2)}=-\frac{2\alpha+\alpha^2}{4}<0, \\ g_2^{(\alpha,2)}=\frac{\alpha^3+\alpha^2-4\alpha+4}{8}>0, \quad g_3^{(\alpha,2)}=\frac{\alpha(\alpha-2)(\alpha^2+\alpha-8)}{6}, \\ 1\geq g_0^{(\alpha,2)}\geq g_2^{(\alpha,2)}\geq g_4^{(\alpha,2)}\geq g_5^{(\alpha,2)}\geq\cdots\geq 0, \\ \sum_{k=0}^{\infty}g_k^{(\alpha,2)}=0, \quad \sum_{k=m}^{\infty}g_k^{(\alpha,2)}<0, \quad m=1 \quad or \quad m\geq 3. \end{cases} \tag{4.160}$$

Lemma 4.3.3 ([146]) *Let* $g_k^{(\alpha,1)}$ *and* $g_k^{(\alpha,2)}$ *be defined by* (4.155) *and* (4.158)*, respectively,* $1<\alpha\leq 2$*, and*

$$S_{N-1}^{(m,\alpha)}=\begin{pmatrix} g_1^{(\alpha,m)} & g_0^{(\alpha,m)} & 0 & \cdots & 0 \\ g_2^{(\alpha,m)} & g_1^{(\alpha,m)} & g_0^{(\alpha,m)} & \cdots & 0 \\ \vdots & \vdots & \vdots & \ddots & \vdots \\ g_{N-2}^{(\alpha,m)} & g_{N-3}^{(\alpha,m)} & g_{N-4}^{(\alpha,m)} & \cdots & g_0^{(\alpha,m)} \\ g_{N-1}^{(\alpha,m)} & g_{N-2}^{(\alpha)} & g_{N-3}^{(\alpha,m)} & \cdots & g_1^{(\alpha,m)} \end{pmatrix}_{(N-1)\times(N-1)}. \tag{4.161}$$

Then the real part of the eigenvalue λ *of* $S_{N-1}^{(m,\alpha)}$ *is negative, and the eigenvalues of* $S_{N-1}^{(m,\alpha)}+(S_{N-1}^{(m,\alpha)})^T$ *are negative.*

From (4.143), (4.153), and (4.156), we can obtain the following finite difference methods for (4.124)

$$\begin{cases} \delta_t u_i^{n+\frac{1}{2}}=d_i\left[(1-\theta)_L\delta_x^{(\alpha,m)}u_i^{n+1}+\theta_L\delta_x^{(\alpha,m)}u_i^n\right]+(1-\theta)g_i^{n+1}+\theta g_i^n, \ i=1,2,\cdots,N-1, \\ u_i^0=\phi_0(x_i), \quad i=0,1,2,\cdots,N, \\ u_0^n=U_a(t_n), \ u_N^n=U_b(t_n), \end{cases} \tag{4.162}$$

where $0\leq\theta\leq 1$, and ${}_L\delta_x^{(\alpha,m)}$ is defined by (4.153) for $m=1$ or by (4.156) for $m=2$.

If $U_a(t) = U_b(t) = 0$, then the method (4.162) has second-order accuracy in space, the matrix representation of which is given by

$$(E - \mu(1-\theta)S^{(m)})\underline{\mathbf{u}}^{n+1} = (E + \theta\mu S^{(m)})\underline{\mathbf{u}}^n + \Delta t\left((1-\theta)\underline{\mathbf{g}}^n + \theta\underline{\mathbf{g}}^{n+1}\right), \tag{4.163}$$

where $\mu = \Delta t/\Delta x^\alpha$, $\underline{\mathbf{u}}^n = (u_1^n, \cdots, u_{N-1}^n)^T$, $\underline{\mathbf{g}}^n = (g_1^n, \cdots, g_{N-1}^n)^T$, E is an $(N-1)\times(N-1)$ identity matrix, $S^{(m)}$ is given by

$$S^{(m)} = diag(d_1, d_2, \cdots, d_{N-1})S_{N-1}^{(m,\alpha)}.$$

in which $S_{N-1}^{(m,\alpha)}$ is defined by (4.161).

From Lemma 4.3.2, one knows that $g_0^{(\alpha,1)} + \sum_{k=2}^{n} g_k^{(\alpha,1)} < -g_1^{(\alpha,1)}$ for $\frac{\sqrt{17}-1}{2} \leq \alpha \leq 2$. In such a case, the matrix $S^{(1)}$ has eigenvalues with negative parts. So we can easily prove that the method (4.124) with $m = 1$ is unconditionally stable for $0 \leq \theta \leq 1/2$, and conditionally stable for $1/2 < \theta \leq 1$.

Assume that $d_i = d$ is a constant. Using Lemma 4.3.3, we can easily prove that weighted finite difference method (4.162) is unconditionally stable for $0 \leq \theta \leq 1/2$, and conditionally stable for $1/2 < \theta \leq 1$ by the energy method. For $\theta = 1/2$, the method (4.162) has second order accuracy both in time and space [146].

4.3.2 Two-Sided Space-Fractional Diffusion Equation

In this subsection, we consider the finite difference methods for two-sided space-fractional partial differential equations. A class of *two-sided space-fractional partial differential equations* can be written as

$$\begin{cases} \partial_t U = c(x,t)_{RL}\mathrm{D}_{a,x}^\alpha U + d(x,t)_{RL}\mathrm{D}_{x,b}^\alpha U + g(x,t), & (x,t) \in (a,b)\times(0,T], \\ U(x,0) = \phi_0(x), \quad x \in (a,b), \\ U(a,t) = U(b,t) = 0, \quad t \in (0,T], \end{cases} \tag{4.164}$$

where $1 < \alpha < 2$ and $c(x,t), d(x,t) \geq 0$.

We can similarly construct the explicit Euler method, the implicit Euler method, the Crank–Nicolson method, and the weighted average method for (4.164) as those for (4.124), see (4.143) and (4.162).

If the left and right Riemann–Liouville fractional derivative operators are respectively discretized by the right and left shifted formulas with one shift, then the weighted average method for (4.164) is given by: Find u_i^n $(i = 1, 2, \cdots, N-1, n = 0, 1, 2, \cdots, n_T - 1)$, such that

$$\begin{cases} \delta_t u_i^{n+\frac{1}{2}} = \left[(1-\theta)c_i^{n+1}{}_L\delta_x^{(\alpha)} u_{i+1}^{n+1} + \theta c_i^n{}_L\delta_x^{(\alpha)} u_{i+1}^n\right] \\ \qquad + \left[(1-\theta)d_i^{n+1}{}_R\delta_x^{(\alpha)} u_{i-1}^{n+1} + \theta d_i^n{}_R\delta_x^{(\alpha)} u_{i-1}^n\right] \\ \qquad + (1-\theta)g_i^{n+1} + \theta g_i^n, \quad i = 1, 2, \cdots, N-1, \\ u_i^0 = \phi_0(x_i), \quad i = 0, 1, 2, \cdots, N, \\ u_0^n = U_a(t_n),\ u_N^n = U_b(t_n), \end{cases} \tag{4.165}$$

where ${}_L\delta_x^{(\alpha)}$ and ${}_R\delta_x^{(\alpha)}$ are defined by (4.8) and (4.9), respectively.

Let $\mu = \frac{\Delta t}{\Delta x^\alpha}$, $\mu_{c,i}^n = \mu c_i^n$, and $\mu_{d,i}^n = \mu d_i^n$. Then (4.165) can be written as

$$\begin{aligned} & u_i^{n+1} - (1-\theta)\left[\mu_{c,i}^{n+1}\,{}_L\delta_x^{(\alpha)} u_i^{n+1} + \mu_{d,i}^{n+1}\,{}_R\delta_x^{(\alpha)} u_i^{n+1}\right] \\ =& u_i^n + \theta\left[\mu_{c,i}^n\,{}_L\delta_x^{(\alpha)} u_i^n + \mu_{d,i}^n\,{}_R\delta_x^{(\alpha)} u_i^n\right] + \Delta t\left[(1-\theta)g_i^{n+1} + \theta g_i^n\right], \quad i = 1,2,\cdots,N-1. \end{aligned} \tag{4.166}$$

Hence, the matrix representation of (4.166) (or (4.165)) can be given below

$$\left[E - (1-\theta)\mu S^{n+1}\right]\underline{\mathbf{u}}^{n+1} = (E + \theta\mu S^n)\underline{\mathbf{u}}^n + \Delta t\left[(1-\theta)\underline{\mathbf{g}}^{n+1} + \theta\underline{\mathbf{g}}^n\right], \tag{4.167}$$

where E is an $(N-1)\times(N-1)$ identity matrix and

$$S^n = diag(c_1^n, c_2^n, \cdots, c_{N-1}^n) S_{N-1}^{(\alpha)} + diag(d_1^n, d_2^n, \cdots, d_{N-1}^n)(S_{N-1}^{(\alpha)})^T. \tag{4.168}$$

Next, we consider the stability of the weighted finite difference methods (4.165). For simplicity, we suppose that $c(x,t)$ and $d(x,t)$ are time independent. And we denote that by $c_{\max} = \max\limits_{0\le i\le N} c(x_i)$, $d_{\max} = \max\limits_{0\le i\le N} d(x_i)$. Therefore, the matrix S^n is independent of n, so we denote it by $S = S^n$.

According to the Gerschgorin's theorem, one has

$$\left|\lambda - \omega_1^{(\alpha)}(c_i + d_i)\right| \le c_i \sum_{j=0, j\ne 1}^{i} |\omega_j^{(\alpha)}| + d_i \sum_{j=0, j\ne 1}^{N-i} |\omega_j^{(\alpha)}|.$$

Noticing that $\omega_j^{(\alpha)} > 0, j \ne 1$, and $\sum\limits_{j=0}^{\infty} \omega_j^{(\alpha)} = 0$, one has $\sum_{j=0,j\ne1}^N \omega_j^{(\alpha)} \le -\omega_1^{(\alpha)}$. Hence,

$$\left|\lambda - \omega_1^{(\alpha)}(c_i + d_i)\right| \le -\omega_1^{(\alpha)}(c_i + d_i).$$

The eigenvalues λ of the matrix S satisfy

$$-2\alpha(c_{\max} + d_{\max}) \le 2\omega_1^{(\alpha)}(c_i + d_i) \le \lambda \le 0.$$

Next, we are in a position to estimate the eigenvalues of the following matrix

$$[E - \mu(1-\theta)A]^{-1}(E + \mu\theta S).$$

Suppose that λ is the eigenvalue of the matrix S. Then the eigenvalue of $[E - \mu(1-\theta)S]^{-1}(E + \mu\theta S)$ is $\frac{1+\mu\theta\lambda}{1-\mu(1-\theta)\lambda}$.

If $0\le\theta\le1/2$, then we always have $|\frac{1+\mu\theta\lambda}{1-\mu(1-\theta)\lambda}| \le 1$, so the weighted finite difference method (4.165) is unconditionally stable. If $1/2<\theta\le1$, we deduce from $-1 \le \frac{1+\mu\theta\lambda}{1-\mu(1-\theta)\lambda} \le 1$ that $\mu = \frac{\Delta t}{\Delta x^\alpha} \le \frac{1}{\alpha(c_{\max}+d_{\max})(2\theta-1)}$. Hence the weighted finite difference method (4.165) is conditionally stable for $1/2<\theta\le1$ and $\frac{\Delta t}{\Delta x^\alpha} \le \frac{1}{\alpha(c_{\max}+d_{\max})(2\theta-1)}$.

Obviously, the first-order method is used in the space discretization in (4.165). As in (4.162), we can use the second-order discretization in space.

Replacing the operators ${}_L\delta_x^{(\alpha)}$ and ${}_R\delta_x^{(\alpha)}$ in (4.165) by ${}_L\delta_x^{(\alpha,m)}$ and ${}_R\delta_x^{(\alpha,m)}$, $m = 1,2,$

respectively, we can get the following difference method for (4.124): Find u_i^n ($i = 1,2,\cdots,N-1, n = 0,1,2,\cdots,n_T-1$), such that

$$\begin{cases} \delta_t u_i^{n+\frac{1}{2}} = \left[(1-\theta)c_i^{n+1}{}_L\delta_x^{(\alpha,m)}u_{i+1}^{n+1} + \theta c_i^n{}_L\delta_x^{(\alpha,m)}u_{i+1}^n\right] \\ \qquad + \left[(1-\theta)d_i^{n+1}{}_R\delta_x^{(\alpha,m)}u_{i-1}^{n+1} + \theta d_i^n{}_R\delta_x^{(\alpha,m)}u_{i-1}^n\right] \\ \qquad + (1-\theta)g_i^{n+1} + \theta g_i^n, \quad i = 1,2,\cdots,N-1, \\ u_i^0 = \phi_0(x_i), \quad i = 0,1,2,\cdots,N, \\ u_0^n = U_a(t_n),\ u_N^n = U_b(t_n), \end{cases} \tag{4.169}$$

where $0\leq\theta\leq 1$, ${}_L\delta_x^{(\alpha,m)}$ is defined by (4.153) for $m = 1$ or by (4.156) for $m = 2$, and ${}_R\delta_x^{(\alpha,m)}$ is defined by (4.154) for $m = 1$ or by (4.157) for $m = 2$.

If $\theta = 1/2$, method (4.169) is reduced to the CN method with second-order accuracy both in time and space.

As in method (4.162), we can easily obtain that for $\frac{\sqrt{17}-1}{2}\leq\alpha\leq 2$, the method (4.169) with $m = 1$ is unconditionally stable for $0\leq\theta\leq 1/2$, and conditionally stable for $1/2<\theta\leq 1$.

If $0\leq\theta\leq 1/2$ and $c(x,t) = d(x,t) = K$, $K > 0$, then method (4.169) is unconditionally stable, which can be proved by the energy method. Readers can refer to [146] for related information.

4.3.3 Riesz Space-Fractional Diffusion Equation

This subsection considers finite difference methods for the fractional differential equations with Riesz space fractional derivatives. For simplicity, we consider the following fractional diffusion equation

$$\begin{cases} \partial_t U = {}_{RZ}\mathrm{D}_x^\alpha U + g(x,t), \quad (x,t)\in(a,b)\times(0,T], \\ U(x,0) = \phi_0(x), \quad x\in(a,b), \\ U(a,t) = U(b,t) = 0, \quad t\in(0,T], \end{cases} \tag{4.170}$$

where $1 < \alpha\leq 2$, $d(x) > 0$, and ${}_{RZ}\mathrm{D}_x^\alpha$ is the Riesz space fractional derivative defined by

$${}_{RZ}\mathrm{D}_x^\alpha U = -c_\alpha({}_{RL}\mathrm{D}_{a,x}^\alpha U + {}_{RL}\mathrm{D}_{x,b}^\alpha U), \tag{4.171}$$

in which $c_\alpha = \frac{1}{2\cos(\alpha\pi/2)}$.

Obviously, the Riesz space fractional derivative can be seen as the linear combination of the left and right Riemann–Liouville derivatives. Therefore, equation (4.170) can be solved by the difference methods (4.165) or (4.169) by letting $c_i^n = d_i^n = c_\alpha$. For the Riesz derivative operator, there exists a special discretization method named the fractional central difference method [14, 37, 38, 39].

The *fractional central difference method* is defined by

$$D_{\Delta x}^\alpha f(x) = \sum_{k=-\infty}^{\infty}\frac{(-1)^k\Gamma(\alpha+1)}{\Gamma(\alpha/2-k+1)\Gamma(\alpha/2+k+1)}f(x-k\Delta x), \quad \alpha > -1. \tag{4.172}$$

Then

$$\lim_{\Delta x\to 0}\frac{D^{\alpha}_{\Delta x}f(x)}{\Delta x^{\alpha}} = \lim_{\Delta x\to 0}\sum_{k=-\infty}^{\infty}\frac{(-1)^k\Gamma(\alpha+1)}{\Gamma(\alpha/2-k+1)\Gamma(\alpha/2+k+1)}f(x-k\Delta x) \qquad (4.173)$$

represents the Riesz derivative (4.171) for the case $1<\alpha\leq 2$ with $a=-\infty$ and $b=\infty$. Hence, (4.173) can be used as the discretization of the Riesz derivative.

Lemma 4.3.4 ([14]) *Let* $g_k = \frac{(-1)^k\Gamma(\alpha+1)}{\Gamma(\alpha/2-k+1)\Gamma(\alpha/2+k+1)}$ *be the coefficients of the centered difference approximation* (4.172) *for* $k=0,\mp1,\mp2,\cdots,$ *and* $\alpha>-1$. *Then*

$$g_0\geq 0,\quad g_{-k}=g_k\leq 0,\quad |k|>0, \qquad (4.174)$$

and

$$\sum_{k=-\infty}^{\infty} g_k e^{ikz} = |2\sin(z/2)|^{\alpha},\quad i^2=-1. \qquad (4.175)$$

Lemma 4.3.5 ([14]) *Let* $f\in C^5(\mathbb{R})$ *and all derivatives up to order five belong to* $L_1(\mathbb{R})$. *Then*

$$-\frac{D^{\alpha}_{\Delta x}f(x)}{\Delta x^{\alpha}} = {}_{RZ}\mathrm{D}^{\alpha}_{x}f(x)+O(\Delta x^2), \qquad (4.176)$$

where ${}_{RZ}\mathrm{D}^{\alpha}_{x}f(x) = -c_{\alpha}({}_{RL}\mathrm{D}^{\alpha}_{-\infty,x}f(x)+{}_{RL}\mathrm{D}^{\alpha}_{x,\infty}f(x))$ *and* $1<\alpha\leq 2$.

Suppose that $f(a)=f(b)=0$. Then the Riesz derivative ${}_{RZ}\mathrm{D}^{\alpha}_{x}f(x)$ can be approximated by

$${}_{RZ}\mathrm{D}^{\alpha}_{x}f(x) = -\frac{1}{\Delta x^{\alpha}}\sum_{k=-\frac{b-x}{\Delta x}}^{\frac{x-a}{\Delta x}} g_k f(x-k\Delta x)+O(\Delta x^2). \qquad (4.177)$$

Similar to (4.169), we can give the following finite difference methods for (4.124): Find u_i^n $(i=1,2,\cdots,N-1, n=0,1,2,\cdots,n_T-1)$, such that

$$\begin{cases} \delta_t u_i^{n+\frac{1}{2}} = (1-\theta)\,{}_{RZ}\delta_x^{(\alpha)}u_i^{n+1}+\theta\,{}_{RZ}\delta_x^{(\alpha)}u_i^n \\ \qquad\qquad +(1-\theta)g_i^{n+1}+\theta g_i^n,\quad i=1,2,\cdots,N-1,\ 0\leq\theta\leq 1, \\ u_i^0=\phi_0(x_i),\quad i=0,1,2,\cdots,N, \\ u_0^n=0,\ u_N^n=0, \end{cases} \qquad (4.178)$$

where

$${}_{RZ}\delta_x^{(\alpha)}u_i^n = -\frac{1}{\Delta x^{\alpha}}\sum_{k=-N+i}^{i} g_k u_{i-k}^n,\quad g_k=\frac{(-1)^k\Gamma(\alpha+1)}{\Gamma(\alpha/2-k+1)\Gamma(\alpha/2+k+1)}. \qquad (4.179)$$

For simplicity, we suppose $u_0^n=u_N^n=0$. In such a case, the matrix representation of (4.178) is given as:

$$[E-(1-\theta)\mu S]\underline{\mathbf{u}}^{n+1} = (E+\mu\theta S)\underline{\mathbf{u}}^n+\Delta t\left[(1-\theta)\underline{\mathbf{g}}^{n+1}+\theta\underline{\mathbf{g}}^n\right], \qquad (4.180)$$

where $\mu = \frac{\Delta t}{\Delta x^\alpha}$, E is an $(N-1)\times(N-1)$ matrix, and $S \in \mathbb{R}^{(N-1)\times(N-1)}$ satisfying

$$S = -\begin{pmatrix} g_0 & g_{-1} & g_{-2} & \cdots & g_{-N+1} \\ g_1 & g_0 & g_{-1} & \cdots & g_{-N+2} \\ \vdots & \vdots & \vdots & \ddots & \vdots \\ g_{N-2} & g_{N-3} & g_{N-4} & \cdots & g_1 \\ g_{N-1} & g_{N-2} & g_{N-3} & \cdots & g_0 \end{pmatrix}. \tag{4.181}$$

By Gerschgorin's circle theorem, Eqs. (4.174) and (4.175), we can get that the eigenvalue λ of S satisfies

$$|\lambda + g_0| \le \mu \sum_{k=-N+i,k\neq 0}^{i-1} |g_k| \le g_0,$$

which yields

$$-2g_0 \le \lambda \le 0.$$

If $0\le\theta\le 1/2$, then one can easily check that the eigenvalues $\frac{1+\mu\theta\lambda}{1-\mu(1-\theta)\lambda}$ of the matrix $[E-(1-\theta)S]^{-1}(E+\theta S)$ satisfy $\left|\frac{1+\mu\theta\lambda}{1-\mu(1-\theta)\lambda}\right| \le 1$. Hence, method (4.178) is unconditionally stable for $0\le\theta\le 1/2$. For $1/2<\theta\le 1$, we can have $\left|\frac{1+\mu\theta\lambda}{1-\mu(1-\theta)\lambda}\right| \le 1$ if $\mu \le \frac{1}{g_0(2\theta-1)}$. Therefore, method (4.178) is conditionally stable for $1/2<\theta\le 1$ and $\mu = \frac{\Delta t}{\Delta x^\alpha} \le \frac{1}{g_0(2\theta-1)}$.

4.3.4 Numerical Examples

Example 6 *Consider the following space-fractional diffusion equation*

$$\begin{cases} \partial_t U = {}_{RL}\mathrm{D}_{0,x}^{\alpha} U + {}_{RL}\mathrm{D}_{x,1}^{\alpha} U + g(x,t), & (x,t) \in (0,1)\times(0,1], \\ U(x,0) = x^4(1-x)^4, & x \in (0,1), \\ U(0,t) = U(1,t) = 0, & t \in (0,1], \end{cases} \tag{4.182}$$

where $1 < \alpha < 2$. Choose the suitable $g(x,t)$ such that Eq. (4.182) *has the exact solution $U(x,t) = \cos(t)x^4(1-x)^4$.*

Propositions 4.3.1 and 4.3.2 show that the explicit Euler method and implicit Euler method based on the standard Grünwald–Letnikov formula are unstable. We test the Crank–Nicolson method (4.178) ($\theta = 1/2$) based on the standard Grünwald–Letnikov formula to solve the problem (4.182), the results are shown in Table 4.14. From the computations, one can find that the numerical solutions blow up.

In Tables 4.15–4.18, we use the Crank–Nicolson method (see (4.165) with $\theta = 1/2$) based on the shifted Grünwald–Letnikov formula, the Crank–Nicolson method (4.169) ($\theta = 1/2$) based on the weighted shifted Grünwald–Letnikov formulas (see (4.153), (4.154), (4.156), and (4.157)), and the Crank–Nicolson method (4.178) ($\theta = 1/2$) based on the fractional central difference method to solve (4.182). The numerical results are shown in Tables 4.15–4.18. We can see that the satisfactory numerical results are obtained, which are in line with the theoretical analysis.

Note that we should rewrite ${}_{RL}\mathrm{D}^{\alpha}_{0,x}U + {}_{RL}\mathrm{D}^{\alpha}_{x,1}U$ in (4.182) in the form ${}_{RL}\mathrm{D}^{\alpha}_{0,x}U + {}_{RL}\mathrm{D}^{\alpha}_{x,1}U = c_{\alpha}{}_{RZ}\mathrm{D}^{\alpha}_{x}U$, so that method (4.178) can be applied properly.

TABLE 4.14: The L^2 error at $t = 1$ for the Crank–Nicolson method (4.178) ($\theta = 1/2$) based on the standard Grünwald–Letnikov formula, $\Delta t = 10^{-3}$.

N	$\alpha = 1.2$	$\alpha = 1.5$	$\alpha = 1.8$
8	7.9306e+017	8.8737e+046	2.7457e+112
16	1.9441e+048	4.6781e+149	NaN
32	9.7458e+119	NaN	NaN
64	NaN	NaN	NaN
128	NaN	NaN	NaN

TABLE 4.15: The L^2 error at $t = 1$ for the Crank–Nicolson method (4.165) ($\theta = 1/2$), $\Delta t = 10^{-3}$.

N	$\alpha = 1.2$	order	$\alpha = 1.5$	order	$\alpha = 1.8$	order
8	6.4217e−4		1.9017e−4		6.5329e−5	
16	4.0174e−4	0.6767	1.1500e−4	0.7256	1.2114e−5	2.4311
32	2.3453e−4	0.7765	6.3839e−5	0.8491	6.2524e−6	0.9542
64	1.2995e−4	0.8519	3.3688e−5	0.9222	3.9517e−6	0.6619
128	6.9087e−5	0.9114	1.7311e−5	0.9605	2.2292e−6	0.8259

TABLE 4.16: The L^2 error at $t = 1$ for the Crank–Nicolson method (4.169) ($\theta = 1/2$) with $m = 1$, $\Delta t = 10^{-3}$.

N	$\alpha = 1.2$	order	$\alpha = 1.5$	order	$\alpha = 1.8$	order
8	5.2424e−5		7.1713e−5		8.8409e−5	
16	1.2433e−5	2.0761	1.6744e−5	2.0986	2.0355e−5	2.1188
32	3.0732e−6	2.0163	4.1248e−6	2.0213	5.0045e−6	2.0240
64	7.6635e−7	2.0037	1.0278e−6	2.0047	1.2464e−6	2.0054
128	1.9155e−7	2.0003	2.5686e−7	2.0006	3.1143e−7	2.0008

TABLE 4.17: The L^2 error at $t = 1$ for the Crank–Nicolson method (4.169) ($\theta = 1/2$) with $m = 2$, $\Delta t = 10^{-3}$.

N	$\alpha = 1.2$	order	$\alpha = 1.5$	order	$\alpha = 1.8$	order
8	1.5557e−4		1.6274e−4		1.3416e−4	
16	3.9381e−5	1.9820	3.9862e−5	2.0295	3.1921e−5	2.0714
32	1.0054e−5	1.9697	1.0043e−5	1.9888	7.9320e−6	2.0087
64	2.5530e−6	1.9776	2.5338e−6	1.9868	1.9849e−6	1.9986
128	6.4422e−7	1.9866	6.3736e−7	1.9911	4.9703e−7	1.9977

TABLE 4.18: The L^2 error at $t = 1$ for the Crank–Nicolson method (4.178) ($\theta = 1/2$), $\Delta t = 10^{-3}$.

N	$\alpha = 1.2$	order	$\alpha = 1.5$	order	$\alpha = 1.8$	order
8	8.9228e−5		1.5660e−4		1.4640e−4	
16	3.5343e−5	1.3361	5.6465e−5	1.4717	4.1501e−5	1.8186
32	1.5504e−5	1.1888	1.6347e−5	1.7883	1.0880e−5	1.9315
64	4.7751e−6	1.6991	4.3544e−6	1.9085	2.7739e−6	1.9717
128	1.3052e−6	1.8713	1.1218e−6	1.9566	6.9973e−7	1.9870

4.4 One-Dimensional Time-Space Fractional Differential Equations

In this section, we numerically investigate the time-space fractional differential equations, where the time derivative and the spatial derivative are both fractional.

4.4.1 Time-Space Fractional Diffusion Equation with Caputo Derivative in Time

We now consider the following *time-space fractional diffusion equation*

$$\begin{cases} {}_C\mathrm{D}_{0,t}^{\gamma} U = (L^{(\alpha)}U)(x,t) + g(x,t), & (x,t) \in (a,b)\times(0,T], \\ U(x,0) = \phi_0(x), & x \in (a,b), \\ U(a,t) = 0,\ U(b,t) = 0, & t \in (0,T], \end{cases} \tag{4.183}$$

where $L^{(\alpha)} = c(x,t)_{RL}\mathrm{D}_{a,x}^{\alpha} + d(x,t)_{RL}\mathrm{D}_{x,b}^{\alpha}$, $0 < \gamma \le 1, 1 < \alpha < 2$, and $c,d > 0$.

Naturally, we can combine the time discretization techniques for the time-fractional equation (4.86) and the space discretization techniques for the space-fractional equation (4.164) to solve (4.183).

In order to illustrate the algorithms clearly and simply, we introduce the notation $L_{\Delta x,q}^{(\alpha,n)}$ defined by

$$L_{\Delta x,q}^{(\alpha,n)} U_i^n = \begin{cases} d_i^n {}_L\delta_x^{(\alpha)} U_{i+1}^n + c_i^n {}_R\delta_x^{(\alpha)} U_{i-1}^n, & q = 1, \\ d_i^n {}_L\delta_x^{(\alpha,1)} U_i^n + c_i^n {}_R\delta_x^{(\alpha,1)} U_i^n, & q = 2, \\ d_i^n {}_L\delta_x^{(\alpha,2)} U_i^n + c_i^n {}_R\delta_x^{(\alpha,2)} U_i^n, & q = 3, \\ {}_{RZ}\delta_x^{(\alpha)} U_i^n, & q = 4, \end{cases} \tag{4.184}$$

where ${}_L\delta_x^{(\alpha)}$, ${}_R\delta_x^{(\alpha)}$, ${}_L\delta_x^{(\alpha,1)}$, ${}_R\delta_x^{(\alpha,1)}$, ${}_L\delta_x^{(\alpha,2)}$, and ${}_R\delta_x^{(\alpha,2)}$ are defined by (4.8), (4.9), (4.153), (4.154), (4.156), and (4.157), respectively; and ${}_{RZ}\delta_x^{(\alpha)}$ is defined by (4.179) with $c = d = -\frac{1}{2\cos(\alpha\pi/2)}$.

It is known from the previous sections that

$$L_{\Delta x,q}^{(\alpha,n)} U_i^n = (L^{(\alpha)}U)(x_i,t_n) + O(\Delta t^p), \tag{4.185}$$

where

$$p = \begin{cases} 1, & q = 1, \\ 2, & q = 2,3, \\ 2, & q = 4 \text{ with } c(x,t) = d(x,t) = -(2\cos(\alpha\pi/2))^{-1}. \end{cases} \tag{4.186}$$

- The time fractional derivative is discretized by the Grünwald–Letnikov formula as in (4.96) and the space operator $L^{(\alpha)}$ in (4.183) is discretized as in (4.185); the fully discrete finite difference method for (4.183) is given by: Find u_i^n $(i = 1,2,\cdots,N-1, n = 1,2,\cdots,n_T)$, such that

$$\begin{cases} \delta_t^{(\gamma)}(u_i^n - u_i^0) = L_{\Delta x,q}^{(\alpha,n)} u_i^n + g_i^n, & i = 1,2,\cdots,N-1, \\ u_i^0 = \phi_0(x_i), & i = 0,1,2,\cdots,N, \\ u_0^n = u_N^n = 0, \end{cases} \tag{4.187}$$

where $\delta_t^{(\gamma)}$ is defined as in (4.96) and $L_{\Delta x,q}^{(\alpha,n)}$ is defined by (4.184).

- The time fractional derivative is discretized as in (4.97) and the space operator $L^{(\alpha)}$ in (4.183) is discretized as in (4.185); the fully discrete finite difference method for (4.183) is given by: Find u_i^n $(i = 1,2,\cdots,N-1, n = 1,2,\cdots,n_T)$, such that

$$\begin{cases} {}_C^{L1}\delta_t^{(\gamma)} u_i^n = L_{\Delta x,q}^{(\alpha,n)} u_i^n + g_i^n, & i = 1,2,\cdots,N-1, \\ u_i^0 = \phi_0(x_i), & i = 0,1,2,\cdots,N, \\ u_0^n = u_N^n = 0, \end{cases} \tag{4.188}$$

where ${}_C^{L1}\delta_t^{(\gamma)}$ is defined by (4.5) and $L_{\Delta x,q}^{(\alpha,n)}$ is defined by (4.184).

If the time direction of (4.183) is discretized by the FLMM as those in the FLMM finite difference methods (4.116), (4.117), (4.120), or (4.121), and the space operator $L^{(\alpha)}$ is discretized by (4.185), then we can similarly derive the corresponding FLMM finite difference methods for (4.183). We only need to replace $\delta_x^2 u_i^n$ in (4.116), (4.117), (4.120), or (4.121) with $L_{\Delta x,q}^{(\alpha,n)} u_i^n$ defined by (4.184) to derive the corresponding algorithms. We list these methods below:

- Find u_i^n $(i = 1,2,\cdots,N-1, n = 1,2,\cdots,n_T)$, such that

$$\begin{cases} \dfrac{1}{\Delta t^\gamma}\displaystyle\sum_{k=0}^{n} \omega_k (u_i^{n-k} - u_i^0) = \dfrac{1}{2^\gamma}\displaystyle\sum_{k=0}^{n} (-1)^k \omega_k L_{\Delta x,q}^{(\alpha,n-k)} u_i^{n-k} + B_n^{(1)} L_{\Delta x,q}^{(\alpha,0)} u_i^0 \\ \qquad\qquad + \dfrac{1}{\Delta t^\gamma}\displaystyle\sum_{k=0}^{n} \omega_{n-k} G_i^{n-k}, \\ u_i^0 = \phi_0(x_i), \quad i = 0,1,2,\cdots,N, \\ u_0^n = u_N^n = 0, \end{cases} \tag{4.189}$$

where $\omega_k = (-1)^k \binom{\gamma}{k}$, $G_i^n = \left[\mathrm{D}_{0,t}^{\gamma} g(x_i,t)\right]_{t=t_n}$, and $B_n^{(1)}$ is defined by (4.114) with $m = 1$.

- Find u_i^n ($i = 1,2,\cdots,N-1, n = 1,2,\cdots,n_T$), such that

$$\begin{cases} \frac{1}{\Delta t^\gamma}\sum_{k=0}^{n}\omega_k(u_i^{n-k}-u_i^0) = (1-\frac{\gamma}{2})L_{\Delta x,q}^{(\alpha,n)}u_i^n + \frac{\gamma}{2}L_{\Delta x,q}^{(\alpha,n-1)}u_i^{n-1} \\ \qquad + B_n^{(2)}L_{\Delta x,q}^{(\alpha,0)}u_i^0 + \frac{1}{\Delta t^\gamma}\left[(1-\frac{\gamma}{2})G_i^n + \frac{\gamma}{2}G_i^{n-1}\right], \\ u_i^0 = \phi_0(x_i), \quad i = 0,1,2,\cdots,N, \\ u_0^n = u_N^n = 0, \end{cases} \tag{4.190}$$

where $\omega_k = (-1)^k\binom{\gamma}{k}$, $G_i^n = \left[\mathrm{D}_{0,t}^{\gamma}g(x_i,t)\right]_{t=t_n}$, and $B_n^{(2)}$ is defined by (4.114) with $m = 2$.

- Find u_i^n ($i = 1,2,\cdots,N-1, n = 1,2,\cdots,n_T$), such that

$$\begin{cases} \frac{1}{\Delta t^\gamma}\sum_{k=0}^{n}\omega_k(u_i^{n-k}-u_i^0) = \frac{1}{2^\gamma}\sum_{k=0}^{n}(-1)^k\omega_k L_{\Delta x,q}^{(\alpha,n-k)}u_i^{n-k} + B_n^{(1)}L_{\Delta x,q}^{(\alpha,0)}u_i^0 \\ \qquad + C_n^{(1)}(L_{\Delta x,q}^{(\alpha,1)}u_i^1 - L_{\Delta x,q}^{(\alpha,0)}u_i^0) + \frac{1}{\Delta t^\gamma}\sum_{k=0}^{n}\omega_{n-k}G_i^{n-k}, \\ u_i^0 = \phi_0(x_i), \quad i = 0,1,2,\cdots,N, \\ u_0^n = u_N^n = 0, \end{cases} \tag{4.191}$$

where $\omega_k = (-1)^k\binom{\gamma}{k}$, $G_i^n = \left[\mathrm{D}_{0,t}^{\gamma}g(x_i,t)\right]_{t=t_n}$, $B_n^{(1)}$ is defined by (4.114) with $m = 1$, and $C_n^{(1)}$ is defined by (4.119) with $m = 1$.

- Find u_i^n ($i = 1,2,\cdots,N-1, n = 1,2,\cdots,n_T$), such that

$$\begin{cases} \frac{1}{\Delta t^\gamma}\sum_{k=0}^{n}\omega_k(u_i^{n-k}-u_i^0) = (1-\frac{\gamma}{2})L_{\Delta x,q}^{(\alpha,n)}u_i^n + \frac{\gamma}{2}L_{\Delta x,q}^{(\alpha,n-1)}u_i^{n-1} \\ \qquad + B_n^{(2)}L_{\Delta x,q}^{(\alpha,0)}u_i^0 + C_n^{(2)}(L_{\Delta x,q}^{(\alpha,1)}u_i^1 - L_{\Delta x,q}^{(\alpha,0)}u_i^0) \\ \qquad + \frac{1}{\Delta t^\gamma}\left[(1-\frac{\gamma}{2})G_i^n + \frac{\gamma}{2}G_i^{n-1}\right], \\ u_i^0 = \phi_0(x_i), \quad i = 0,1,2,\cdots,N, \\ u_0^n = u_N^n = 0, \end{cases} \tag{4.192}$$

where $\omega_k = (-1)^k\binom{\gamma}{k}$, $G_i^n = \left[\mathrm{D}_{0,t}^{\gamma}g(x_i,t)\right]_{t=t_n}$, $B_n^{(2)}$ is defined by (4.114) with $m = 2$, and $C_n^{(2)}$ is defined by (4.119) with $m = 2$.

Next, we present the stability analysis for the methods (4.187)–(4.188). For simplicity, we suppose that $c(x,t) = d(x,t) = constant$. We first focus on the stability for (4.188). The matrix representation of (4.188) is given by:

$$\left(b_0^{(\gamma)}E - \mu S\right)\underline{\mathbf{u}}^n = \sum_{k=1}^{n-1}(b_{n-k-1}^{(\gamma)} - b_{n-k}^{(\gamma)})\underline{\mathbf{u}}^k + b_n^{(\gamma)}\underline{\mathbf{u}}^0 + \Delta t^\gamma\underline{\mathbf{g}}^n, \tag{4.193}$$

where $\mu = \frac{\Delta t^\gamma}{\Delta x^\alpha}$, the matrix S is defined as that in (4.167), and E is an $(N-1)\times(N-1)$ identity matrix. In the following, $\mathbf{u}^n$ and $\underline{\mathbf{u}}^n$ means

$$\mathbf{u}^n = (u_0^n, u_1^n, \cdots, u_N^n)^T, \quad \underline{\mathbf{u}}^n = (u_1^n, u_2^n, \cdots, u_{N-1}^n)^T, \quad n = 0, 1, \cdots.$$

$\mathbf{g}^n, \underline{\mathbf{g}}^n, \mathbf{e}^n, \underline{\mathbf{e}}^n, \underline{\mathbf{G}}$ and $\mathbf{R}^n$ with $g_0^n = g_N^n = e_0^n = e_N^n = R_0^n = R_N^n = 0$ have the same meaning.

It is known that all the eigenvalues of the matrix S defined in (4.193) (see also (4.167)) have negative real parts. Therefore, for any vector $\mathbf{u}\in\mathbb{R}^{N-1}$, we have $(S\mathbf{u},\mathbf{u}) = \mathbf{u}^T S\mathbf{u}\le 0$. Hence, we have from $u_0^n = u_N^0 = 0$ and (4.193) that

$$\begin{aligned} b_0^{(\gamma)}\|\mathbf{u}^n\|_N^2 &\le b_0^{(\gamma)}\|\mathbf{u}^n\|_N^2 + \mu\Delta x(-S\underline{\mathbf{u}}^n, \underline{\mathbf{u}}^n) \\ &= \sum_{k=1}^{n-1}(b_{n-k-1}^{(\gamma)} - b_{n-k}^{(\gamma)})(\mathbf{u}^k, \mathbf{u}^n)_N + b_n^{(\gamma)}(\mathbf{u}^0, \mathbf{u}^n)_N + \Delta t^\gamma(\mathbf{g}^n, \mathbf{u}^n)_N. \end{aligned} \tag{4.194}$$

Applying Lemma 4.2.5 (see Eq. (4.101)) yields

$$\|\mathbf{u}^n\|_N^2 \le 2\|\mathbf{u}^0\|_N^2 + C\max_{1\le n\le n_T}\|\mathbf{g}^n\|_N^2. \tag{4.195}$$

For method (4.187), one can similarly obtain that the numerical solution of (4.187) satisfies (4.195).

Next, we consider the convergence analysis. Let $e_i^n = U(x_i, t_n) - u_i^n$. Then one gets the error equation for (4.188) as

$${}_C^{L1}\delta_t^{(\gamma)} e_i^n = L_{\Delta x,q}^{(\alpha,n)} e_i^n + R_i^n, \tag{4.196}$$

where R_i^n is the truncation error satisfying $|R_i^n|\le C(\Delta t^{2-\gamma} + \Delta x^p)$. From (4.195), we get

$$\|\mathbf{e}^n\|_N^2 \le 2\|\mathbf{e}^0\|_N^2 + C\max_{1\le n\le n_T}\|\mathbf{R}^n\|_N^2 \le C(\Delta t^{2-\gamma} + \Delta x^p).$$

The error bounds for the method (4.187) can be similarly obtained, which is given by

$$\|\mathbf{e}^n\|_N^2 \le C(\Delta t + \Delta x^p).$$

The stability and convergence analysis of methods (4.189)–(4.192) are a little different from (4.187) and (4.188). If $c(x,t) = d(x,t) = constant$, the stability and convergence analysis of methods (4.189)–(4.192) are similar to those of the schemes (4.116)–(4.117) and (4.120)–(4.121).

We just analyze the stability and convergence for (4.189), which is the same as those for (4.190)–(4.192). If $c(x,t) = d(x,t) = K_\gamma > 0$, then the matrix representation of method (4.189) is given by:

$$\sum_{k=0}^{n}\omega_k(\underline{\mathbf{u}}^{n-k} - \underline{\mathbf{u}}^0) = \Delta t^\gamma\mu\sum_{k=0}^{n}(-1)^k\omega_k S\underline{\mathbf{u}}^{n-k} + \Delta t^\gamma B_n^{(1)} S\underline{\mathbf{u}}^0 + \sum_{k=0}^{n}\omega_{n-k}\underline{\mathbf{G}}^{n-k}, \tag{4.197}$$

where $\mu = \frac{1}{2^\gamma\Delta x^\alpha}$ and S is a negative symmetric matrix. Hence, we can define the inner product

$$(\mathbf{u},\mathbf{v})_S = -\mathbf{v}^T S\mathbf{u}, \qquad \mathbf{u},\mathbf{v}\in\mathbb{R}^{N-1}.$$

with the norm $||\mathbf{u}||_S = \sqrt{(\mathbf{u},\mathbf{u})_S}$. We also have

$$(\mathbf{u},\mathbf{v})_S \leq ||\mathbf{u}||_S\,||\mathbf{v}||_S \leq \epsilon||\mathbf{u}||_S^2 + \frac{1}{4\epsilon}||\mathbf{v}||_S^2, \quad \epsilon > 0.$$

Denote by $|||\mathbf{u}|||_1 = \sqrt{||\mathbf{u}||^2 + \Delta t^{\gamma}\mu||\mathbf{u}||_S^2}$, $||\mathbf{u}|| = \sqrt{(\mathbf{u},\mathbf{u})}$, with

$$(\mathbf{u},\mathbf{v}) = \sum_{i=1}^{N-1} u_i v_i, \quad \mathbf{u},\mathbf{v} \in \mathbb{R}^{(N-1)\times 1}.$$

Then we have from (4.197) that

$$\begin{aligned}|||\mathbf{u}|||_1^2 &= (\underline{\mathbf{u}}^n,\underline{\mathbf{u}}^n) + \Delta t^{\gamma}\mu(\underline{\mathbf{u}}^n,\underline{\mathbf{u}}^n)_S \\ &= \sum_{k=1}^{n}\omega_k\left[(\underline{\mathbf{u}}^{n-k},\underline{\mathbf{u}}^n) - \mu\Delta t^{\gamma}(-1)^k(\underline{\mathbf{u}}^{n-k},\underline{\mathbf{u}}^n)_S\right] \\ &\quad + b_n(\underline{\mathbf{u}}^0,\underline{\mathbf{u}}^n) - \mu\Delta t^{\gamma}B_n^{(1)}(\underline{\mathbf{u}}^0,\underline{\mathbf{u}}^n)_S + \sum_{k=0}^{n}\omega_{n-k}(\underline{\mathbf{G}}^{n-k},\underline{\mathbf{u}}^n).\end{aligned} \tag{4.198}$$

Similar to Theorem 29, we can obtain

$$|||\underline{\mathbf{u}}^n|||_1^2 \leq C_1(|||\underline{\mathbf{u}}^0|||_1^2 + \Delta t^{\gamma}||\underline{\mathbf{u}}^0||_S^2) + C_2 \max_{0\leq k\leq n_T}||\underline{\mathbf{g}}^n||^2, \tag{4.199}$$

where C_1 is independent of $n, \Delta t, \Delta x$ and T, and C_2 is independent of $n, \Delta t, \Delta x$.

Using the properties $||\mathbf{u}^n||_N^2 = \Delta x||\underline{\mathbf{u}}||^2$ and $||\mathbf{g}^n||_N^2 = \Delta x||\underline{\mathbf{g}}||^2$ gives

$$||\mathbf{u}^n||_N^2 \leq C_1\left(||\mathbf{u}^0||_N^2 + \Delta t^{\gamma}\Delta x||\underline{\mathbf{u}}^0||_S^2\right) + C_2 \max_{0\leq k\leq n_T}||\mathbf{g}^n||_N^2. \tag{4.200}$$

Next we consider the convergence. Let $e_i^n = U(x_i,t_n) - u_i^n$, u_i^n be the solution of (4.189). Then the error equation of (4.189) is given by

$$\frac{1}{\Delta t^{\gamma}}\sum_{k=0}^{n}\omega_k(e_i^{n-k} - e_i^0) = \frac{1}{2^{\gamma}}\sum_{k=0}^{n}(-1)^k\omega_k L_{\Delta x,q}^{(\alpha,n-k)}e_i^{n-k} + \frac{1}{\Delta t^{\gamma}}R_i^n, \tag{4.201}$$

where $R_i^n = O(\Delta t + \Delta x^p)$.

Hence, we have from (4.200) and (4.201) that

$$\begin{aligned}||\mathbf{e}^n||_N^2 &\leq C_1\left(||\mathbf{e}^0||_N^2 + \Delta t^{\gamma}\Delta x||\underline{\mathbf{e}}^0||_S^2\right) + C_2 \max_{0\leq k\leq n_T}||\mathbf{R}^n||_N^2 = C_2 \max_{0\leq k\leq n_T}||\mathbf{R}^n||_N^2 \\ &\leq C(\Delta t + \Delta x^p).\end{aligned} \tag{4.202}$$

We can similarly derive that the solution u_i^n to (4.190), (4.191), or (4.192) satisfies (4.200), the convergence order of (4.190) is $(\Delta t + \Delta x^p)$, and the convergence orders of (4.191) and (4.192) are $(\Delta t^2 + \Delta x^p)$.

4.4.2 Time-Space Fractional Diffusion Equation with Riemann–Liouville Derivative in Time

Next, we consider the finite difference methods for the following *time-space fractional diffusion equation*

$$\begin{cases} \partial_t U = {}_{RL}\mathrm{D}_{0,t}^{1-\gamma}\left(L^{(\alpha)}U\right) + g(x,t), & (x,t)\in(a,b)\times(0,T], \\ U(x,0)=\phi_0(x), & x\in(a,b), \\ U(a,t)=U(b,t)=0, & t\in(0,T], \end{cases} \tag{4.203}$$

where $L^{(\alpha)} = c(x,t){}_{RL}\mathrm{D}_{a,x}^{\alpha} + d(x,t){}_{RL}\mathrm{D}_{x,b}^{\alpha}$, $0<\gamma\le 1, 1<\alpha<2$, and $c,d>0$.

One can find that the subdiffusion equation (4.203) is similar to (4.10), except that the second-order space derivative operator ∂_x^2 is replaced by the fractional derivative operator $L^{(\alpha)}$. Hence, the time discretization of (4.10) can be used for (4.203). The space derivative is discretized as that of (4.183).

Next, we directly list several finite difference methods for (4.203).

- **Explicit Euler type methods:** The time direction is discretized as that in (4.16), the space operator $L^{(\alpha)}$ at $t=t_n$ is approximated by $L_{\Delta x,q}^{(\alpha,n)}$ which is defined as (4.184), and the fully discrete finite difference method for (4.203) is given by: Find u_i^n $(i=1,2,\cdots,N-1, n=0,1,2,\cdots,n_T-1)$, such that

$$\begin{cases} \delta_t u_i^{n+\frac{1}{2}} = {}^{GL}\delta_t^{(1-\gamma)}\left(L_{\Delta x,q}^{(\alpha,n)}u_i^n\right) + f_i^n, & i=1,2,\cdots,N-1, \\ u_i^0=\phi_0(x_i), & i=0,1,2,\cdots,N, \\ u_0^n=u_N^n=0, \end{cases} \tag{4.204}$$

  where ${}^{GL}\delta_t^{(1-\gamma)}$ is defined by (4.184).

  The time fractional derivative in (4.203) can be discretized by the L1 method or the fractional backward difference formula; we just need to replace ${}^{GL}\delta_t^{(1-\gamma)}$ in (4.204) by ${}_{RL}^{L1}\delta_t^{(1-\gamma)}$ which is defined by (4.4) or ${}_p^B\delta_t^{(1-\gamma)}$ defined by (4.6) to obtain the corresponding algorithms.

- **Implicit Euler type methods:** The time direction is discretized as in (4.30), the space is discretized as in (4.204), the fully implicit Euler type method for (4.203) is given by: Find u_i^n $(i=1,2,\cdots,N-1, n=1,2,\cdots,n_T)$, such that

$$\begin{cases} \delta_t u_i^{n-\frac{1}{2}} = {}^{GL}\delta_t^{(1-\gamma)}\left(L_{\Delta x,q}^{(\alpha,n)}u_i^n\right) + f_i^n, & i=1,2,\cdots,N-1, \\ u_i^0=\phi_0(x_i), & i=0,1,2,\cdots,N, \\ u_0^n=u_N^n=0, \end{cases} \tag{4.205}$$

  where ${}^{GL}\delta_t^{(1-\gamma)}$ is defined by (4.3) and $L_{\Delta x,q}^{(\alpha,n-k)}$ is defined by (4.184). The operator ${}^{GL}\delta_t^{(1-\gamma)}$ in (4.205) can be replaced by ${}_{RL}^{L1}\delta_t^{(1-\gamma)}$ or ${}_p^B\delta_t^{(1-\gamma)}$ when the L1 method (see also (4.49)) or the fractional BDF method is used in the discretization of the time fractional derivative, which yields various Euler type methods.

- **Crank–Nicolson type methods:** The time direction is discretized as that in the CN method (4.59), the space operator $L^{(\alpha)}$ at $t=t_n$ is approximated by $L_{\Delta x,q}^{(\alpha,n-\frac{1}{2})}$ given in (4.184), the fully discrete Crank–Nicolson type method for (4.203) is given by: Find u_i^n $(i=1,2,\cdots,N-1, n=1,2,\cdots,n_T)$, such that

$$\begin{cases} \delta_t u_i^{n-\frac{1}{2}} = \delta_t^{(1-\gamma)}\left(L_{\Delta x,q}^{(\alpha,n-\frac{1}{2})} u_i^{n-\frac{1}{2}}\right) + f_i^n, \quad i=1,2,\cdots,N-1, \\ u_i^0 = \phi_0(x_i), \quad i=0,1,2,\cdots,N, \\ u_0^n = u_N^n = 0, \end{cases} \tag{4.206}$$

where $\delta_t^{(1-\gamma)}(L_{\Delta x,q}^{(\alpha,n-\frac{1}{2})} u_i^{n-\frac{1}{2}})$ is defined by

$$\begin{aligned} \delta_t^{(1-\gamma)}\left(L_{\Delta x,q}^{(\alpha,n-\frac{1}{2})} u_i^{n-\frac{1}{2}}\right) = & \frac{1}{\Delta t^{1-\gamma}}\Big[b_0 L_{\Delta x,q}^{(\alpha,n-\frac{1}{2})} u_i^{n-\frac{1}{2}} - \sum_{k=1}^{n-1}(b_{n-1-k}-b_{n-k}) L_{\Delta x,q}^{(\alpha,k-\frac{1}{2})} u_i^{k-\frac{1}{2}} \\ & - (b_n - B_n) L_{\Delta x,q}^{(\alpha,\frac{1}{2})} u_i^{\frac{1}{2}} - A_n L_{\Delta x,q}^{(\alpha,0)} u_i^0\Big], \end{aligned}$$

in which $A_n = B_n - \frac{\gamma(n+1/2)^{\gamma-1}}{\Gamma(1+\gamma)\Delta t^{1-\gamma}}$, $B_n = \frac{2\Delta t^{\gamma-1}}{\Gamma(1+\gamma)}[(n+1/2)^\gamma - n^\gamma]$, $b_l = \frac{1}{\Gamma(1+\gamma)}[(l+1)^\gamma - l^\gamma]$, and $L_{\Delta x,q}^{(\alpha,n)}$ is defined by (4.184).

If the time direction is discretized as that in (4.60) or (4.67), we can obtain different CN type methods which are not listed here.

- **Integration methods:** The time direction is discretized the same as that in (4.82), the space operator $L^{(\alpha)}$ at $t=t_n$ is approximated by $L_{\Delta x,q}^{(\alpha,n)}$ which is defined in (4.184), the explicit method for (4.203) is given by: Find u_i^n $(i=1,2,\cdots,N-1, n=1,2,\cdots,n_T)$, such that

$$\begin{cases} \delta_t u_i^{n-\frac{1}{2}} = \delta_t^{(1-\gamma)}\left(L_{\Delta x,q}^{(\alpha,n)} u_i^n\right) + f_i^n, \quad i=1,2,\cdots,N-1, \\ u_i^0 = \phi_0(x_i), \quad i=0,1,2,\cdots,N, \\ u_0^n = u_N^n = 0, \end{cases} \tag{4.207}$$

where $L_{\Delta x,q}^{(\alpha,n)}$ is defined by (4.184) and $\delta_t^{(1-\gamma)}$ is defined by

$$\delta_t^{(1-\gamma)}(L_{\Delta x,q}^{(\alpha,n)} u_i^n) = \frac{1}{\Delta t^{1-\gamma}}\left[\sum_{k=1}^{n} b_{n-k}^{(1-\gamma)}(L_{\Delta x,q}^{(\alpha,k-1)} u_i^{k-1}) - \sum_{l=1}^{n-1} b_{n-k-1}^{(1-\gamma)}(L_{\Delta x,q}^{(\alpha,k-1)} u_i^{k-1})\right],$$

$$b_k^{(1-\gamma)} = \frac{1}{\Gamma(1+\gamma)}[(k+1)^\gamma - k^\gamma].$$

If the time direction is discretized the same as that in (4.83), then we can obtain the implicit method given by:

$$\begin{cases} \delta_t u_i^{n-\frac{1}{2}} = \delta_t^{(1-\gamma)}\left(L_{\Delta x,q}^{(\alpha,n)} u_i^n\right) + f_i^n, \quad i=1,2,\cdots,N-1, \\ u_i^0 = \phi_0(x_i), \quad i=0,1,2,\cdots,N, \\ u_0^n = u_N^n = 0, \end{cases} \tag{4.208}$$

where $L_{\Delta x,q}^{(\alpha,n)}$ is defined by (4.184) and $\delta_t^{(1-\gamma)}$ is defined by

$$\delta_t^{(1-\gamma)}(L_{\Delta x,q}^{(\alpha,n)}u_i^n) = \frac{1}{\Delta t^{1-\gamma}}\left[\sum_{k=1}^{n} b_{n-k}^{(1-\gamma)}(L_{\Delta x,q}^{(\alpha,k)}u_i^k) - \sum_{l=1}^{n-1} b_{n-k-1}^{(1-\gamma)}(L_{\Delta x,q}^{(\alpha,k)}u_i^k)\right],$$

$$b_k^{(1-\gamma)} = \frac{1}{\Gamma(1+\gamma)}[(k+1)^\gamma - k^\gamma].$$

If the time direction is discretized as that in (4.84), we can also derive the corresponding implicit method, which is not listed here.

If $\gamma \to 1$ and $\alpha \to 2$, the explicit methods (4.204) and (4.207) are reduced to the classical forward Euler method, the implicit methods (4.205) and (4.208) are reduced to the classical backward Euler method, and the Crank–Nicolson type method (4.206) is reduced to the classical CN method.

The stability and convergence analyses of the methods (4.204)–(4.208) are more complicated than their counterparts of the classical equations.

If $c(x,t) = d(x,t) = K_\gamma > 0$, then the implicit method (4.205), the CN type method (4.206), and the integration method (4.208) are all unconditionally stable and are convergent to order $(\Delta t + \Delta x^p)$, $(\Delta t^{2-\gamma} + \Delta x^p)$, and $(\Delta t + \Delta x^p)$, respectively.

4.4.3 Numerical Examples

Example 7 *Consider the following time-space fractional diffusion equation*

$$\begin{cases} {}_CD_{0,t}^{\gamma}U = {}_{RL}\mathrm{D}_{0,x}^{\alpha}U + {}_{RL}\mathrm{D}_{x,1}^{\alpha}U + g(x,t), & (x,t) \in (0,1)\times(0,1], \\ U(x,0) = 2x^4(1-x)^4, & x \in (0,1), \\ U(0,t) = U(1,t) = 0, & t \in (0,1], \end{cases} \tag{4.209}$$

where $0 < \gamma < 1, 1 < \alpha < 2$. Choose the suitable $g(x,t)$ such that Eq. (4.209) *has the exact solution $U(x,t) = (t^{2+\gamma} + t + 2)x^4(1-x)^4$.*

We first test method (4.188). The L^2 error at $t = 1$ is shown in Table 4.19. We can see that first-order accuracy for $q = 1$ in space and second-order accuracy for $q = 2, 3, 4$ in space are observed, which are in line with the theoretical analysis. For $q = 4$ in Table 4.19, ${}_{RL}\mathrm{D}_{0,x}^{\alpha}U + {}_{RL}\mathrm{D}_{x,1}^{\alpha}U$ in (4.209) is written in the form of $c_{\alpha RZ}\mathrm{D}_x^{\alpha}U$ so that method (4.188) can be used properly, which is the same as in Tables 4.20–4.23. Tables 4.20–4.21 give the L^2 errors at $t = 1$ of methods (4.191) and (4.192), respectively. Obviously, we get satisfactory numerical results.

Example 8 *Consider the following space-fractional diffusion equation*

$$\begin{cases} \partial_t U = {}_{RL}\mathrm{D}_{0,t}^{1-\gamma}\left({}_{RL}\mathrm{D}_{0,x}^{\alpha}U + {}_{RL}\mathrm{D}_{x,1}^{\alpha}U\right) + g(x,t), & (x,t) \in (0,1)\times(0,1], \\ U(x,0) = 0, & x \in (0,1), \\ U(0,t) = U(1,t) = 0, & t \in (0,1], \end{cases} \tag{4.210}$$

where $0 < \gamma < 1, 1 < \alpha < 2$. Choose the suitable $g(x,t)$ such that Eq. (4.209) *has the exact solution $U(x,t) = (t^{2.5} + t)x^4(1-x)^4$.*

TABLE 4.19: The L^2 error at $t = 1$ for method (4.188), $\gamma = 0.8, \Delta t = 10^{-3}$.

q	N	$\alpha = 1.2$	order	$\alpha = 1.5$	order	$\alpha = 1.8$	order
1	8	2.6218e−3		7.5458e−4		2.6271e−4	
	16	1.6674e−3	0.6529	4.6301e−4	0.7046	4.8195e−5	2.4465
	32	9.7591e−4	0.7728	2.5856e−4	0.8405	2.4504e−5	0.9759
	64	5.3754e−4	0.8604	1.3678e−4	0.9186	1.5648e−5	0.6470
	128	2.8401e−4	0.9204	7.0337e−5	0.9595	8.8464e−6	0.8228
2	8	2.0521e−4		2.8394e−4		3.5194e−4	
	16	4.8876e−5	2.0699	6.6752e−5	2.0887	8.1674e−5	2.1074
	32	1.2163e−5	2.0066	1.6502e−5	2.0161	2.0134e−5	2.0203
	64	3.1149e−6	1.9653	4.1429e−6	1.9940	5.0316e−6	2.0005
	128	8.9300e−7	1.8024	1.0696e−6	1.9535	1.2734e−6	1.9823
3	8	6.0945e−4		6.4575e−4		5.3546e−4	
	16	1.5473e−4	1.9777	1.5906e−4	2.0214	1.2818e−4	2.0626
	32	3.9597e−5	1.9663	4.0155e−5	1.9859	3.1909e−5	2.0061
	64	1.0126e−5	1.9672	1.0160e−5	1.9827	8.0018e−6	1.9956
	128	2.6389e−6	1.9401	2.5858e−6	1.9742	2.0194e−6	1.9864
4	8	3.7015e−4		6.3728e−4		5.9009e−4	
	16	1.4568e−4	1.3453	2.2853e−4	1.4795	1.6719e−4	1.8194
	32	6.1691e−5	1.2396	6.5678e−5	1.7989	4.3810e−5	1.9322
	64	1.8924e−5	1.7048	1.7470e−5	1.9105	1.1180e−5	1.9704
	128	5.2385e−6	1.8530	4.5247e−6	1.9490	2.8350e−6	1.9795

TABLE 4.20: The L^2 error at $t = 1$ for method (4.191), $\gamma = 0.5, \Delta t = 10^{-3}$.

q	N	$\alpha = 1.2$	order	$\alpha = 1.5$	order	$\alpha = 1.8$	order
1	8	2.6125e−3		7.5259e−4		2.6266e−4	
	16	1.6639e−3	0.6509	4.6202e−4	0.7039	4.8193e−5	2.4463
	32	9.7474e−4	0.7714	2.5810e−4	0.8400	2.4530e−5	0.9743
	64	5.3713e−4	0.8597	1.3659e−4	0.9181	1.5686e−5	0.6451
	128	2.8389e−4	0.9199	7.0287e−5	0.9585	8.8899e−6	0.8192
2	8	2.0451e−4		2.8378e−4		3.5187e−4	
	16	4.8650e−5	2.0717	6.6681e−5	2.0894	8.1640e−5	2.1077
	32	1.2035e−5	2.0152	1.6456e−5	2.0187	2.0110e−5	2.0214
	64	3.0014e−6	2.0036	4.1015e−6	2.0044	5.0102e−6	2.0049
	128	7.5000e−7	2.0007	1.0247e−6	2.0009	1.2516e−6	2.0011
3	8	6.0724e−4		6.4530e−4		5.3535e−4	
	16	1.5419e−4	1.9776	1.5893e−4	2.0216	1.2813e−4	2.0628
	32	3.9392e−5	1.9687	4.0093e−5	1.9870	3.1882e−5	2.0069
	64	1.0002e−5	1.9776	1.0115e−5	1.9868	7.9797e−6	1.9983
	128	2.5234e−6	1.9868	2.5438e−6	1.9915	1.9979e−6	1.9978
4	8	3.6988e−4		6.3601e−4		5.8985e−4	
	16	1.4535e−4	1.3476	2.2826e−4	1.4784	1.6712e−4	1.8194
	32	6.1410e−5	1.2430	6.5588e−5	1.7992	4.3778e−5	1.9326
	64	1.8765e−5	1.7104	1.7418e−5	1.9128	1.1156e−5	1.9723
	128	5.1185e−6	1.8743	4.4816e−6	1.9585	2.8133e−6	1.9875

TABLE 4.21: The L^2 error at $t = 1$ for method (4.192), $\gamma = 0.2, \Delta t = 10^{-3}$.

q	N	$\alpha = 1.2$	order	$\alpha = 1.5$	order	$\alpha = 1.8$	order
1	8	2.6122e−3		7.5168e−4		2.6259e−4	
	16	1.6646e−3	0.6501	4.6155e−4	0.7036	4.8156e−5	2.4470
	32	9.7537e−4	0.7712	2.5785e−4	0.8399	2.4505e−5	0.9746
	64	5.3746e−4	0.8598	1.3647e−4	0.9180	1.5673e−5	0.6448
	128	2.8405e−4	0.9200	7.0224e−5	0.9585	8.8836e−6	0.8191
2	8	2.0424e−4		2.8363e−4		3.5180e−4	
	16	4.8590e−5	2.0716	6.6645e−5	2.0894	8.1622e−5	2.1077
	32	1.2020e−5	2.0151	1.6447e−5	2.0187	2.0105e−5	2.0214
	64	2.9976e−6	2.0036	4.0992e−6	2.0044	5.0089e−6	2.0050
	128	7.4894e−7	2.0009	1.0240e−6	2.0011	1.2512e−6	2.0012
3	8	6.0646e−4		6.4491e−4		5.3522e−4	
	16	1.5400e−4	1.9775	1.5884e−4	2.0215	1.2810e−4	2.0628
	32	3.9344e−5	1.9687	4.0071e−5	1.9869	3.1874e−5	2.0069
	64	9.9895e−6	1.9777	1.0110e−5	1.9868	7.9776e−6	1.9983
	128	2.5202e−6	1.9869	2.5423e−6	1.9916	1.9973e−6	1.9979
4	8	3.7056e−4		6.3545e−4		5.8965e−4	
	16	1.4551e−4	1.3486	2.2812e−4	1.4780	1.6708e−4	1.8194
	32	6.1359e−5	1.2457	6.5551e−5	1.7991	4.3766e−5	1.9326
	64	1.8745e−5	1.7108	1.7409e−5	1.9128	1.1153e−5	1.9723
	128	5.1123e−6	1.8744	4.4791e−6	1.9585	2.8125e−6	1.9875

If acting $\mathrm{D}_{0,t}^{\gamma-1}$ on both sides of equation (4.210), then it can be changed into a time-space fractional equation. In this example, we test methods (4.206) and (4.208); the L^2 errors are shown in Tables 4.22 and 4.23. In Table 4.22, the L^2 errors and the corresponding convergence orders in space for method (4.206) are displayed. Table 4.23 shows the L^2 errors and the corresponding convergence rates in time for method (4.208). We can see that the numerical results fit well with the theoretical analysis.

4.5 Fractional Differential Equations in Two Space Dimensions

In this section, we introduce the finite difference methods for the fractional partial differential equations in two spatial dimensions.

We focus on the discretization of several two–dimensional models, such as the two-dimensional time-fractional diffusion equation [4, 5], two-dimensional space-fractional diffusion equation [8], two-dimensional fractional advection-dispersion equation [122], and some other models [150, 174].

TABLE 4.22: The L^2 error at $t = 1$ for method (4.206), $\Delta t = 10^{-3}$.

q	N	$(\gamma,\alpha) = (0.2,1.2)$	order	$(\gamma,\alpha) = (0.5,1.5)$	order	$(\gamma,\alpha) = (0.8,1.8)$	order
1	8	1.2716e−3		3.6152e−4		1.3011e−4	
	16	8.0981e−4	0.6510	2.2321e−4	0.6957	2.3371e−5	2.4769
	32	4.7454e−4	0.7710	1.2499e−4	0.8366	1.1761e−5	0.9908
	64	2.6145e−4	0.8600	6.6209e−5	0.9167	7.5874e−6	0.6323
	128	1.3805e−4	0.9214	3.4072e−5	0.9584	4.3156e−6	0.8140
2	8	1.0064e−4		1.4026e−4		1.7471e−4	
	16	2.4149e−5	2.0592	3.2964e−5	2.0891	4.0502e−5	2.1089
	32	6.1760e−6	1.9672	8.1538e−6	2.0154	9.9729e−6	2.0219
	64	1.7577e−6	1.8130	2.0527e−6	1.9899	2.4853e−6	2.0046
	128	6.9293e−7	1.3429	5.3414e−7	1.9422	6.2181e−7	1.9989
3	8	2.9761e−4		3.1824e−4		2.6556e−4	
	16	7.5926e−5	1.9708	7.8473e−5	2.0198	6.3540e−5	2.0633
	32	1.9605e−5	1.9534	1.9822e−5	1.9851	1.5808e−5	2.0070
	64	5.1805e−6	1.9201	5.0220e−6	1.9808	3.9573e−6	1.9980
	128	1.5276e−6	1.7618	1.2836e−6	1.9680	9.9178e−7	1.9964
4	8	1.7941e−4		3.0921e−4		2.9140e−4	
	16	7.1000e−5	1.3374	1.1228e−4	1.4615	8.2738e−5	1.8164
	32	3.0397e−5	1.2239	3.2374e−5	1.7942	2.1694e−5	1.9313
	64	9.4881e−6	1.6798	8.6250e−6	1.9082	5.5311e−6	1.9716
	128	2.7937e−6	1.7640	2.2403e−6	1.9448	1.3959e−6	1.9863

TABLE 4.23: The L^2 error at $t = 1$ for method (4.208), $N = 1000$.

q	$1/\Delta t$	$(\gamma,\alpha) = (0.2,1.2)$	order	$(\gamma,\alpha) = (0.5,1.5)$	order	$(\gamma,\alpha) = (0.8,1.8)$	order
1	8	6.9113e−4		7.8814e−4		8.3716e−4	
	16	3.6568e−4	0.9184	4.1360e−4	0.9302	4.3487e−4	0.9449
	32	1.9555e−4	0.9030	2.1437e−4	0.9481	2.2202e−4	0.9699
	64	1.0782e−4	0.8589	1.1097e−4	0.9499	1.1244e−4	0.9815
	128	6.3099e−5	0.7730	5.8082e−5	0.9340	5.6807e−5	0.9850
2	8	6.7782e−4		7.8484e−4		8.3669e−4	
	16	3.5114e−4	0.9489	4.0996e−4	0.9369	4.3435e−4	0.9458
	32	1.8028e−4	0.9618	2.1054e−4	0.9614	2.2148e−4	0.9717
	64	9.2008e−5	0.9704	1.0704e−4	0.9760	1.1188e−4	0.9852
	128	4.6767e−5	0.9763	5.4088e−5	0.9848	5.6242e−5	0.9923
3	8	6.7781e−4		7.8483e−4		8.3669e−4	
	16	3.5113e−4	0.9489	4.0995e−4	0.9369	4.3435e−4	0.9458
	32	1.8027e−4	0.9619	2.1054e−4	0.9614	2.2147e−4	0.9717
	64	9.1999e−5	0.9704	1.0703e−4	0.9760	1.1188e−4	0.9852
	128	4.6759e−5	0.9764	5.4081e−5	0.9849	5.6238e−5	0.9923
4	8	6.7781e−4		7.8482e−4		8.3669e−4	
	16	3.5112e−4	0.9489	4.0994e−4	0.9369	4.3434e−4	0.9459
	32	1.8026e−4	0.9619	2.1053e−4	0.9614	2.2147e−4	0.9717
	64	9.1987e−5	0.9705	1.0702e−4	0.9761	1.1187e−4	0.9852
	128	4.6746e−5	0.9766	5.4072e−5	0.9850	5.6234e−5	0.9924

4.5.1 Time-Fractional Diffusion Equation with Riemann–Liouville Derivative in Time

First consider the following *time-fractional diffusion equation*

$$\begin{cases} \partial_t U = {}_{RL}\mathrm{D}_{0,t}^{1-\gamma}\left(K_1\partial_x^2 U + K_2\partial_y^2 U\right) + f(x,y,t), & (x,y,t)\in(a,b)\times(c,d)\times(0,T],\\ U(x,y,0)=\phi_0(x,y), \quad (x,y)\in(a,b)\times(c,d), \\ U(a,y,t)=U_a(y,t),\ U(b,y,t)=U_b(y,t), \quad (y,t)\in(c,d)\times(0,T],\\ U(x,c,t)=U_c(x,t),\ U(x,d,t)=U_d(x,t), \quad (x,t)\in(a,b)\times(0,T], \end{cases} \tag{4.211}$$

where $K_1, K_2 > 0$ and $0 < \gamma < 1$.

Before giving the discretization of the subdiffusion equation (4.211), we introduce some notations. Let $\Delta t = T/n_T$, $\Delta x = (b-a)/N_x$ and $\Delta y = (d-c)/N_y$ be the step sizes in time, x direction, and y direction, respectively, where n_T, N_x and N_y are positive integers. The grid points t_k, x_i and y_j are defined as $t_k = k\Delta t$, $x_i = a + i\Delta x$ and $y_j = c + j\Delta y$, respectively with $t_{k+\frac{1}{2}} = (t_k + t_{k+1})/2$. For the function $U(x,y,t)$ defined on the domain $\Omega = (a,b)\times(c,d)\times[0,T]$, denote by $U^n = U^n(\cdot) = U(x,y,t_n)$, $U_{i,j}^n = U(x_i,y_j,t_n)$, and

$$\delta_x U^n_{i+\frac{1}{2},j} = \frac{U^n_{i+1,j}-U^n_{i,j}}{\Delta x}, \qquad \delta_y U^n_{i,j+\frac{1}{2}} = \frac{U^n_{i,j+1}-U^n_{i,j}}{\Delta y}, \tag{4.212}$$

$$\delta_x^2 U^n_{i,j} = \frac{U^n_{i+1,j}-2U^n_{i,j}+U^n_{i-1,j}}{\Delta x^2}, \quad \delta_y^2 U^n_{i,j} = \frac{U^n_{i,j+1}-2U^n_{i,j}+U^n_{i,j-1}}{\Delta y^2}, \tag{4.213}$$

$$\delta_t U^{n+\frac{1}{2}}_{i,j} = \frac{U^{n+1}_{i,j}-U^n_{i,j}}{\Delta t}, \qquad U^{n+\frac{1}{2}}_{i,j} = \frac{U^n_{i,j}+U^{n+1}_{i,j}}{2}. \tag{4.214}$$

All the finite difference methods for (4.10) can be directly extended to (4.211), and the stability and convergence analyses are almost the same. We introduce the first method for (4.211) that can be seen as an extension of (4.30) to a two-dimensional problem.

- **The Implicit Method**

Letting $(x,y,t) = (x_i,y_j,t_n)$ in (4.211) yields

$$\partial_t U(x_i,y_j,t_n) = \left[{}_{RL}\mathrm{D}_{0,t}^{1-\gamma}\left(K_1\partial_x^2 U + K_2\partial_y^2 U\right)\right]_{(x,y,t)=(x_i,y_j,t_n)} + f(x_i,y_j,t_n). \tag{4.215}$$

The first-order time derivative, the time-fractional derivatives and the space derivatives in (4.215) at $(x,y,t) = (x_i,y_j,t_n)$ are discretized by the backward Euler scheme, the Grunwald scheme, and the central difference methods, respectively, i.e.,

$$\begin{aligned}
&\partial_t U(x_i,y_j,t_n) = \frac{U(x_i,y_j,t_n)-U(x_i,y_j,t_{n-1})}{\Delta t} + O(\Delta t) = \delta_t U_{i,j}^{n-\frac{1}{2}} + O(\Delta t),\\
&\left({}_{RL}\mathrm{D}_{0,t}^{1-\gamma}\partial_x^2 U\right)(x_i,y_j,t_n) = {}^{GL}\delta_t^{(1-\gamma)}(\partial_x^2 U^n(x_i,y_j)) + O(\Delta t),\\
&\left({}_{RL}\mathrm{D}_{0,t}^{1-\gamma}\partial_y^2 U\right)(x_i,y_j,t_n) = {}^{GL}\delta_t^{(1-\gamma)}(\partial_y^2 U^n(x_i,y_j)) + O(\Delta t),\\
&\partial_x^2 U(x_i,y_j,t_n) = \delta_x^2 U_{i,j}^n + O(\Delta x^2),\\
&\partial_y^2 U(x_i,y_j,t_n) = \delta_y^2 U_{i,j}^n + O(\Delta y^2),
\end{aligned}$$

where ${}^{GL}\delta_t^{(1-\gamma)}$ is defined by (4.3). Inserting the above equations into (4.215), we can get

$$\delta_t U_{i,j}^{n-\frac{1}{2}} = {}^{GL}\delta_t^{(1-\gamma)}(K_1\delta_x^2 U_{i,j}^n + K_2\delta_y^2 U_{i,j}^n) + f_{i,j}^n + O(\Delta t + \Delta x^2 + \Delta y^2). \tag{4.216}$$

Dropping the truncation error $O(\Delta t + \Delta x^2 + \Delta y^2)$ in (4.216) and replacing $U_{i,j}^k$ with $u_{i,j}^k$, we can obtain the finite difference scheme for (4.211) as: Find $u_{i,j}^n$ ($i = 1,2,\cdots,N_x-1, j = 1,2,\cdots,N_y-1, n = 1,2,\cdots,n_T$), such that

$$\begin{cases}
\delta_t u_{i,j}^{n-\frac{1}{2}} = {}^{GL}\delta_t^{(1-\gamma)}(K_1\delta_x^2 u_{i,j}^n + K_2\delta_y^2 u_{i,j}^n) + f_{i,j}^n,\\
u_{i,j}^0 = \phi_0(x_i,y_j), \quad i=0,1,2,\cdots,N_x, \; j=0,1,2,\cdots,N_y,\\
u_{0,j}^n = U_a(y_j,t_n), \quad u_{N_x,j}^n = U_b(y_j,t_n), \quad j=0,1,2,\cdots,N_y,\\
u_{i,0}^n = U_c(x_i,t_n), \quad u_{i,N_y}^n = U_d(x_i,t_n), \quad i=0,1,2,\cdots,N_x,
\end{cases} \tag{4.217}$$

where ${}^{GL}\delta_t^{(1-\gamma)}$is defined by (4.3).

Next, we present the matrix representation of the method (4.217). Rewrite the scheme (4.217) as the following form

$$u_{i,j}^n = u_{i,j}^{n-1} + \sum_{k=0}^{n}\omega_{n-k}^{(1-\gamma)}\left[\mu_1(u_{i+1,j}^k - 2u_{i,j}^k + u_{i-1,j}^k) + \mu_2(u_{i,j+1}^k - 2u_{i,j}^k + u_{i,j-1}^k)\right] + \Delta t f_i^n, \tag{4.218}$$

where $\mu_1 = K_1\Delta t^\gamma/\Delta x^2$ and $\mu_2 = K_2\Delta t^\gamma/\Delta y^2$.

Denote by

$$\underline{\mathbf{u}}^n = \begin{bmatrix}
u_{1,1}^n & u_{1,2}^n & \cdots & u_{1,N_y-2}^n & u_{1,N_y-1}^n\\
u_{2,1}^n & u_{2,2}^n & \cdots & u_{2,N_y-2}^n & u_{2,N_y-1}^n\\
\vdots & \vdots & \ddots & \vdots & \vdots\\
u_{N_x-1,1}^n & u_{N_x-1,2}^n & \cdots & u_{N_x-1,N_y-2}^n & u_{N_x-1,N_y-1}^n
\end{bmatrix}_{(N_x-1)\times(N_y-1)},$$

$$\mathbf{u}_{ab}^n = \begin{bmatrix}
u_{0,1}^n & u_{0,2}^n & \cdots & u_{0,N_y-1}^n\\
u_{N_x,1}^n & u_{N_x,2}^n & \cdots & u_{N_x,N_y-1}^n
\end{bmatrix}_{2\times(N_y-1)}, \quad
\mathbf{u}_{cd}^n = \begin{bmatrix}
u_{1,0}^n & u_{1,N_y}^n\\
u_{2,0}^n & u_{2,N_y}^n\\
\vdots & \vdots\\
u_{N_x-1,0}^n & u_{N_x-1,N_y}^n
\end{bmatrix}_{(N_x-1)\times 2}.$$

Let $E_N \in \mathbb{R}^{N\times N}$ be an identity matrix. The matrices $B_{N,2}$ and S_N are defined as

$$B_{N,2} = \begin{bmatrix} 1 & 0 \\ 0 & 0 \\ \vdots & \vdots \\ 0 & 0 \\ 0 & 1 \end{bmatrix}_{N\times 2}, \quad S_N = \begin{bmatrix} -2 & 1 & 0 & \cdots & 0 & 0 \\ 1 & -2 & 1 & \cdots & 0 & 0 \\ 0 & 1 & -2 & \cdots & 0 & 0 \\ \vdots & \vdots & \vdots & \ddots & \vdots & \vdots \\ 0 & 0 & 0 & \cdots & -2 & 1 \\ 0 & 0 & 0 & \cdots & 1 & -2 \end{bmatrix}_{N\times N}.$$

Then, the matrix representation of (4.218) can be rewritten as

$$\underline{\mathbf{u}}^n - \left(\mu_1 S_{N_x-1}\underline{\mathbf{u}}^n + \mu_2 \underline{\mathbf{u}}^n S^T_{N_y-1}\right) = RHS^n, \tag{4.219}$$

where

$$\begin{aligned} RHS^n =& \underline{\mathbf{u}}^{n-1} + \sum_{k=0}^{n-1} \omega_{n-k}^{(1-\gamma)}\left(\mu_1 S_{N_x-1}\underline{\mathbf{u}}^k + \mu_2 \underline{\mathbf{u}}^k S^T_{N_y-1}\right) \\ &+ \sum_{k=0}^{n} \omega_{n-k}^{(1-\gamma)}\left(\mu_1 B_{N_x-1,2}\mathbf{u}^k_{ab} + \mu_2 \mathbf{u}^k_{cd} B^T_{N_y-1,2}\right) + \Delta t \mathbf{F}^n, \\ (\mathbf{F}^n)_{i,j} =& f(x_i, y_j, t_n), \quad \mathbf{F}^n \in \mathbb{R}^{(N_x-1)\times(N_y-1)}. \end{aligned}$$

The matrix equation (4.219) can be solved by the iteration method or by using the Kronecker product to transform (4.219) into the following equivalent system

$$A\,\mathrm{vec}(\underline{\mathbf{u}}^n) = \mathrm{vec}(RHS^n). \tag{4.220}$$

Here

$$A = E_{N_y-1}\otimes E_{N_x-1} - \omega_0^{(1-\gamma)}\left(\mu_1 E_{N_y-1}\otimes S_{N_x-1} + \mu_2 S_{N_y-1}\otimes E_{N_x-1}\right),$$

in which the vec operator creates a column vector from a matrix $M \in \mathbb{R}^{I\times J}$, i.e.,

$$\mathrm{vec}(M) = \begin{pmatrix} m_1 \\ m_2 \\ \vdots \\ m_J \end{pmatrix}, \quad M = (m_1, m_2, \cdots, m_J).$$

Next, we consider the stability and convergence for (4.217). We first introduce some notations. Denote by $N = (N_x, N_y)$, and define the discrete inner product $(\cdot,\cdot)_N$ and norm $\|\cdot\|_N$ as

$$(\mathbf{u},\mathbf{v})_N = \Delta x \Delta y \sum_{i=0}^{N_x-1}\sum_{j=0}^{N_y-1} u_{i,j}v_{i,j}, \quad \|\mathbf{u}\|_N = \sqrt{(\mathbf{u},\mathbf{u})_N}, \tag{4.221}$$

where $\mathbf{u},\mathbf{v} \in \mathbb{R}^{(N_x+1)\times(N_y+1)}$, satisfying $(\mathbf{u})_{i,j} = u_{i,j}$ and $(\mathbf{v})_{i,j} = v_{i,j}$.

For convenience, we introduce also the following notations

$$
\begin{aligned}
(\delta_x \mathbf{u}, \delta_x \mathbf{v})_N &= \Delta x \Delta y \sum_{i=0}^{N_x-1} \sum_{j=0}^{N_y-1} \delta_x u_{i+\frac{1}{2},j} \delta_x v_{i+\frac{1}{2},j}, \\
(\delta_y \mathbf{u}, \delta_y \mathbf{v})_N &= \Delta x \Delta y \sum_{i=0}^{N_x-1} \sum_{j=0}^{N_y-1} \delta_y u_{i,j+\frac{1}{2}} \delta_y v_{i,j+\frac{1}{2}}, \\
(\delta_x^2 \mathbf{u}, \mathbf{v})_N &= \Delta x \Delta y \sum_{i=1}^{N_x-1} \sum_{j=0}^{N_y-1} v_{i,j} \delta_x^2 u_{i,j}, \\
(\delta_y^2 \mathbf{u}, \mathbf{v})_N &= \Delta x \Delta y \sum_{i=0}^{N_x-1} \sum_{j=1}^{N_y-1} v_{i,j} \delta_y^2 u_{i,j}, \\
||\mathbf{u}||_{1,N} &= \sqrt{(\delta_x \mathbf{u}, \delta_x \mathbf{u})_N + (\delta_y \mathbf{u}, \delta_y \mathbf{u})_N}.
\end{aligned}
$$

If $u_{i,0} = u_{i,N_y} = u_{j,0} = u_{j,N_x} = 0$ and $v_{i,0} = v_{i,N_y} = v_{j,0} = v_{j,N_x} = 0$, then one has

$$
(\delta_x^2 \mathbf{u}, \mathbf{v})_N = -(\delta_x \mathbf{u}, \delta_x \mathbf{v})_N, \tag{4.222}
$$

$$
(\delta_y^2 \mathbf{u}, \mathbf{v})_N = -(\delta_y \mathbf{u}, \delta_y \mathbf{v})_N. \tag{4.223}
$$

Similar to Theorem 23, we can easily get the following theorem.

Theorem 32 *Let $(\mathbf{u}^n)_{i,j} = u^n_{i,j}$ $(i = 0, 1, \cdots, N_x, j = 0, 1, \cdots, N_y)$ be the solution to the finite difference scheme (4.217), $u^n_{0,j} = u^n_{N_x,j} = u^n_{i,0} = u^n_{i,N_y} = 0$, $(\mathbf{F}^n)_{i,j} = f^n_{i,j}$ $(i = 0, 1, \cdots, N_x, j = 0, 1, \cdots, N_y)$. Then there exists a positive constant C independent of n, Δt and Δx, such that*

$$
||\mathbf{u}^n||_N^2 \le ||\mathbf{u}^0||_N^2 + C \max_{0 \le k \le n_T} ||\mathbf{F}^k||_N^2.
$$

The proof of Theorem 32 is almost the same as that of Theorem 23 with the help of (4.222)–(4.223), which is omitted here.

From (4.216), one can easily get that the truncation error of the scheme (4.217) is $(\mathbf{R}^n)_{i,j} = R^n_{i,j} = O(\Delta t + \Delta x^2 + \Delta y^2)$. Denote by $(\mathbf{e}^n)_{i,j} = e^n_{i,j} = U(x_i, y_j, t_n) - u^n_{i,j}$. Then the error equation of (4.217) is given by

$$
\delta_t e^{n-\frac{1}{2}}_{i,j} = {}^{GL}\delta_t^{(1-\gamma)} (K_1 \delta_x^2 e^n_{i,j} + K_2 \delta_y^2 e^n_{i,j}) + R^n_{i,j}.
$$

From Theorem 32, we can get

$$
||\mathbf{e}^n||_N^2 \le ||\mathbf{e}^0||_N^2 + C \max_{0 \le k \le n_T} ||\mathbf{R}^k||_N^2 \le C(\Delta t + \Delta x^2 + \Delta y^2).
$$

We know that the numerical solution of (4.211) can be obtained by solving the matrix equation (4.219) whose equivalent linear algebraic system is (4.220) with large coefficient matrix A of size $(N_x - 1)(N_y - 1) \times (N_x - 1)(N_y - 1)$.

Next, we mainly focus on the alternating direction implicit (ADI) finite difference methods for the discretization of (4.211). The ADI technique can transform the computation of a two-dimensional problem to a series of one dimensional problems that can be solved in parallel.

- **Review of the classical ADI method**

We recall the construction of the *ADI finite difference methods* for the classical equation in the following form

$$\begin{cases} \partial_t U = K_1\partial_x^2 U + K_2\partial_y^2 U + f(x,y,t), & (x,y,t)\in(a,b)\times(c,d)\times(0,T], \\ U(x,y,0) = \phi_0(x,y), & (x,y)\in(a,b)\times(c,d), \\ U(a,y,t) = U_a(y,t),\ U(b,y,t) = U_b(y,t), & (y,t)\in(c,d)\times(0,T], \\ U(x,c,t) = U_c(x,t),\ U(x,d,t) = U_d(x,t), & (x,t)\in(a,b)\times(0,T], \end{cases} \tag{4.224}$$

where $K_1, K_2 > 0$.

Denote by $L_x U = K_1\partial_x^2 U$ and $L_y U = K_2\partial_y^2 U$. Then it follows from (4.224) that

$$\partial_t U = (L_x + L_y)U(x,y,t) + f(x,y,t), \tag{4.225}$$

Letting $t = t_{n-\frac{1}{2}}$ in (4.225) yields

$$\partial_t U(t_{n-\frac{1}{2}}) = (L_x + L_y)U(t_{n-\frac{1}{2}}) + f(t_{n-\frac{1}{2}}). \tag{4.226}$$

By $\partial_t U(t_{n-\frac{1}{2}}) = \delta_t U^{n-\frac{1}{2}} + O(\Delta t^2)$ and $U(t_{n-\frac{1}{2}}) = U^{n-\frac{1}{2}} + O(\Delta t^2)$, one has

$$\delta_t U^{n-\frac{1}{2}} = (L_x + L_y)U^{n-\frac{1}{2}} + f(t_{n-\frac{1}{2}}) + O(\Delta t^2). \tag{4.227}$$

In order to derive the ADI method, we add the perturbation term $\left(\frac{\Delta t}{2}\right)^2 L_x L_y \delta_t U^{n-\frac{1}{2}} = O(\Delta t^2)$ to the left-hand side of (4.227), which yields

$$\delta_t U^{n-\frac{1}{2}} + \left(\frac{\Delta t}{2}\right)^2 L_x L_y \delta_t U^{n-\frac{1}{2}} = (L_x + L_y)U^{n-\frac{1}{2}} + f(t_{n-\frac{1}{2}}) + O(\Delta t^2). \tag{4.228}$$

Rewrite (4.228) as the following form

$$(1 - \frac{\Delta t}{2}L_x)(1 - \frac{\Delta t}{2}L_y)U^n = (1 + \frac{\Delta t}{2}L_x)(1 + \frac{\Delta t}{2}L_y)U^{n-1} + \Delta t f(t_{n-\frac{1}{2}}) + O(\Delta t^3). \tag{4.229}$$

Denote by

$$L_{\Delta x}U_{i,j}^n = K_1\delta_x^2 U_{i,j}^n, \quad L_{\Delta y}U_{i,j}^n = K_2\delta_y^2 U_{i,j}^n. \tag{4.230}$$

Then

$$(L_x U^n)(x_i, y_j) = L_{\Delta x}U_{i,j}^n + O(\Delta x^2), \quad (L_y U^n)(x_i, y_j) = L_{\Delta y}U_{i,j}^n + O(\Delta y^2).$$

Hence, we can obtain

$$\begin{aligned}(1-\frac{\Delta t}{2}L_{\Delta x})(1-\frac{\Delta t}{2}L_{\Delta y})U_{i,j}^{n} =&(1+\frac{\Delta t}{2}L_{\Delta x})(1+\frac{\Delta t}{2}L_{\Delta y})U_{i,j}^{n-1}\\ &+\Delta t f(x_i,y_j,t_{n-\frac{1}{2}})+O(\Delta t(\Delta t^2+\Delta x^2)).\end{aligned} \tag{4.231}$$

Removing the truncation error $O(\Delta t(\Delta t+\Delta x^2))$ and replacing $U_{i,j}^n$ with $u_{i,j}^n$ in (4.231), we can get the ADI difference method for (4.224) as

$$(1-\frac{\Delta t}{2}L_{\Delta x})(1-\frac{\Delta t}{2}L_{\Delta y})u_{i,j}^{n} = (1+\frac{\Delta t}{2}L_{\Delta x})(1+\frac{\Delta t}{2}L_{\Delta y})u_{i,j}^{n-1}+\Delta t f(x_i,y_j,t_{n-\frac{1}{2}}). \tag{4.232}$$

There are two methods commonly used to solve (4.232), the first one that is called *PR factorization* [123] which is given by

$$(1-\frac{\Delta t}{2}L_{\Delta x})u_{i,j}^{*} = (1+\frac{\Delta t}{2}L_{\Delta y})u_{i,j}^{n-1}+\frac{\Delta t}{2}f(x_i,y_j,t_{n-\frac{1}{2}}), \tag{4.233}$$

$$(1-\frac{\Delta t}{2}L_{\Delta y})u_{i,j}^{n} = (1+\frac{\Delta t}{2}L_{\Delta x})u_{i,j}^{*}+\frac{\Delta t}{2}f(x_i,y_j,t_{n-\frac{1}{2}}). \tag{4.234}$$

Eliminating the intermediate term $u_{i,j}^*$ from (4.233) and (4.234) yields (4.232).

From (4.233), we can find that if j is given, then we can solve the linear system (4.233) to obtain $u_j^* = (u_{1,j}^*, u_{2,j}^*, \cdots, u_{N_x-1,j}^*)^T$, where the size of the coefficient matrix derived from (4.233) is $(N_x-1)\times(N_x-1)$, which is much smaller than that of the system (4.220) (The size of the coefficient matrix of (4.220) is $(N_x-1)(N_y-1)\times(N_x-1)(N_y-1)$). Obviously, $u_{j_1}^*$ and $u_{j_2}^*$ for $j_1 \neq j_2$ can be computed in parallel. We can similarly obtain $u_i^{n+1} = (u_{i,1}^{n+1}, u_{i,2}^{n+1}, \cdots, u_{i,N_y-1}^{n+1})$ from (4.234) for a fixed i.

Eliminating $f(x_i,y_j,t_{n-\frac{1}{2}})$ from (4.233) and (4.234), we can get

$$u_{i,j}^{*} = u_{i,j}^{n-\frac{1}{2}} - \frac{\Delta t^2}{4}L_{\Delta y}\delta_t u_{i,j}^{n-\frac{1}{2}}, \tag{4.235}$$

Hence, the boundary conditions of $u_{i,j}^*$ needed in (4.233) can be taken as

$$u_{0,j}^{*} = u_{0,j}^{n-\frac{1}{2}} - \frac{\Delta t^2}{4}L_{\Delta y}\delta_t u_{0,j}^{n-\frac{1}{2}}, \quad u_{N_x,j}^{*} = u_{0,j}^{n-\frac{1}{2}} - \frac{\Delta t^2}{4}L_{\Delta y}\delta_t u_{N_x,j}^{n-\frac{1}{2}}.$$

Another factorization, called the *D'Yakonov factorization*, is given by

$$\begin{aligned}(1-\frac{\Delta t}{2}L_{\Delta x})u_{i,j}^{*} &= (1-\frac{\Delta t}{2}L_{\Delta x})(1+\frac{\Delta t}{2}L_{\Delta y})u_{i,j}^{n-1}+\Delta t f(x_i,y_j,t_{n-\frac{1}{2}}),\\ &\qquad i=1,2,\cdots,N_x-1,\end{aligned} \tag{4.236}$$

$$(1-\frac{\Delta t}{2}L_{\Delta y})u_{i,j}^{n} = u_{i,j}^{*}, \quad j=1,2,\cdots,N_y-1. \tag{4.237}$$

The system (4.236)–(4.237) can be similarly solved as (4.233)–(4.234). From (4.237), we can obtain the boundary conditions for $u_{i,j}^*$ needed in (4.236) as

$$u_{0,j}^{*} = (1-\frac{\Delta t}{2}L_{\Delta y})u_{0,j}^{n}, \quad u_{N_x,j}^{*} = (1-\frac{\Delta t}{2}L_{\Delta y})u_{N_x,j}^{n}.$$

Generally speaking, the factorization (4.236)–(4.237) can be easily extended to the three-dimensional or much higher dimensional fractional differential equations.

- **The first ADI method for** (4.211)

Next, we introduce the first ADI finite difference method for (4.211). From (4.216), we can get

$$\begin{aligned}\delta_t U_{i,j}^{n-\frac{1}{2}} =& {}^{GL}\delta_t^{(1-\gamma)}(K_1\delta_x^2 U_{i,j}^n + K_2\delta_y^2 U_{i,j}^n) + f_{i,j}^n + O(\Delta t + \Delta x^2 + \Delta y^2)\\ =&\Delta t^{\gamma-1}\sum_{k=0}^{n}\omega_{n-k}^{(1-\gamma)}(L_{\Delta x}+L_{\Delta y})U_{i,j}^k + f_{i,j}^n + O(\Delta t + \Delta x^2 + \Delta y^2)\\ =&-(\omega_0^{(1-\gamma)}\Delta t^{\gamma})^2 L_{\Delta x}L_{\Delta y}\delta_t U_{i,j}^{n-\frac{1}{2}} + O(\Delta t^{2\gamma})\\ &+\Delta t^{\gamma-1}\sum_{k=0}^{n}\omega_{n-k}^{(1-\gamma)}(L_{\Delta x}+L_{\Delta y})U_{i,j}^k + f_{i,j}^n + O(\Delta t + \Delta x^2 + \Delta y^2).\end{aligned} \tag{4.238}$$

In fact, we add a perturbation term $-(\omega_0^{(1-\gamma)}\Delta t^{\gamma})^2 L_{\Delta x}L_{\Delta y}\delta_t U_{i,j}^{n-\frac{1}{2}} = O(\Delta t^{2\gamma})$ to the right-hand side of (4.216) to obtain (4.238). Dropping the truncation error $O(\Delta t + \Delta x^2 + \Delta y^2) + O(\Delta t^{2\gamma})$ in the above equation (4.238) and replacing $U_{i,j}^n$ with $u_{i,j}^n$, we get the following ADI finite difference method for (4.211) as: Find $u_{i,j}^n$ $(i = 1,2,\cdots,N_x - 1, j = 1,2,\cdots,N_y, n = 1,2,\cdots,n_T)$, such that

$$\begin{cases}\delta_t u_{i,j}^{n-\frac{1}{2}} + K_1K_2(\omega_0^{(1-\gamma)}\Delta t^{\gamma})^2\delta_x^2\delta_y^2\delta_t u_{i,j}^{n-\frac{1}{2}} = {}^{GL}\delta_t^{(1-\gamma)}(K_1\delta_x^2 u_{i,j}^n + K_2\delta_y^2 u_{i,j}^n) + f_{i,j}^n,\\ u_{i,j}^0 = \phi_0(x_i,y_j), \quad i = 0,1,2,\cdots,N_x,\ j = 0,1,2,\cdots,N_y,\\ u_{0,j}^n = U_a(y_j,t_n), \quad u_{N_x,j}^n = U_b(y_j,t_n), \quad j = 0,1,2,\cdots,N_y,\\ u_{i,0}^n = U_c(x_i,t_n), \quad u_{i,N_y}^n = U_d(x_i,t_n), \quad i = 0,1,2,\cdots,N_x,\end{cases} \tag{4.239}$$

where ${}^{GL}\delta_t^{(1-\gamma)}$ is defined by (4.3).

Next, we give a brief illustration that (4.239) is the ADI algorithm. Rewrite (4.239) in the following form

$$\begin{aligned}&u_{i,j}^n - \omega_0^{(1-\gamma)}\Delta t^{\gamma}(K_1\delta_x^2 u_{i,j}^n + K_2\delta_y^2 u_{i,j}^n) + K_1K_2(\omega_0^{(1-\gamma)}\Delta t^{\gamma})^2\delta_x^2\delta_y^2 u_{i,j}^n\\ =&u_{i,j}^{n-1} + K_1K_2(\omega_0^{(1-\gamma)}\Delta t^{\gamma})^2\delta_x^2\delta_y^2 u_{i,j}^{n-1} + \Delta t^{\gamma}\sum_{k=0}^{n-1}\omega_{n-k}^{(1-\gamma)}(K_1\delta_x^2 u_{i,j}^k + K_2\delta_y^2 u_{i,j}^k) + \Delta t f_{i,j}^n.\end{aligned} \tag{4.240}$$

Notice that

$$\begin{aligned}&u_{i,j}^n - \omega_0^{(1-\gamma)}\Delta t^{\gamma}(K_1\delta_x^2 u_{i,j}^n + K_2\delta_y^2 u_{i,j}^n) + K_1K_2(\omega_0^{(1-\gamma)}\Delta t^{\gamma})^2(\delta_x^2\delta_y^2 u_{i,j}^n)\\ =&(1 - K_1\omega_0^{(1-\gamma)}\Delta t^{\gamma}\delta_x^2)(1 - K_2\omega_0^{(1-\gamma)}\Delta t^{\gamma}\delta_y^2)u_{i,j}^n.\end{aligned} \tag{4.241}$$

Hence,

$$(1 - K_1\omega_0^{(1-\gamma)}\Delta t^{\gamma}\delta_x^2)(1 - K_2\omega_0^{(1-\gamma)}\Delta t^{\gamma}\delta_y^2)u_{i,j}^n = (RHS)_{i,j}^n, \tag{4.242}$$

where

$$
\begin{aligned}
(RHS)_{i,j}^n =& u_{i,j}^{n-1} + K_1K_2(\omega_0^{(1-\gamma)}\Delta t^\gamma)^2(\delta_x^2\delta_y^2 u_{i,j}^{n-1})\\
&+\Delta t^\gamma \sum_{k=0}^{n-1}\omega_{n-k}^{(1-\gamma)}(K_1\delta_x^2 u_{i,j}^k + K_2\delta_y^2 u_{i,j}^k) + \Delta t f_{i,j}^n.
\end{aligned}
$$

Eq. (4.242) is equivalent to the following form

$$(1-K_1\omega_0^{(1-\gamma)}\Delta t^\gamma\delta_x^2)u_{i,j}^* = (RHS)_{i,j}^n, \quad i=1,2,\cdots,N_x-1, \tag{4.243}$$

$$(1-K_2\omega_0^{(1-\gamma)}\Delta t^\gamma\delta_y^2)u_{i,j}^n = u_{i,j}^*, \quad j=1,2,\cdots,N_y-1. \tag{4.244}$$

From (4.244), we can get the boundary conditions for $u_{i,j}^*$ needed in (4.243), which are taken as

$$u_{i,0}^* = (1-K_2\omega_0^{(1-\gamma)}\Delta t^\gamma\delta_y^2)u_{i,0}^n, \quad u_{i,N_y}^* = (1-K_2\omega_0^{(1-\gamma)}\Delta t^\gamma\delta_y^2)u_{i,N_y}^n. \tag{4.245}$$

Next, we consider the stability and convergence for (4.239).

Lemma 4.5.1 *Let $(\mathbf{u})_{i,j} = u_{i,j}$ $(i=0,1,\cdots,N_x, j=0,1,\cdots,N_y)$ be the grid functions with $u_{0,j}=u_{N_x,j}=u_{i,0}=u_{i,N_y}=0$. Then there exists a positive constant C such that*

$$||\mathbf{u}||_N \le C||\mathbf{u}||_{1,N}.$$

The following lemma illustrates that the ADI scheme (4.239) is unconditionally stable.

Theorem 33 *Let $(\mathbf{u})_{i,j} = u_{i,j}^n$ $(i=0,1,\cdots,N_x, j=0,1,\cdots,N_y)$ be the solutions to the ADI finite difference scheme (4.239), $u_{0,j}^n = u_{N_x,j}^n = u_{i,0}^n = u_{i,N_y}^n = 0$, $(\mathbf{f}^n)_{i,j} = f_{i,j}^n$ $(i=0,1,\cdots,N_x, j=0,1,\cdots,N_y)$. Then there exists a positive constant independent of n, Δt and Δx, such that*

$$||\mathbf{u}||_N^2 \le ||\mathbf{u}^0||_N^2 + \Delta t^\gamma\left(K_1||\delta_x\mathbf{u}^0||_N^2 + K_2||\delta_y\mathbf{u}^0||_N^2\right) + \Delta t^{2\gamma}||\delta_x\delta_y\mathbf{u}^0||_N^2 + C\max_{0\le k\le n_T}||\mathbf{f}^k||_N^2.$$

Proof. From (4.239), one has

$$
\begin{aligned}
&\sum_{i=1}^{N_x-1}\sum_{j=1}^{N_y-1}\delta_t u_{i,j}^n u_{i,j}^{n-\frac{1}{2}} + K_1K_2(\omega_0^{(1-\gamma)}\Delta t^\gamma)^2\sum_{i=1}^{N_x-1}\sum_{j=1}^{N_y-1}u_{i,j}^n(\delta_x^2\delta_y^2\delta_t u_{i,j}^{n-\frac{1}{2}})\\
=&{}^{GL}\delta_t^{(1-\gamma)}\left(K_1\sum_{i=1}^{N_x-1}\sum_{j=1}^{N_y-1}u_{i,j}^n\delta_x^2u_{i,j}^n + K_2\sum_{i=1}^{N_x-1}\sum_{j=1}^{N_y-1}u_{i,j}^n\delta_y^2u_{i,j}^n\right) + \sum_{i=1}^{N_x-1}\sum_{j=1}^{N_y-1}u_{i,j}^n f_{i,j}^n,
\end{aligned}
\tag{4.246}
$$

which implies

$$
\begin{aligned}
&(\mathbf{u}^n,\mathbf{u}^n)_N + K_1K_2(\omega_0^{(1-\gamma)}\Delta t^\gamma)^2(\delta_x^2\delta_y^2\mathbf{u}^n,\mathbf{u}^n)_N\\
=&(\mathbf{u}^n,\mathbf{u}^{n-1})_N + K_1K_2(\omega_0^{(1-\gamma)}\Delta t^\gamma)^2(\delta_x\delta_y\mathbf{u}^n,\delta_x\delta_y\mathbf{u}^{n-1})_N\\
&+\Delta t^\gamma\sum_{k=0}^{n-1}\omega_{n-k}^{(1-\gamma)}\left(K_1(\delta_x^2\mathbf{u}^k,\mathbf{u}^n)_N + K_2(\delta_y^2\mathbf{u}^k,\mathbf{u}^n)_N\right) + \Delta t(\mathbf{f}^n,\mathbf{u}^n)_N.
\end{aligned}
\tag{4.247}
$$

Using (4.222), (4.223), and Cauchy inequality one has

$$\begin{aligned}
&||\mathbf{u}^n||_N^2 + \Delta t^\gamma \omega_0^{(1-\gamma)} \left(K_1||\delta_x \mathbf{u}^n||_N^2 + K_2||\delta_y \mathbf{u}^n||_N^2\right) + K_1K_2(\omega_0^{(1-\gamma)}\Delta t^\gamma)^2||\delta_x\delta_y\mathbf{u}^n||_N^2 \\
=&(\mathbf{u}^n, \mathbf{u}^{n-1})_N + K_1K_2(\omega_0^{(1-\gamma)}\Delta t^\gamma)^2(\delta_x\delta_y\mathbf{u}^n, \delta_x\delta_y\mathbf{u}^{n-1})_N \\
&-\Delta t^\gamma \sum_{k=0}^{n-1} \omega_{n-k}^{(1-\gamma)} \left(K_1(\delta_x\mathbf{u}^k, \delta_x\mathbf{u}^n)_N + K_2(\delta_y\mathbf{u}^k, \delta_y\mathbf{u}^n)_N\right) + \Delta t(\mathbf{f}^n, \mathbf{u}^n)_N \\
\leq&\frac{1}{2}(||\mathbf{u}^n||_N^2 + ||\mathbf{u}^{n-1}||_N^2) + \frac{1}{2}K_1K_2(\omega_0^{(1-\gamma)}\Delta t^\gamma)^2(||\delta_x\delta_y\mathbf{u}^n||_N^2 + ||\delta_x\delta_y\mathbf{u}^{n-1}||_N^2) \\
&+\frac{1}{2}\Delta t^\gamma \sum_{k=0}^{n-1} \omega_{n-k}^{(1-\gamma)} \left[K_1(||\delta_x\mathbf{u}^k||_N^2 + ||\delta_x\mathbf{u}^n||_N^2) + K_2(||\delta_y\mathbf{u}^k||_N^2 + ||\delta_y\mathbf{u}^n||_N^2\right] \\
&+\Delta t(\epsilon||\mathbf{u}^n||_N^2 + \frac{1}{4\epsilon}||\mathbf{f}^n||_N^2).
\end{aligned} \tag{4.248}$$

For simplicity, we denote by

$$E^n = ||\mathbf{u}^n||_N^2 + K_1K_2(b_0\Delta t^\gamma)^2||\delta_x\delta_y\mathbf{u}^n||_N^2 + \Delta t^\gamma \sum_{k=0}^{n} b_{n-k}\left(K_1||\delta_x\mathbf{u}^k||_N^2 + K_2||\delta_y\mathbf{u}^k||_N^2\right),$$

where $b_n = \sum_{k=0}^{n} \omega_k^{(1-\gamma)} = \frac{\Gamma(n+\gamma)}{\Gamma(\gamma)\Gamma(n+1)} = O((n+1)^{\gamma-1})$. Then one has $b_0 = \omega_0^{(1-\gamma)}, \omega_n^{(1-\gamma)} = b_{n-1} - b_n < 0, n > 0$, and

$$E^n + \Delta t^\gamma b_n\left(K_1||\delta_x\mathbf{u}^n||_N^2 + K_2||\delta_y\mathbf{u}^n||_N^2\right) \leq E^{n-1} + \Delta t(2C_2\epsilon||\mathbf{u}^n||_{1,N}^2 + \frac{1}{2\epsilon}||\mathbf{f}^n||_N^2), \tag{4.249}$$

where we have used Lemma 4.5.1. It easy to check that $\Delta t \leq C_1 b_n \Delta t^\gamma$ $(C_1 > 0)$ is independent of n and Δt. Hence, we can choose a suitable $\epsilon = \frac{\min\{K_1, K_2\}}{2C_2C_1}$ such that

$$2\epsilon C_2\Delta t||\mathbf{u}^n||_{1,N}^2 \leq \Delta t^\gamma b_n\left(K_1||\delta_x\mathbf{u}^n||_N^2 + K_2||\delta_y\mathbf{u}^n||_N^2\right).$$

Therefore,

$$E^n \leq E^{n-1} + C\Delta t||\mathbf{f}^n||_N^2 \leq E^0 + C\Delta t\sum_{k=1}^{n} ||\mathbf{f}^k||_N^2, \tag{4.250}$$

which yields the desired result. The proof is completed. □

Let $(\mathbf{e}^n)_{i,j}^n = e_{i,j}^n = U(x_i, y_j, t_n) - u_{i,j}^n$. Then we can get the error equation of the ADI difference method (4.239) as

$$\delta_t e_{i,j}^{n-\frac{1}{2}} + K_1K_2(\omega_0^{(1-\gamma)}\Delta t^\gamma)^2(\delta_x^2\delta_y^2\delta_t e_{i,j}^{n-\frac{1}{2}}) = {}^{GL}\delta_t^{(1-\gamma)}(K_1\delta_x^2 e_{i,j}^n + K_2\delta_y^2 e_{i,j}^n) + R_{i,j}^n,$$

where $(\mathbf{R}^n)_{i,j} = R_{i,j}^n$ is the truncation error satisfying $|R_{i,j}^n| \leq C(\Delta t + \Delta t^{2\gamma} + \Delta x^2 + \Delta y^2)$.

From Theorem 33, we derive

$$\begin{aligned}
||\mathbf{e}^n||_N^2 \leq& ||\mathbf{e}^0||_N^2 + \Delta t^\gamma\left(K_1||\delta_x\mathbf{e}^0||_N^2 + K_2||\delta_y\mathbf{e}^0||_N^2\right) + \Delta t^{2\gamma}||\delta_x\delta_y\mathbf{e}^0||_N^2 + C\max_{0\leq k\leq n_T}||\mathbf{R}^k||_N^2 \\
\leq& C(\Delta t + \Delta t^{2\gamma} + \Delta x^2 + \Delta y^2).
\end{aligned}$$

When $0 < \gamma < 1/2$, the convergence order of the ADI scheme (4.239) is less than one. We can use an extrapolation technique to improve the convergence order in time. Let $u_{i,j}^{n,1}$ be the numerical solution at time level n based on time step Δt. We use the time step $\Delta t/2$ to compute the numerical solution $u_{i,j}^{2n,2}$ at time level $2n$. Then one has

$$U(x_i, y_j, t_n) - u_{i,j}^{n,1}(\Delta t) = C\Delta t^{2\gamma} + O(\Delta t + \Delta x^2 + \Delta y^2), \tag{4.251}$$

$$U(x_i, y_j, t_n) - u_{i,j}^{n,2}(\Delta t) = C(\Delta t/2)^{2\gamma} + O(\Delta t + \Delta x^2 + \Delta y^2). \tag{4.252}$$

Eliminating $\Delta t^{2\gamma}$ from the above two equations yields

$$U(x_i, y_j, t_n) = \frac{u_{i,j}^{n,2} - 2^{-2\gamma} u_{i,j}^{n,1}}{1 - 2^{-2\gamma}} + O(\Delta t + \Delta x^2 + \Delta y^2).$$

Hence, we can use

$$u_{i,j}^{n} = \frac{u_{i,j}^{n,2} - 2^{-2\gamma} u_{i,j}^{n,1}}{1 - 2^{-2\gamma}}$$

as the numerical solution of the ADI scheme (4.239), which has first-order accuracy in time.

Another remedy procedure is to use $-(\omega_0^{(1-\gamma)}\Delta t^{\gamma})^2 L_{\Delta x} L_{\Delta y}(\delta_t U_{i,j}^{n-\frac{1}{2}} - \delta_t U_{i,j}^{n-\frac{3}{2}}) = O(\Delta t^{2\gamma+1})$ to replace the perturbation term $-(\omega_0^{(1-\gamma)}\Delta t^{\gamma})^2 L_{\Delta x} L_{\Delta y}(\delta_t U_{i,j}^{n-\frac{1}{2}}) = O(\Delta t^{2\gamma})$ in (4.238). Thus the term $K_1 K_2 (\omega_0^{(1-\gamma)}\Delta t^{\gamma})^2 (\delta_x^2 \delta_y^2 \delta_t u_{i,j}^{n-\frac{1}{2}})$ in (4.239) is replaced by $K_1 K_2 (\omega_0^{(1-\gamma)}\Delta t^{\gamma})^2 \delta_x^2 \delta_y^2 (\delta_t u_{i,j}^{n-\frac{1}{2}} - \delta_t u_{i,j}^{n-\frac{3}{2}})$ to get the improved ADI algorithm with higher order local accuracy in time.

- **More ADI Algorithms**

We introduce a technique to derive the ADI methods from the non-ADI methods. We know that almost all the numerical methods for the one-dimensional problem (4.10) can be directly extended to a two-dimensional problem (4.211), which has the following form

$$\delta_t u_{i,j}^{n-\frac{1}{2}} = \delta_t^{(1-\gamma)}(L_{\Delta x} + L_{\Delta y}) u_{i,j}^{n} + F_{i,j}^{n} = (L_{\Delta x} + L_{\Delta y}) \delta_t^{(1-\gamma)} u_{i,j}^{n} + F_{i,j}^{n}, \tag{4.253}$$

where $\delta_t^{(1-\gamma)}$ is defined as the form

$$\delta_t^{(1-\gamma)} u_{i,j}^{n} = \frac{1}{\Delta t^{1-\gamma}} \sum_{k=0}^{n} a_{k,n} u_{i,j}^{k}. \tag{4.254}$$

For example, $\delta_t^{(1-\gamma)} = {}_{RL}^{L1}\delta_t^{(1-\gamma)}$ and $F_{i,j}^{n} = f_{i,j}^{n}$ when the L1 method is used to discretize the time fractional derivative in (4.211) (see also (4.49) and (4.4)). If we apply the time discretization technique used in (4.59) to the time discretization of (4.211), we have $F^n_{i,j} = f(x_i, y_j, t_{n-\frac{1}{2}})$ and

$$\delta_t^{(1-\gamma)} u^n_{i,j} = b_0 u^{n-\frac{1}{2}}_{i,j} - \sum_{j=1}^{n-1} (b_{n-1-j} - b_{n-j}) u^{j-\frac{1}{2}}_{i,j} - (b_n - B_n) u^{\frac{1}{2}}_{i,j} - A_n u^0_{i,j},$$

where $A_n = B_n - \frac{\gamma(n+1/2)^{\gamma-1}}{\Gamma(1+\gamma)\Delta t^{1-\gamma}}$, b_n and B_n are defined by

$$b_n = \frac{\Delta t^{\gamma-1}}{\Gamma(1+\gamma)} [(n+1)^\gamma - n^\gamma], \quad B_n = \frac{2\Delta t^{\gamma-1}}{\Gamma(1+\gamma)} [(n+1/2)^\gamma - n^\gamma]. \tag{4.255}$$

We can choose other time discretization techniques such as (4.67), (4.79), (4.83), or (4.84) to derive the corresponding methods, which are not listed here.

Next, we consider how to construct the ADI algorithms from (4.253). From (4.253) and (4.254), we have

$$u^n_{i,j} - u^{n-1}_{i,j} = \Delta t^\gamma \sum_{k=0}^{n} a_{k,n} (L_{\Delta x} + L_{\Delta y}) u^k_{i,j} + \Delta t F^n_{i,j}, \tag{4.256}$$

Rewrite the above equation as

$$\begin{aligned} & u^n_{i,j} - \Delta t^\gamma a_{n,n} (L_{\Delta x} + L_{\Delta y}) u^n_{i,j} + (\Delta t^\gamma a_{n,n})^2 L_{\Delta x} L_{\Delta y} u^n_{i,j} \\ =& u^{n-1}_{i,j} + (\Delta t^\gamma a_{n,n})^2 L_{\Delta x} L_{\Delta y} u^n_{i,j} + \Delta t^\gamma \sum_{k=0}^{n-1} a_{k,n} (L_{\Delta x} + L_{\Delta y}) u^k_{i,j} + \Delta t F^n_{i,j} \\ & + (\Delta t^\gamma a_{n,n})^2 L_{\Delta x} L_{\Delta y} (u^n_{i,j} - u^{n-1}_{i,j}). \end{aligned} \tag{4.257}$$

Dropping the last term $(\Delta t^\gamma a_{n,n})^2 L_{\Delta x} L_{\Delta y} (u^n_{i,j} - u^{n-1}_{i,j})$ in the above equation yields the desired ADI method

$$\begin{aligned} & u^n_{i,j} - \Delta t^\gamma a_{n,n} (L_{\Delta x} + L_{\Delta y}) u^n_{i,j} + (\Delta t^\gamma a_{n,n})^2 L_{\Delta x} L_{\Delta y} u^n_{i,j} \\ =& u^{n-1}_{i,j} + (\Delta t^\gamma a_{n,n})^2 L_{\Delta x} L_{\Delta y} u^n_{i,j} + \Delta t^\gamma \sum_{k=0}^{n-1} a_{k,n} (L_{\Delta x} + L_{\Delta y}) u^k_{i,j} + \Delta t F^n_{i,j}. \end{aligned} \tag{4.258}$$

Rewriting the above equation into the following equivalent form

$$\delta_t u^{n-\frac{1}{2}}_{i,j} + (\Delta t^\gamma a_{n,n})^2 L_{\Delta x} L_{\Delta y} \delta_t u^{n-\frac{1}{2}}_{i,j} = (L_{\Delta x} + L_{\Delta y}) \delta_t^{(1-\gamma)} u^n_{i,j} + F^n_{i,j}. \tag{4.259}$$

One can find the ADI method (4.259) can be derived from the non-ADI method (4.253) by adding the perturbation term $(\Delta t^\gamma a_{n,n})^2 L_{\Delta x} L_{\Delta y} \delta_t u^{n-\frac{1}{2}}_{i,j}$ to the left of (4.253).

Next, we illustrate that the method (4.258) or (4.259) is indeed the ADI method. We also have the following equivalent form of (4.258) as

$$(1 - \Delta t^\gamma a_{n,n} L_{\Delta x})(1 - \Delta t^\gamma a_{n,n} L_{\Delta y}) u^n_{i,j} = (RHS)^n_{i,j}, \tag{4.260}$$

where

$$(RHS)_{i,j}^{n} = u_{i,j}^{n-1} + (\Delta t^{\gamma} a_{n,n})^{2} L_{\Delta x} L_{\Delta y} u_{i,j}^{n} + \Delta t^{\gamma} \sum_{k=0}^{n-1} a_{k,n} (L_{\Delta x} + L_{\Delta y}) u_{i,j}^{k} + \Delta t F_{i,j}^{n}.$$

The scheme (4.260) can be solved by the following two steps:
Stage 1: For each $j, j = 1,2,\cdots,N_y - 1$, solve

$$(1 - \Delta t^{\gamma} a_{n,n} L_{\Delta x}) u_{i,j}^{*} = (RHS)_{i,j}^{n}, \quad i = 1,2,\cdots,N_x - 1, \tag{4.261}$$

to obtain $u_{i,j}^{*}$ with $u_{0,j}^{*} = (1 - \Delta t^{\gamma} a_{n,n} L_{\Delta y}) u_{0,j}^{n}$ and $u_{N_x,j}^{*} = (1 - \Delta t^{\gamma} a_{n,n} L_{\Delta y}) u_{N_x,j}^{n}$.
Stage 2: For each $i, i = 1,2,\cdots,N_x - 1$, solve

$$(1 - \Delta t^{\gamma} a_{n,n} L_{\Delta y}) u_{i,j}^{n} = u_{i,j}^{*}, \quad j = 1,2,\cdots,N_y - 1, \tag{4.262}$$

to obtain $u_{i,j}^{n}$ with $u_{i,0}^{n} = U_c(x_i, t_n)$ and $u_{i,N_y}^{n} = U_d(x_i, t_n)$.

Next, we just list some non-ADI and ADI algorithms for (4.211) as follows:

- **Non-ADI method (1)**: The time is discretized the same as that in (4.49); the space is discretized by the central difference method in (4.217). So the finite difference method for (4.211) is given by: Find $u_{i,j}^{n}$ ($i = 1,2,\cdots,N_x - 1, j = 1,2,\cdots,N_y, n = 1,2,\cdots,n_T$), such that

$$\begin{cases} \delta_t u_{i,j}^{n-\frac{1}{2}} = {}_{RL}^{L1}\delta_t^{(1-\gamma)} (K_1 \delta_x^2 u_{i,j}^{n} + K_2 \delta_y^2 u_{i,j}^{n}) + f_{i,j}^{n}, \\ u_{i,j}^{0} = \phi_0(x_i, y_j), \quad i = 0,1,2,\cdots,N_x, \ j = 0,1,2,\cdots,N_y, \\ u_{0,j}^{n} = U_a(y_j, t_n), \quad u_{N_x,j}^{n} = U_b(y_j, t_n), \quad j = 0,1,2,\cdots,N_y, \\ u_{i,0}^{n} = U_c(x_i, t_n), \quad u_{i,N_y}^{n} = U_d(x_i, t_n), \quad i = 0,1,2,\cdots,N_x, \end{cases} \tag{4.263}$$

  where ${}_{RL}^{L1}\delta_t^{(1-\gamma)}$ is defined by (4.4).
  ADI method (1): From (4.259) and (4.263), we derive the corresponding ADI method for (4.211) as: Find $u_{i,j}^{n}$ ($i = 1,2,\cdots,N_x - 1, j = 1,2,\cdots,N_y, n = 1,2,\cdots,n_T$), such that

$$\begin{cases} \delta_t u_{i,j}^{n-\frac{1}{2}} + (\Delta t^{\gamma} a_{n,n})^2 K_1 K_2 \delta_x^2 \delta_y^2 \delta_t u_{i,j}^{n-\frac{1}{2}} = {}_{RL}^{L1}\delta_t^{(1-\gamma)} (K_1 \delta_x^2 u_{i,j}^{n} + K_2 \delta_y^2 u_{i,j}^{n}) + f_{i,j}^{n}, \\ u_{i,j}^{0} = \phi_0(x_i, y_j), \quad i = 0,1,2,\cdots,N_x, \ j = 0,1,2,\cdots,N_y, \\ u_{0,j}^{n} = U_a(y_j, t_n), \quad u_{N_x,j}^{n} = U_b(y_j, t_n), \quad j = 0,1,2,\cdots,N_y, \\ u_{i,0}^{n} = U_c(x_i, t_n), \quad u_{i,N_y}^{n} = U_d(x_i, t_n), \quad i = 0,1,2,\cdots,N_x, \end{cases} \tag{4.264}$$

  where $a_{n,n} = b_0^{(1-\gamma)} = \frac{1}{\Gamma(1+\gamma)}$, see (4.4).

- **Non-ADI method (2)**: The time is discretized the same as that in (4.59), the space is discretized by the central difference methods as in (4.217). Then the finite difference method for (4.211) is given by: Find $u_{i,j}^n$ $(i=1,2,\cdots,N_x-1, j=1,2,\cdots,N_y, n=1,2,\cdots,n_T)$, such that

$$\begin{cases} \delta_t u_{i,j}^{n-\frac{1}{2}} = \delta_t^{(1-\gamma)}\left(K_1\delta_x^2 u_{i,j}^{n-\frac{1}{2}} + K_2\delta_y^2 u_{i,j}^{n-\frac{1}{2}}\right) + f(x_i,y_j,t_{n-\frac{1}{2}}), \quad n=2,3,\cdots,n_T, \\ u_{i,j}^0 = \phi_0(x_i,y_j), \quad i=0,1,2,\cdots,N_x,\ j=0,1,2,\cdots,N_y, \\ u_{0,j}^n = U_a(y_j,t_n), \quad u_{N_x,j}^n = U_b(y_j,t_n), \quad j=0,1,2,\cdots,N_y, \\ u_{i,0}^n = U_c(x_i,t_n), \quad u_{i,N_y}^n = U_d(x_i,t_n), \quad i=0,1,2,\cdots,N_x, \end{cases} \tag{4.265}$$

where $\delta_t^{(1-\gamma)}$ is defined by (4.57).

ADI method (2): From (4.265) and (4.259), we derive the corresponding ADI method for (4.211) as: Find $u_{i,j}^n$ $(i=1,2,\cdots,N_x-1, j=1,2,\cdots,N_y, n=1,2,\cdots,n_T)$, such that

$$\begin{cases} \delta_t u_{i,j}^{n-\frac{1}{2}} + (\Delta t^\gamma a_{n,n})^2 K_1K_2\delta_x^2\delta_y^2\delta_t u_{i,j}^{n-\frac{1}{2}} = \delta_t^{(1-\gamma)}\left(K_1\delta_x^2 u_{i,j}^{n-\frac{1}{2}} + K_2\delta_y^2 u_{i,j}^{n-\frac{1}{2}}\right) \\ \qquad + f(x_i,y_j,t_{n-\frac{1}{2}}), \quad n=2,3,\cdots,n_T, \\ u_{i,j}^0 = \phi_0(x_i,y_j), \quad i=0,1,2,\cdots,N_x,\ j=0,1,2,\cdots,N_y, \\ u_{0,j}^n = U_a(y_j,t_n), \quad u_{N_x,j}^n = U_b(y_j,t_n), \quad j=0,1,2,\cdots,N_y, \\ u_{i,0}^n = U_c(x_i,t_n), \quad u_{i,N_y}^n = U_d(x_i,t_n), \quad i=0,1,2,\cdots,N_x, \end{cases} \tag{4.266}$$

where $a_{n,n} = \frac{b_0}{2} = \frac{1}{2\Gamma(1+\gamma)}$ for $n>1$ and $a_{n,n} = \frac{1}{2}B_0 = \frac{2^{1-\gamma}}{\Gamma(1+\gamma)}$ for $n=1$, see (4.58).

- **Non-ADI method (3)**: The time is discretized as in (4.83), the space is discretized by the central difference methods as in (4.217). Hence the finite difference method for (4.211) is given by: Find $u_{i,j}^n$ $(i=1,2,\cdots,N_x-1, j=1,2,\cdots,N_y, n=1,2,\cdots,n_T)$, such that

$$\begin{cases} u_{i,j}^n = u_{i,j}^{n-1} + \Delta t^\gamma\Big[\sum_{k=1}^{n} b_{n-k}^{(1-\gamma)}(K_1\delta_x^2 u_{i,j}^k + K_2\delta_y^2 u_{i,j}^k) \\ \qquad - \sum_{k=1}^{n-1} b_{n-k-1}^{(1-\gamma)}(K_1\delta_x^2 u_{i,j}^k + K_2\delta_y^2 u_{i,j}^k)\Big] + \Delta t f_{i,j}^n, \\ u_{i,j}^0 = \phi_0(x_i,y_j), \quad i=0,1,2,\cdots,N_x,\ j=0,1,2,\cdots,N_y, \\ u_{0,j}^n = U_a(y_j,t_n), \quad u_{N_x,j}^n = U_b(y_j,t_n), \quad j=0,1,2,\cdots,N_y, \\ u_{i,0}^n = U_c(x_i,t_n), \quad u_{i,N_y}^n = U_d(x_i,t_n), \quad i=0,1,2,\cdots,N_x, \end{cases} \tag{4.267}$$

where $b_k^{(1-\gamma)} = \frac{1}{\Gamma(1+\gamma)}[(k+1)^\gamma - k^\gamma]$.

ADI method (3): From (4.267) and (4.259), we derive the corresponding ADI method for (4.211) as: Find $u_{i,j}^n$ $(i=1,2,\cdots,N_x-1, j=1,2,\cdots,N_y, n=$

$1,2,\cdots,n_T$), such that

$$\begin{cases} u_{i,j}^{n} + (\Delta t^{\gamma} a_{n,n})^2 K_1 K_2 \delta_x^2 \delta_y^2 \delta_t u_{i,j}^{n-\frac{1}{2}} = u_{i,j}^{n-1} + \Delta t^{\gamma} \Big[\sum_{k=1}^{n} b_{n-k}^{(1-\gamma)} (K_1 \delta_x^2 u_{i,j}^{k} + K_2 \delta_y^2 u_{i,j}^{k}) \\ \quad - \sum_{k=1}^{n-1} b_{n-k-1}^{(1-\gamma)} (K_1 \delta_x^2 u_{i,j}^{k} + K_2 \delta_y^2 u_{i,j}^{k}) \Big] + \Delta t f_{i,j}^{n}, \\ u_{i,j}^{0} = \phi_0(x_i, y_j), \quad i = 0,1,2,\cdots,N_x,\ j = 0,1,2,\cdots,N_y, \\ u_{0,j}^{n} = U_a(y_j, t_n), \quad u_{N_x,j}^{n} = U_b(y_j, t_n), \quad j = 0,1,2,\cdots,N_y, \\ u_{i,0}^{n} = U_c(x_i, t_n), \quad u_{i,N_y}^{n} = U_d(x_i, t_n), \quad i = 0,1,2,\cdots,N_x, \end{cases} \tag{4.268}$$

where $b_k^{(1-\gamma)} = \frac{1}{\Gamma(1+\gamma)}[(k+1)^{\gamma} - k^{\gamma}]$ and $a_{n,n} = b_0^{(1-\gamma)}$.

The stability and convergence of the above three non-ADI methods (4.263), (4.265), and (4.267) are similar to those of their corresponding one-dimensional problems, see (4.49), (4.59), and (4.83). The convergence orders of methods (4.263), (4.265), and (4.267) are $O(\Delta t + \Delta x^2 + \Delta y^2)$, $O(\Delta t^{1+\gamma} + \Delta x^2 + \Delta y^2)$, and $O(\Delta t + \Delta x^2 + \Delta y^2)$, respectively.

The above three ADI methods (4.264), (4.266), and (4.268) are also unconditionally stable, the proofs are similar to those of the first ADI method (4.239), the convergence of which are $O(\Delta t + \Delta t^{2\gamma} + \Delta x^2 + \Delta y^2)$, $O(\Delta t^{1+\gamma} + \Delta t^{2\gamma} + \Delta x^2 + \Delta y^2)$, and $O(\Delta t + \Delta t^{2\gamma} + \Delta x^2 + \Delta y^2)$, respectively.

4.5.2 Time-Fractional Diffusion Equation with Caputo Derivative in Time

In this subsection, we consider the ADI finite difference methods for the following equation

$$\begin{cases} {}_{C}\mathrm{D}_{0,t}^{\gamma} U = K_1 \partial_x^2 U + K_2 \partial_y^2 U + g(x,y,t), \quad (x,y,t) \in (a,b)\times(c,d)\times(0,T], \\ U(x,y,0) = \phi_0(x,y), \quad (x,y) \in (a,b)\times(c,d), \\ U(a,y,t) = U_a(y,t),\ U(b,y,t) = U_b(y,t), \quad (y,t) \in (c,d)\times(0,T], \\ U(x,c,t) = U_c(x,t),\ U(x,d,t) = U_d(x,t), \quad (x,t) \in (a,b)\times(0,T], \end{cases} \tag{4.269}$$

where $K_1, K_2 > 0$ and $0 < \gamma < 1$.

Next, we first consider the time discretization of (4.269). Rewrite (4.269) in the following form

$${}_{C}\mathrm{D}_{0,t}^{\gamma} U = (L_x + L_y)U + g(x,y,t), \tag{4.270}$$

where $L_x U = K_1 \partial_x^2 U$ and $L_y U = K_1 \partial_y^2 U$.

If we use the time discretization in (4.96), (4.97), (4.116), (4.117), (4.120), or (4.121) to discretize (4.270) in time, then at each time level n, the time discretization of (4.270) has the following general form

$$\sum_{k=0}^{n}\alpha_{n,k}U^k = \Delta t^\gamma \sum_{k=0}^{n}\theta_{n,k}(L_x+L_y)U^k + \Delta t^\gamma G^n + \Delta t^\gamma R^n, \tag{4.271}$$

where G^n is related to $g^0, g^1, \cdots, g^n$, and R^n is the truncation error.

Rearranging the above equation gives

$$U^n - \frac{\theta_{n,n}\Delta t^\gamma}{\alpha_{n,n}}(L_x+L_y)U^n = -\sum_{k=0}^{n-1}\frac{\alpha_{n,k}}{\alpha_{n,n}}U^k + \frac{\Delta t^\gamma}{\alpha_{n,n}}\sum_{k=0}^{n-1}\theta_{n,k}(L_x+L_y)U^k + \frac{\Delta t^\gamma G^n}{\alpha_{n,n}} + \frac{\Delta t^\gamma R^n}{\alpha_{n,n}}. \tag{4.272}$$

Adding the perturbation term

$$\left(\frac{\theta_{n,n}\Delta t^\gamma}{\alpha_{n,n}}\right)^2 L_xL_y(U^n - U^{n-1}) = O(\Delta t^{1+2\gamma})$$

to the left-hand side of (4.272) leads to

$$\begin{aligned} &U^n - \frac{\theta_{n,n}\Delta t^\gamma}{\alpha_{n,n}}(L_x+L_y)U^n + \left(\frac{\theta_{n,n}\Delta t^\gamma}{\alpha_{n,n}}\right)^2 L_xL_y(U^n - U^{n-1}) \\ =& -\sum_{k=0}^{n-1}\frac{\alpha_{n,k}}{\alpha_{n,n}}U^k + \frac{\Delta t^\gamma}{\alpha_{n,n}}\sum_{k=0}^{n-1}\theta_{n,k}(L_x+L_y)U^k + \frac{\Delta t^\gamma G^n}{\alpha_{n,n}} + \frac{\Delta t^\gamma R^n}{\alpha_{n,n}} + O(\Delta t^{1+2\gamma}). \end{aligned} \tag{4.273}$$

The above equation is equivalent to the following form

$$\begin{aligned} &\left(1 - \frac{\theta_{n,n}\Delta t^\gamma}{\alpha_{n,n}}L_x\right)\left(1 - \frac{\theta_{n,n}\Delta t^\gamma}{\alpha_{n,n}}L_y\right)U^n \\ =& \left(\frac{\theta_{n,n}\Delta t^\gamma}{\alpha_{n,n}}\right)^2 L_xL_yU^{n-1} - \sum_{k=0}^{n-1}\frac{\alpha_{n,k}}{\alpha_{n,n}}U^k + \frac{\Delta t^\gamma}{\alpha_{n,n}}\sum_{k=0}^{n-1}\theta_{n,k}(L_x+L_y)U^k \\ &+ \frac{\Delta t^\gamma G^n}{\alpha_{n,n}} + \frac{\Delta t^\gamma R^n}{\alpha_{n,n}} + O(\Delta t^{1+2\gamma}). \end{aligned} \tag{4.274}$$

The space derivative of (4.274) is discretized by the central difference, i.e.,

$$\begin{aligned} (L_xU^n)(x_i,y_j) &= L_{\Delta x}U^n_{i,j} + O(\Delta x^2) = K_1\delta_x^2U^n_{i,j} + O(\Delta x^2), \\ (L_yU^n)(x_i,y_j) &= L_{\Delta y}U^n_{i,j} + O(\Delta y^2) = K_2\delta_y^2U^n_{i,j} + O(\Delta x^2), \end{aligned}$$

we can get

$$\begin{aligned} &\left(1 - \frac{\theta_{n,n}\Delta t^\gamma}{\alpha_{n,n}}L_{\Delta x}\right)\left(1 - \frac{\theta_{n,n}\Delta t^\gamma}{\alpha_{n,n}}L_{\Delta y}\right)U^n_{i,j} \\ =& \left(\frac{\theta_{n,n}\Delta t^\gamma}{\alpha_{n,n}}\right)^2 L_{\Delta x}L_{\Delta y}U^{n-1}_{i,j} - \sum_{k=0}^{n-1}\frac{\alpha_{n,k}}{\alpha_{n,n}}U^k_{i,j} + \frac{\Delta t^\gamma}{\alpha_{n,n}}\sum_{k=0}^{n-1}\theta_{n,k}(L_{\Delta x}+L_{\Delta y})U^k_{i,j} \\ &+ \frac{\Delta t^\gamma G^n_{i,j}}{\alpha_{n,n}} + \frac{\Delta t^\gamma R^n_{i,j}}{\alpha_{n,n}} + O(\Delta t^\gamma(\Delta t^{1+\gamma} + \Delta x^2 + \Delta y^2)). \end{aligned} \tag{4.275}$$

Removing the truncation error $\frac{\Delta t^{\gamma} R_{i,j}^{n}}{\alpha_{n,n}} + O(\Delta t^{\gamma}(\Delta t^{1+\gamma} + \Delta x^2 + \Delta y^2))$ and replacing $U_{i,j}^{k}$ by $u_{i,j}^{k}$ in the above equation gives

$$\left(1 - \frac{\theta_{n,n}\Delta t^{\gamma}}{\alpha_{n,n}} L_{\Delta x}\right)\left(1 - \frac{\theta_{n,n}\Delta t^{\gamma}}{\alpha_{n,n}} L_{\Delta y}\right) u_{i,j}^{n} = (RHS)_{i,j}^{n}, \tag{4.276}$$

where

$$\begin{aligned}(RHS)_{i,j}^{n} = &\left(\frac{\theta_{n,n}\Delta t^{\gamma}}{\alpha_{n,n}}\right)^{2} L_{\Delta x} L_{\Delta y} u_{i,j}^{n-1} - \sum_{k=0}^{n-1} \frac{\alpha_{n,k}}{\alpha_{n,n}} u_{i,j}^{k} \\ &+ \frac{\Delta t^{\gamma}}{\alpha_{n,n}} \sum_{k=0}^{n-1} \theta_{n,k}(L_{\Delta x} + L_{\Delta y}) u_{i,j}^{k} + \frac{\Delta t^{\gamma} G_{i,j}^{n}}{\alpha_{n,n}}.\end{aligned}$$

Eq. (4.276) has the following factorization

$$\left(1 - \frac{\theta_{n,n}\Delta t^{\gamma}}{\alpha_{n,n}} L_{\Delta x}\right) u_{i,j}^{*} = (RHS)_{i,j}^{n}, \quad i = 1, 2, \cdots, N_x - 1, \tag{4.277}$$

$$\left(1 - \frac{\theta_{n,n}\Delta t^{\gamma}}{\alpha_{n,n}} L_{\Delta y}\right) u_{i,j}^{n} = u_{i,j}^{*}, \quad j = 1, 2, \cdots, N_y - 1. \tag{4.278}$$

From (4.278), we can obtain that the boundary conditions for $u_{i,j}^{*}$ are taken as

$$u_{0,j}^{*} = \left(1 - \frac{\theta_{n,n}\Delta t^{\gamma}}{\alpha_{n,n}} L_{\Delta y}\right) u_{0,j}^{n}, \quad u_{N_x,j}^{*} = \left(1 - \frac{\theta_{n,n}\Delta t^{\gamma}}{\alpha_{n,n}} L_{\Delta y}\right) u_{N_x,j}^{n}. \tag{4.279}$$

In order to illustrate the relationships between the ADI difference methods (4.276) for two-dimensional subdiffusion equation (4.269) and the finite difference methods for the one-dimensional subdiffusion equation (4.86), (4.276) can be rewritten as below

$$\frac{1}{\Delta t^{\gamma}} \sum_{k=0}^{n} \alpha_{n,k} u_{i,j}^{k} + \frac{\theta_{n,n}^{2}\Delta t^{1+\gamma}}{\alpha_{n,n}} L_{\Delta x} L_{\Delta y} \delta_t u_{i,j}^{n-\frac{1}{2}} = \sum_{k=0}^{n} \theta_{n,k}(L_{\Delta x} + L_{\Delta y}) u_{i,j}^{k} + G_{i,j}^{n}. \tag{4.280}$$

We find that if we remove $\frac{\theta_{n,n}^{2}\Delta t^{1+\gamma}}{\alpha_{n,n}} L_{\Delta x} L_{\Delta y} u_{i,j}^{n-\frac{1}{2}}$ in (4.280), we can get the non-ADI algorithms for (4.269) as follows

$$\frac{1}{\Delta t^{\gamma}} \sum_{k=0}^{n} \alpha_{n,k} u_{i,j}^{k} = \sum_{k=0}^{n} \theta_{n,k}(L_{\Delta x} + L_{\Delta y}) u_{i,j}^{k} + G_{i,j}^{n}. \tag{4.281}$$

Next, we list some non-ADI and ADI algorithms. The non-ADI algorithms for (4.269) can be seen as the direct extensions of the corresponding algorithms for a one-dimensional problem (4.86) (see (4.96), (4.97), (4.116), (4.117), (4.120), and (4.121)).

- **Non-ADI method (1)**: The time is discretized as in (4.96), the non-ADI method for (4.269) is given by: Find $u_{i,j}^n$ $(i = 1,2,\cdots,N_x-1, j = 1,2,\cdots,N_y, n = 1,2,\cdots,n_T)$, such that

$$\frac{1}{\Delta t^\gamma}\sum_{k=0}^{n}\omega_{n-k}^{(\gamma)}(u_{i,j}^k - u_{i,j}^0) = (K_1\delta_x^2 + K_2\delta_y^2)u_{i,j}^n + g_{i,j}^n, \tag{4.282}$$

where $\omega_k^{(\gamma)} = (-1)^k\binom{\gamma}{k}$.
ADI method (1): From (4.280) and (4.282), the corresponding ADI method for (4.269) is given by: Find $u_{i,j}^n$ $(i = 1,2,\cdots,N_x-1, j = 1,2,\cdots,N_y, n = 1,2,\cdots,n_T)$, such that

$$\frac{1}{\Delta t^\gamma}\sum_{k=0}^{n}\omega_k^{(\gamma)}(u_{i,j}^k - u_{i,j}^0) + K_1K_2\Delta t^{1+\gamma}\delta_x^2\delta_y^2\delta_t u_{i,j}^{n-\frac{1}{2}} = (K_1\delta_x^2 + K_2\delta_y^2)u_{i,j}^n + g_{i,j}^n. \tag{4.283}$$

In such a case, $\alpha_{n,n}$ and $\theta_{n,n}$ in (4.281) are chosen as: $\alpha_{n,n} = \omega_0^{(\gamma)} = 1$, $\theta_{n,n} = 1$.

- **Non-ADI method (2)**: The time is discretized as in (4.97), the non-ADI method for (4.269) is given by: Find $u_{i,j}^n$ $(i = 1,2,\cdots,N_x-1, j = 1,2,\cdots,N_y, n = 1,2,\cdots,n_T)$, such that

$$\frac{1}{\Delta t^\gamma}\sum_{k=0}^{n-1}b_{n-k}^{(\gamma)}(u_{i,j}^{k+1} - u_{i,j}^k) = (K_1\delta_x^2 + K_2\delta_y^2)u_{i,j}^n + g_{i,j}^n, \tag{4.284}$$

where $b_k^{(\gamma)} = \frac{1}{\Gamma(2-\gamma)}\left[(k+1)^{1-\gamma} - k^{1-\gamma}\right]$.
ADI method (2): From (4.280) and (4.282), the corresponding ADI method for (4.269) is given by: Find $u_{i,j}^n$ $(i = 1,2,\cdots,N_x-1, j = 1,2,\cdots,N_y, n = 1,2,\cdots,n_T)$, such that

$$\frac{1}{\Delta t^\gamma}\sum_{k=0}^{n-1}b_{n-k}^{(\gamma)}(u_{i,j}^{k+1} - u_{i,j}^k) + \frac{K_1K_2\Delta t^{1+\gamma}}{b_0^{(\gamma)}}\delta_x^2\delta_y^2\delta_t u_{i,j}^{n-\frac{1}{2}} = (K_1\delta_x^2 + K_2\delta_y^2)u_{i,j}^n + g_{i,j}^n. \tag{4.285}$$

In such a case, $\alpha_{n,n}$ and $\theta_{n,n}$ in (4.281) are chosen as: $\alpha_{n,n} = b_0^{(\gamma)}$, $\theta_{n,n} = 1$.

- **Non-ADI method (3)**: The time is discretized as in (4.116), the non-ADI method for (4.269) is given by: Find $u_{i,j}^n$ $(i = 1,2,\cdots,N_x-1, j = 1,2,\cdots,N_y, n = 1,2,\cdots,n_T)$, such that

$$\begin{aligned}\frac{1}{\Delta t^\gamma}\sum_{k=0}^{n}\omega_k^{(\gamma)}(u_{i,j}^{n-k} - u_{i,j}^0) = {}& \frac{1}{2^\gamma}\sum_{k=0}^{n}(-1)^k\omega_k^{(\gamma)}(K_1\delta_x^2 + K_2\delta_y^2)u_{i,j}^{n-k} \\ & + B_n^{(1)}(K_1\delta_x^2 + K_2\delta_y^2)u_{i,j}^0 + \sum_{k=0}^{n}\omega_k^{(\gamma)}G_{i,j}^{n-k},\end{aligned} \tag{4.286}$$

where $\omega_k^{(\gamma)} = (-1)^k\binom{\gamma}{k}$, $G_{i,j}^n = \frac{1}{\Delta t^\gamma}\left[\mathrm{D}_{0,t}^{-\gamma}g(x_i,y_i,t)\right]_{t=t_n}$, and $B_n^{(1)}$ is defined by

(4.114) with $m = 1$.

ADI method (3): From (4.280) and (4.286), the corresponding ADI method for (4.269) is given by: Find $u_{i,j}^n$ ($i = 1,2,\cdots,N_x-1, j = 1,2,\cdots,N_y, n = 1,2,\cdots,n_T$), such that

$$\begin{aligned}&\frac{1}{\Delta t^\gamma}\sum_{k=0}^{n}\omega_k^{(\gamma)}(u_{i,j}^{n-k}-u_{i,j}^0)+2^{-2\gamma}K_1K_2\Delta t^{1+\gamma}\delta_x^2\delta_y^2\delta_t u_{i,j}^{n-\frac{1}{2}}\\&=\frac{1}{2^\gamma}\sum_{k=0}^{n}(-1)^k\omega_k^{(\gamma)}(K_1\delta_x^2+K_2\delta_y^2)u_{i,j}^{n-k}+B_n^{(1)}(K_1\delta_x^2+K_2\delta_y^2)u_{i,j}^0+\sum_{k=0}^{n}\omega_k^{(\gamma)}G_{i,j}^{n-k}.\end{aligned} \tag{4.287}$$

In such a case, $\alpha_{n,n}$ and $\theta_{n,n}$ in (4.281) are chosen as: $\alpha_{n,n} = 1$, $\theta_{n,n} = 2^{-\gamma}$.

- **Non-ADI method (4)**: The time is discretized as in (4.117), the non-ADI method for (4.269) is given by: Find $u_{i,j}^n$ ($i = 1,2,\cdots,N_x-1, j = 1,2,\cdots,N_y, n = 1,2,\cdots,n_T$), such that

$$\begin{aligned}\frac{1}{\Delta t^\gamma}\sum_{k=0}^{n}\omega_k^{(\gamma)}(u_{i,j}^{n-k}-u_{i,j}^0)=&(1-\frac{\gamma}{2})(K_1\delta_x^2+K_2\delta_y^2)u_{i,j}^n+\frac{\gamma}{2}(K_1\delta_x^2+K_2\delta_y^2)u_{i,j}^{n-1}\\&+B_n^{(2)}(K_1\delta_x^2+K_2\delta_y^2)u_{i,j}^0+(1-\frac{\gamma}{2})G_{i,j}^n+\frac{\gamma}{2}G_{i,j}^{n-1},\end{aligned} \tag{4.288}$$

where $\omega_k^{(\gamma)} = (-1)^k\binom{\gamma}{k}$, $G_{i,j}^n = \frac{1}{\Delta t^\gamma}\left[\mathrm{D}_{0,t}^{-\gamma}g(x_i,y_i,t)\right]_{t=t_n}$, and $B_n^{(1)}$ is defined by (4.114) with $m = 2$.

ADI method (4): From (4.280) and (4.288), the corresponding ADI method for (4.269) is given by: Find $u_{i,j}^n$ ($i = 1,2,\cdots,N_x-1, j = 1,2,\cdots,N_y, n = 1,2,\cdots,n_T$), such that

$$\begin{aligned}&\frac{1}{\Delta t^\gamma}\sum_{k=0}^{n}\omega_k^{(\gamma)}(u_{i,j}^{n-k}-u_{i,j}^0)+\left(1-\frac{\gamma}{2}\right)^2K_1K_2\Delta t^{1+\gamma}\delta_x^2\delta_y^2\delta_t u_{i,j}^{n-\frac{1}{2}}\\&=(1-\frac{\gamma}{2})(K_1\delta_x^2+K_2\delta_y^2)u_{i,j}^n+\frac{\gamma}{2}(K_1\delta_x^2+K_2\delta_y^2)u_{i,j}^{n-1}\\&\quad+B_n^{(2)}(K_1\delta_x^2+K_2\delta_y^2)u_{i,j}^0+(1-\frac{\gamma}{2})G_{i,j}^n+\frac{\gamma}{2}G_{i,j}^{n-1}.\end{aligned} \tag{4.289}$$

$\alpha_{n,n}$ and $\theta_{n,n}$ in (4.281) are chosen as: $\alpha_{n,n} = 1$, $\theta_{n,n} = 1-\frac{\gamma}{2}$.

- **Non-ADI method (5)**: The time is discretized as in (4.120), the non-ADI method for (4.269) is given by: Find $u_{i,j}^n$ ($i = 1,2,\cdots,N_x-1, j = 1,2,\cdots,N_y, n = 1,2,\cdots,n_T$), such that

$$\begin{aligned}\frac{1}{\Delta t^\gamma}\sum_{k=0}^{n}\omega_k^{(\gamma)}(u_{i,j}^{n-k}-u_{i,j}^0)=&\frac{1}{2^\gamma}\sum_{k=0}^{n}(-1)^k\omega_k^{(\gamma)}(K_1\delta_x^2+K_2\delta_y^2)u_{i,j}^{n-k}\\&+B_n^{(1)}(K_1\delta_x^2+K_2\delta_y^2)u_{i,j}^0+C_n^{(1)}(K_1\delta_x^2+K_2\delta_y^2)(u_{i,j}^1-u_{i,j}^0)+\sum_{k=0}^{n}\omega_k^{(\gamma)}G_{i,j}^{n-k},\end{aligned} \tag{4.290}$$

where $\omega_k^{(\gamma)} = (-1)^k\binom{\gamma}{k}$, $G_{i,j}^n = \frac{1}{\Delta t^\gamma}\left[\mathrm{D}_{0,t}^{-\gamma}g(x_i,y_i,t)\right]_{t=t_n}$, $B_n^{(1)}$ is defined by (4.114) with $m=1$, and $C_n^{(1)}$ is defined by (4.119) with $m=1$.

ADI method (5): From (4.280) and (4.290), the corresponding ADI method for (4.269) is given by: Find $u_{i,j}^n$ $(i=1,2,\cdots,N_x-1, j=1,2,\cdots,N_y, n=1,2,\cdots,n_T)$, such that

$$\begin{aligned}
&\frac{1}{\Delta t^\gamma}\sum_{k=0}^{n}\omega_k^{(\gamma)}(u_{i,j}^{n-k}-u_{i,j}^0)+(2^{-\gamma}+\delta_{1n}C_n^{(1)})^2K_1K_2\Delta t^{1+\gamma}\delta_x^2\delta_y^2\delta_t u_{i,j}^{n-\frac{1}{2}}\\
=&\frac{1}{2^\gamma}\sum_{k=0}^{n}(-1)^k\omega_k^{(\gamma)}(K_1\delta_x^2+K_2\delta_y^2)u_{i,j}^{n-k}+B_n^{(1)}(K_1\delta_x^2+K_2\delta_y^2)u_{i,j}^0\\
&+C_n^{(1)}(K_1\delta_x^2+K_2\delta_y^2)(u_{i,j}^1-u_{i,j}^0)+\sum_{k=0}^{n}\omega_k^{(\gamma)}G_{i,j}^{n-k},
\end{aligned}\tag{4.291}$$

where δ_{1n} is the Kronecker delta, i.e., $\delta_{1n}=1$ if $n=1$ and $\delta_{1n}=0$ if $n>1$. In this case, $\alpha_{n,n}$ and $\theta_{n,n}$ in (4.281) are chosen as: $\alpha_{n,n}=1$, $\theta_{n,n}=2^{-\gamma}$ for $n>1$. If $n=1$, then $\theta_{n,n}=2^{-\gamma}+C_1^{(1)}$.

- **Non-ADI method (6)**: The time is discretized as in (4.121), the non-ADI method for (4.269) is given by: Find $u_{i,j}^n$ $(i=1,2,\cdots,N_x-1, j=1,2,\cdots,N_y, n=1,2,\cdots,n_T)$, such that

$$\begin{aligned}
\frac{1}{\Delta t^\gamma}\sum_{k=0}^{n}\omega_k^{(\gamma)}(u_{i,j}^{n-k}-u_{i,j}^0)=&(1-\frac{\gamma}{2})(K_1\delta_x^2+K_2\delta_y^2)u_{i,j}^n+\frac{\gamma}{2}(K_1\delta_x^2+K_2\delta_y^2)u_{i,j}^{n-1}\\
&+B_n^{(2)}(K_1\delta_x^2+K_2\delta_y^2)u_{i,j}^0+C_n^{(2)}(K_1\delta_x^2+K_2\delta_y^2)(u_{i,j}^1-u_{i,j}^0)\\
&+(1-\frac{\gamma}{2})G_{i,j}^n+\frac{\gamma}{2}G_{i,j}^{n-1},
\end{aligned}\tag{4.292}$$

where $\omega_k^{(\gamma)} = (-1)^k\binom{\gamma}{k}$, $G_{i,j}^n = \frac{1}{\Delta t^\gamma}\left[\mathrm{D}_{0,t}^{-\gamma}g(x_i,y_i,t)\right]_{t=t_n}$, and $B_n^{(2)}$ is defined by (4.114) with $m=2$, $C_n^{(2)}$ is defined by (4.119) with $m=2$.

ADI method (6): From (4.280) and (4.292), the corresponding ADI method for (4.269) is given by: Find $u_{i,j}^n$ $(i=1,2,\cdots,N_x-1, j=1,2,\cdots,N_y, n=1,2,\cdots,n_T)$, such that

$$\begin{aligned}
&\frac{1}{\Delta t^\gamma}\sum_{k=0}^{n}\omega_k^{(\gamma)}(u_{i,j}^{n-k}-u_{i,j}^0)+\left(1-\frac{\gamma}{2}+\delta_{1n}C_n^{(1)}\right)^2K_1K_2\Delta t^{1+\gamma}\delta_x^2\delta_y^2\delta_t u_{i,j}^{n-\frac{1}{2}}\\
=&(1-\frac{\gamma}{2})(K_1\delta_x^2+K_2\delta_y^2)u_{i,j}^n+\frac{\gamma}{2}(K_1\delta_x^2+K_2\delta_y^2)u_{i,j}^{n-1}+B_n^{(2)}(K_1\delta_x^2+K_2\delta_y^2)u_{i,j}^0\\
&+C_n^{(2)}(K_1\delta_x^2+K_2\delta_y^2)(u_{i,j}^1-u_{i,j}^0)+(1-\frac{\gamma}{2})G_{i,j}^n+\frac{\gamma}{2}G_{i,j}^{n-1}.
\end{aligned}\tag{4.293}$$

In such a case, $\alpha_{n,n}$ and $\theta_{n,n}$ in (4.281) are chosen as: $\alpha_{n,n}=1$, $\theta_{n,n}=1-\frac{\gamma}{2}$. If $n=1$, then $\theta_{n,n}=1-\frac{\gamma}{2}+C_1^{(2)}$.

The initial and boundary conditions for the ADI finite difference methods (4.282)-(4.293) are taken as

$$\begin{cases} u_{i,j}^0 = \phi_0(x_i, y_j), \quad i = 0,1,2,\cdots,N_x,\ j = 0,1,2,\cdots,N_y, \\ u_{0,j}^n = U_a(y_j,t_n), \quad u_{N_x,j}^n = U_b(y_j,t_n), \quad j = 0,1,2,\cdots,N_y, \\ u_{i,0}^n = U_c(x_i,t_n), \quad u_{i,N_y}^n = U_d(x_i,t_n), \quad i = 0,1,2,\cdots,N_x. \end{cases} \tag{4.294}$$

4.5.3 Space-Fractional Diffusion Equation

Next, we study the ADI finite difference methods for the following *space-fractional differential equation*

$$\begin{cases} \partial_t U = d_+(x,y,t)_{RL}\mathrm{D}_{a,x}^{\alpha} U + d_-(x,y,t)_{RL}\mathrm{D}_{x,b}^{\alpha} U \\ \qquad + c_+(x,y,t)_{RL}\mathrm{D}_{c,y}^{\alpha} U + c_-(x,y,t)_{RL}\mathrm{D}_{y,d}^{\alpha} U + g(x,y,t), \\ \qquad (x,y,t) \in (a,b)\times(c,d)\times(0,T], \\ U(x,y,0) = \phi_0(x,y), \quad (x,y) \in (a,b)\times(c,d), \\ U(a,y,t) = U(b,y,t) = 0, \quad (y,t) \in (c,d)\times(0,T], \\ U(x,c,t) = U(x,d,t) = 0, \quad (x,t) \in (a,b)\times(0,T], \end{cases} \tag{4.295}$$

where $1 < \alpha < 2$ and $c_+(x,y,t), c_-(x,y,t), d_+(x,y,t), d_-(x,y,t) \geq 0$.

For simplicity, we introduce the notations

$$L_x^{(\alpha)} = d_+(x,y,t)_{RL}\mathrm{D}_{a,x}^{\alpha} + d_-(x,y,t)_{RL}\mathrm{D}_{x,b}^{\alpha}, \tag{4.296}$$

$$L_y^{(\alpha)} = c_+(x,y,t)_{RL}\mathrm{D}_{c,y}^{\alpha} + c_-(x,y,t)_{RL}\mathrm{D}_{y,d}^{\alpha}. \tag{4.297}$$

Then Eq. (4.295) can be written as

$$\partial_t U = (L_x^{(\alpha)} + L_y^{(\alpha)})U + g(x,y,t). \tag{4.298}$$

For simplicity, we denote by

$$L_x^{(\alpha,k)} = d_+(x,y,t_k)_{RL}\mathrm{D}_{a,x}^{\alpha} + d_-(x,y,t_k)_{RL}\mathrm{D}_{x,b}^{\alpha}, \quad k = n, n-\frac{1}{2},$$

$$L_y^{(\alpha,k)} = c_+(x,y,t_k)_{RL}\mathrm{D}_{c,y}^{\alpha} + c_-(x,y,t_k)_{RL}\mathrm{D}_{y,d}^{\alpha}, \quad k = n, n-\frac{1}{2}.$$

Letting $t = t_{n-\frac{1}{2}}$ in (4.298) yields

$$\partial_t U(t_{n-\frac{1}{2}}) = (L_x^{(\alpha,n-\frac{1}{2})} + L_y^{(\alpha,n-\frac{1}{2})})U(t_{n-\frac{1}{2}}) + g(x,y,t_{n-\frac{1}{2}}). \tag{4.299}$$

Similar to (4.229), we can derive

$$\begin{aligned}(1-\frac{\Delta t}{2}L_x^{(\alpha,n-\frac{1}{2})})(1-\frac{\Delta t}{2}L_y^{(\alpha,n-\frac{1}{2})})U^n =& (1+\frac{\Delta t}{2}L_x^{(\alpha,n-\frac{1}{2})})(1+\frac{\Delta t}{2}L_y^{(\alpha,n-\frac{1}{2})})U^{n-1} \\ &+ \Delta t g(x,y,t_{n-\frac{1}{2}}) + O(\Delta t^3).\end{aligned} \tag{4.300}$$

Next, we consider the discretization in space. Similar to (4.184), we can also introduce the following operators

$$L_{\Delta x,q}^{(\alpha,k)}U_{i,j}^{n}=\begin{cases} d_{+}(x_i,y_j,t_k)_L\delta_x^{(\alpha)}U_{i+1,j}^{n}+d_{-}(x_i,y_j,t_k)_R\delta_x^{(\alpha)}U_{i-1,j}^{n}, & q=1,\\ d_{+}(x_i,y_j,t_k)_L\delta_x^{(\alpha,1)}U_{i,j}^{n}+d_{-}(x_i,y_j,t_k)_R\delta_x^{(\alpha,1)}U_{i,j}^{n}, & q=2,\\ d_{+}(x_i,y_j,t_k)_L\delta_x^{(\alpha,2)}U_{i,j}^{n}+d_{-}(x_i,y_j,t_k)_R\delta_x^{(\alpha,2)}U_{i,j}^{n}, & q=3,\\ {}_{RZ}\delta_x^{(\alpha)}U_{i,j}^{n}, \quad q=4,\ d_{+}=d_{-}=c_{+}=c_{-}=-\dfrac{1}{2\cos(\alpha\pi/2)}=c_\alpha, \end{cases} \tag{4.301}$$

where

$${}_L\delta_x^{(\alpha)}U_{i,j}^{n}=\frac{1}{\Delta x^\alpha}\sum_{k=0}^{i}\omega_k^{(\alpha)}U_{i-k,j},\quad {}_R\delta_x^{(\alpha)}U_{i,j}=\frac{1}{\Delta x^\alpha}\sum_{k=0}^{N_x-i}\omega_k^{(\alpha)}U_{i+k,j},\quad \omega_k^{(\alpha)}=(-1)^k\binom{\alpha}{k},$$

$${}_L\delta_x^{(\alpha,1)}U_{i,j}^{n}=\frac{\alpha}{2}{}_L\delta_x^{(\alpha)}U_{i+1,j}^{n}+\frac{2-\alpha}{2}{}_L\delta_x^{(\alpha)}U_{i,j}^{n},$$

$${}_R\delta_x^{(\alpha,1)}U_{i,j}^{n}=\frac{\alpha}{2}{}_R\delta_x^{(\alpha)}U_{i-1,j}^{n}+\frac{2-\alpha}{2}{}_R\delta_x^{(\alpha)}U_{i,j}^{n},$$

$${}_L\delta_x^{(\alpha,2)}U_{i,j}^{n}=\frac{2+\alpha}{4}{}_L\delta_x^{(\alpha)}U_{i+1,j}^{n}+\frac{2-\alpha}{4}{}_L\delta_x^{(\alpha)}U_{i-1,j}^{n},$$

$${}_R\delta_x^{(\alpha,2)}U_{i,j}^{n}=\frac{2+\alpha}{4}{}_R\delta_x^{(\alpha)}U_{i-1,j}^{n}+\frac{2-\alpha}{4}{}_R\delta_x^{(\alpha)}U_{i+1,j}^{n},$$

$${}_{RZ}\delta_x^{(\alpha)}U_{i,j}^{n}=-\frac{1}{\Delta x^\alpha}\sum_{k=-N_x+i}^{i}g_k u_{i-k}^{n},\quad g_k=\frac{(-1)^k\Gamma(\alpha+1)}{\Gamma(\alpha/2-k+1)\Gamma(\alpha/2+k+1)}.$$

We can similarly define the operator $L_{\Delta y,q}^{(\alpha,k)}, q=1,2,3,4$.

From (4.185), we can similarly have

$$L_{\Delta x,q}^{(\alpha,k)}U_{i,j}^{n}-(L_x^{(\alpha,k)}U^n)(x_i,y_j)=O(\Delta x^p),\quad L_{\Delta y,q}^{(\alpha,k)}U_{i,j}^{n}-(L_y^{(\alpha,k)}U^n)(x_i,y_j)=O(\Delta y^p),$$

where p is given as in (4.186)

Replacing $L_x^{(\alpha)}$, $L_y^{(\alpha)}$, and U^n in (4.300) with $L_{\Delta x,q}^{(\alpha,n-\frac{1}{2})}$ and $L_{\Delta y,q}^{(\alpha,n-\frac{1}{2})}$, and $U_{i,j}^{n}$, respectively, we have

$$\begin{aligned}(1-\frac{\Delta t}{2}L_{\Delta x,q}^{(\alpha,n-\frac{1}{2})})(1-\frac{\Delta t}{2}L_{\Delta y,q}^{(\alpha,n-\frac{1}{2})})U_{i,j}^{n}=&(1+\frac{\Delta t}{2}L_{\Delta x,q}^{(\alpha,n-\frac{1}{2})})(1+\frac{\Delta t}{2}L_{\Delta y,q}^{(\alpha,n-\frac{1}{2})})U_{i,j}^{n-1}\\ &+\Delta t g(x_i,y_j,t_{n-\frac{1}{2}})+\Delta t R_{i,j}^{n-\frac{1}{2}},\end{aligned} \tag{4.302}$$

where $R_{i,j}^{n-\frac{1}{2}}$ is the truncation error satisfying $|R_{i,j}^{n-\frac{1}{2}}|\le C(\Delta t^2+\Delta x+\Delta y)$ when $q=1$, and $|R_{i,j}^{n-\frac{1}{2}}|\le C(\Delta t^2+\Delta x^2+\Delta y^2)$ when $q=2,3,4$.

Neglecting the truncation error $R_{i,j}^{n}$ in (4.302) and replacing $U_{i,j}^{n}$ with $u_{i,j}^{n}$, we have

$$\begin{aligned}(1-\frac{\Delta t}{2}L_{\Delta x,q}^{(\alpha,n-\frac{1}{2})})(1-\frac{\Delta t}{2}L_{\Delta y,q}^{(\alpha,n-\frac{1}{2})})u_{i,j}^{n} =&(1+\frac{\Delta t}{2}L_{\Delta x,q}^{(\alpha,n-\frac{1}{2})})(1+\frac{\Delta t}{2}L_{\Delta y,q}^{(\alpha,n-\frac{1}{2})})u_{i,j}^{n-1}\\ &+\Delta t g(x_i,y_j,t_{n-\frac{1}{2}}).\end{aligned} \tag{4.303}$$

Similar to (4.233)-(4.234) and (4.236)-(4.237), we have two ways to solve (4.303).

PR factorization:

$$(1-\frac{\Delta t}{2}L_{\Delta x,q}^{(\alpha,n-\frac{1}{2})})u_{i,j}^{*} = (1+\frac{\Delta t}{2}L_{\Delta y,q}^{(\alpha,n-\frac{1}{2})})u_{i,j}^{n-1}+\frac{\Delta t}{2}g(x_i,y_j,t_{n-\frac{1}{2}}), \tag{4.304}$$

$$(1-\frac{\Delta t}{2}L_{\Delta y,q}^{(\alpha,n-\frac{1}{2})})u_{i,j}^{n} = (1+\frac{\Delta t}{2}L_{\Delta x,q}^{(\alpha,n-\frac{1}{2})})u_{i,j}^{*}+\frac{\Delta t}{2}g(x_i,y_j,t_{n-\frac{1}{2}}). \tag{4.305}$$

From (4.304) and (4.305), we can also have

$$u_{i,j}^{*} = u_{i,j}^{n-\frac{1}{2}} - \frac{\Delta t^2}{4}L_{\Delta x,q}^{(\alpha,n-\frac{1}{2})}L_{\Delta y,q}^{(\alpha,n-\frac{1}{2})}\delta_t u_{i,j}^{n-\frac{1}{2}}. \tag{4.306}$$

So the boundary conditions for (4.304) are given by

$$u_{0,j}^{*} = u_{0,j}^{n-\frac{1}{2}} - \frac{\Delta t^2}{4}L_{\Delta x,q}^{(\alpha,n-\frac{1}{2})}L_{\Delta y,q}^{(\alpha,n)}\delta_t u_{0,j}^{n-\frac{1}{2}}, \tag{4.307}$$

$$u_{N_x,j}^{*} = u_{N_x,j}^{n-\frac{1}{2}} - \frac{\Delta t^2}{4}L_{\Delta x,q}^{(\alpha,n-\frac{1}{2})}L_{\Delta y,q}^{(\alpha,n)}\delta_t u_{N_x,j}^{n-\frac{1}{2}}. \tag{4.308}$$

D'Yakonov factorization:

$$\begin{aligned}(1-\frac{\Delta t}{2}L_{\Delta x,q}^{(\alpha,n-\frac{1}{2})})u_{i,j}^{*} =& (1+\frac{\Delta t}{2}L_{\Delta x,q}^{(\alpha,n-\frac{1}{2})})(1+\frac{\Delta t}{2}L_{\Delta y,q}^{(\alpha,n-\frac{1}{2})})u_{i,j}^{n-1}\\ &+\Delta t g(x_i,y_j,t_{n-\frac{1}{2}}),\end{aligned} \tag{4.309}$$

$$(1-\frac{\Delta t}{2}L_{\Delta y,q}^{(\alpha,n-\frac{1}{2})})u_{i,j}^{n} = u_{i,j}^{*}. \tag{4.310}$$

So the boundary conditions for (4.309) are given by

$$u_{0,j}^{*} = (1-\frac{\Delta t}{2}L_{\Delta y,q}^{(\alpha,n-\frac{1}{2})})u_{0,j}^{n}, \quad u_{N_x,j}^{*} = (1-\frac{\Delta t}{2}L_{\Delta y,q}^{(\alpha,n-\frac{1}{2})})u_{N_x,j}^{n}. \tag{4.311}$$

Next, we consider the stability of (4.303). We first consider the case of $q=1$ in (4.303). For simplicity, we also suppose that $d_+ = d_- = K_1$, $c_+ = c_- = K_2$, K_1 and K_2 are positive constants, and $u_{0,j}^n = u_{N_x,j}^n = u_{i,0}^n = u_{i,N_y}^n = 0$. Let $\underline{\mathbf{u}}^n$ be defined as in (4.219) and $\underline{\mathbf{g}}^{n-\frac{1}{2}} \in \mathbb{R}^{(N_x-1)\times(N_y-1)}$ with $(\underline{\mathbf{g}}^{n-\frac{1}{2}})_{i-1,j-1} = g(x_i,y_j,t_{n-\frac{1}{2}})$, $i=1,2,\cdots,N_x-1$, $j=1,2,\cdots,N_y-1$.

Then, for $q=1,2,3,4$, we always have the matrix representation of (4.303) as

$$\begin{aligned}&(E_{N_x-1}+\mu_1 S_{N_x-1}^{(\alpha)})\underline{\mathbf{u}}^n(E_{N_y-1}+\mu_2 S_{N_y-1}^{(\alpha)})^T\\ =&(E_{N_x-1}-\mu_1 S_{N_x-1}^{(\alpha)})\underline{\mathbf{u}}^{n-1}(E_{N_y-1}-\mu_2 S_{N_y-1}^{(\alpha)})^T+\Delta t\underline{\mathbf{g}}^{n-\frac{1}{2}},\end{aligned} \tag{4.312}$$

where $\mu_1 = \frac{K_1\Delta t}{2\Delta x^\alpha}$, $\mu_2 = \frac{K_2\Delta t}{2\Delta y^\alpha}$, E_N is an $N\times N$ identity matrix, $S^{(\alpha)}_{N-1}$ is a symmetric positive definite matrix. For example, if $q = 1$, then $S^{(\alpha)}_{N-1} = S + S^T$, where S is defined by

$$S = -\begin{bmatrix} \omega_1^{(\alpha)} & \omega_0^{(\alpha)} & 0 & \cdots & 0 \\ \omega_2^{(\alpha)} & \omega_1^{(\alpha)} & \omega_0^{(\alpha)} & \cdots & 0 \\ \vdots & \vdots & \vdots & \ddots & \vdots \\ \omega_{N-2}^{(\alpha)} & \omega_{N-3}^{(\alpha)} & \omega_{N-4}^{(\alpha)} & \cdots & \omega_0^{(\alpha)} \\ \omega_{N-1}^{(\alpha)} & \omega_{N-2}^{(\alpha)} & \omega_{N-3}^{(\alpha)} & \cdots & \omega_1^{(\alpha)} \end{bmatrix}_{(N-1)\times(N-1)}, \omega_k^{(\alpha)} = (-1)^k\binom{\alpha}{k}. \quad (4.313)$$

Let $A = (E_{N_x-1} + \mu_1 S^{(\alpha)}_{N_x-1})^{-1}(E_{N_x-1} - \mu_1 S^{(\alpha)}_{N_x-1})$ and $B = (E_{N_y-1} + \mu_2 S^{(\alpha)}_{N_y-1})^{-1}$ $(E_{N_y-1} - \mu_2 S^{(\alpha)}_{N_y-1})$. Then A and B are symmetric matrices with spectral radius $\rho(A^k) < 1$ and $\rho(B^k) < 1$. We can rewrite (4.312) into

$$\begin{aligned} \underline{\mathbf{u}}^n =& A\underline{\mathbf{u}}^{n-1}B + \Delta t A\underline{\mathbf{g}}^{n-\frac{1}{2}}B = A^2\underline{\mathbf{u}}^{n-2}B^2 + \Delta t A^2\underline{\mathbf{g}}^{n-\frac{3}{2}}B^2 + \Delta t A\underline{\mathbf{g}}^{n-\frac{1}{2}}B \\ =& A^n\underline{\mathbf{u}}^0B^n + \Delta t\sum_{k=1}^{n} A^k\underline{\mathbf{g}}^{n+\frac{1}{2}-k}B^k. \end{aligned} \quad (4.314)$$

As in (4.220), we also have

$$\mathtt{vec}(\underline{\mathbf{u}}^n) = (B^n\otimes A^n)\mathtt{vec}(\underline{\mathbf{u}}^0) + \Delta t\sum_{k=1}^{n}(B^k\otimes A^k)\mathtt{vec}(\underline{\mathbf{g}}^{n+\frac{1}{2}-k}). \quad (4.315)$$

It immediately follows from (4.315) that

$$||\mathtt{vec}(\mathbf{u}^n)||^2 \le ||\mathtt{vec}(\mathbf{u}^0)||^2 + \Delta t\sum_{k=1}^{n}||\mathtt{vec}(\mathbf{g}^{n+\frac{1}{2}-k})||^2, \quad (4.316)$$

where $(\mathbf{u}^n)_{i,j} = u^n_{i,j}$ and $(\mathbf{g}^{n-\frac{1}{2}})_{i,j} = f(x_i, y_j, t_{n-\frac{1}{2}})$.

Therefore, the ADI method (4.303) is unconditionally stable.

Let $(\mathbf{e}^n)_{i,j} = U(x_i, y_j, t_n) = u^n_{i,j}$ and $(\mathbf{R}^{n-\frac{1}{2}})_{i,j} = R^{n-\frac{1}{2}}_{i,j} = O(\Delta t^2 + \Delta x^p + \Delta y^p)$. Then one has

$$||\mathtt{vec}(\mathbf{e}^n)||^2 \le ||\mathtt{vec}(\mathbf{e}^0)||^2 + \Delta t\sum_{k=1}^{n}||\mathtt{vec}(\mathbf{R}^{n+\frac{1}{2}-k})||^2 \le C(\Delta t^2 + \Delta x^p + \Delta y^p). \quad (4.317)$$

4.5.4 Time-Space Fractional Diffusion Equation with Caputo Derivative in Time

This subsection considers the non-ADI and ADI finite difference methods for a two-dimensional *time-space fractional diffusion equation* in the following form

$$\begin{cases} {}_C\mathrm{D}_{0,t}^{\gamma}U = d_+(x,y,t)_{RL}\mathrm{D}_{a,x}^{\alpha}U + d_-(x,y,t)_{RL}\mathrm{D}_{x,b}^{\alpha}U \\ \qquad + c_+(x,y,t)_{RL}\mathrm{D}_{c,y}^{\alpha}U + c_-(x,y,t)_{RL}\mathrm{D}_{y,d}^{\alpha}U + g(x,y,t), \\ \qquad (x,y,t) \in (a,b)\times(c,d)\times(0,T], \\ U(x,y,0) = \phi_0(x,y), \quad (x,y) \in (a,b)\times(c,d), \\ U(a,y,t) = U(b,y,t) = 0, \quad (y,t) \in (c,d)\times(0,T], \\ U(x,c,t) = U(x,d,t) = 0, \quad (x,t) \in (a,b)\times(0,T], \end{cases} \tag{4.318}$$

where $0 < \gamma \leq 1, 1 < \alpha < 2$ and $c_+(x,y,t), c_-(x,y,t), d_+(x,y,t), d_-(x,y,t) \geq 0$.

Let $L_x^{(\alpha)} = d_+(x,y,t)_{RL}\mathrm{D}_{a,x}^{\alpha} + d_-(x,y,t)_{RL}\mathrm{D}_{x,b}^{\alpha}$ and $L_y^{(\alpha)} = c_+(x,y,t)_{RL}\mathrm{D}_{c,y}^{\alpha} + c_-(x,y,t)_{RL}\mathrm{D}_{y,d}^{\alpha}$. Then Eq. (4.318) can be written as

$${}_C\mathrm{D}_{0,t}^{\gamma}U = (L_x^{(\alpha)} + L_y^{(\alpha)})U + g(x,y,t). \tag{4.319}$$

The time in (4.319) is discretized as in Eq. (4.269) (see (4.276) or (4.280)), the space derivatives are discretized as those in Eq. (4.295) (see the space discretization $L_{\Delta x,q}^{(\alpha,n)}$ defined by (4.301) and $L_{\Delta y,q}^{(\alpha,n)}$)), we can obtain the ADI finite difference schemes for (4.318) as

$$\frac{1}{\Delta t^{\gamma}}\sum_{k=0}^{n}\alpha_{n,k}u_{i,j}^{k} + \frac{\theta_{n,n}^2\Delta t^{1+\gamma}}{\alpha_{n,n}}L_{\Delta x,q}^{(\alpha,n)}L_{\Delta y,q}^{(\alpha,n)}\delta_t u_{i,j}^{n-\frac{1}{2}} = \sum_{k=0}^{n}\theta_{n,k}(L_{\Delta x,q}^{(\alpha,k)} + L_{\Delta y,q}^{(\alpha,k)})u_{i,j}^{k} + G_{i,j}^{n}. \tag{4.320}$$

Eq. (4.320) is equivalent to the following form

$$\left(1 - \frac{\theta_{n,n}\Delta t^{\gamma}}{\alpha_{n,n}}L_{\Delta x,q}^{(\alpha,n)}\right)\left(1 - \frac{\theta_{n,n}\Delta t^{\gamma}}{\alpha_{n,n}}L_{\Delta y,q}^{(\alpha,n)}\right)u_{i,j}^{n} = (RHS)_{i,j}^{n}, \tag{4.321}$$

where

$$\begin{aligned}(RHS)_{i,j}^{n} =& \left(\frac{\theta_{n,k}\Delta t^{\gamma}}{\alpha_{n,n}}\right)^2 L_{\Delta x,q}^{(\alpha,n)}L_{\Delta y,q}^{(\alpha,n)}u_{i,j}^{n-1} - \sum_{k=0}^{n-1}\frac{\alpha_{n,k}}{\alpha_{n,n}}u_{i,j}^{k} \\ &+ \frac{\Delta t^{\gamma}}{\alpha_{n,n}}\sum_{k=0}^{n-1}\theta_{n,k}(L_{\Delta x,q}^{(\alpha,n)} + L_{\Delta y,q}^{(\alpha,n)})u_{i,j}^{k} + \frac{\Delta t^{\gamma}G_{i,j}^{n}}{\alpha_{n,n}}.\end{aligned}$$

Eq. (4.321) has the following factorization

$$\left(1 - \frac{\theta_{n,n}\Delta t^{\gamma}}{\alpha_{n,n}}L_{\Delta x,q}^{(\alpha,n)}\right)u_{i,j}^{*} = (RHS)_{i,j}^{n}, \quad i = 1,2,\cdots,N_x - 1, \tag{4.322}$$

$$\left(1 - \frac{\theta_{n,n}\Delta t^{\gamma}}{\alpha_{n,n}}L_{\Delta y,q}^{(\alpha,n)}\right)u_{i,j}^{n} = u_{i,j}^{*}, \quad j = 1,2,\cdots,N_y - 1. \tag{4.323}$$

Similarly to (4.282)–(4.293), we can also derive the corresponding non-ADI and ADI algorithms for (4.318) below.

- **Non-ADI method (1)**: The time derivative is discretized as in (4.282), the non-ADI method for (4.318) is given by: Find $u_{i,j}^n$ $(i = 1,2,\cdots,N_x-1, j = 1,2,\cdots,N_y, n = 1,2,\cdots,n_T)$, such that

$$\frac{1}{\Delta t^{\gamma}}\sum_{k=0}^{n}\omega_{n-k}^{(\gamma)}(u_{i,j}^k - u_{i,j}^0) = (L_{\Delta x,q}^{(\alpha,n)} + L_{\Delta y,q}^{(\alpha,n)})u_{i,j}^n + g_{i,j}^n, \tag{4.324}$$

where $\omega_k^{(\gamma)} = (-1)^k\binom{\gamma}{k}$.
ADI method (1): From (4.320) and (4.324), the corresponding ADI method for (4.318) is given by: Find $u_{i,j}^n$ $(i = 1,2,\cdots,N_x-1, j = 1,2,\cdots,N_y, n = 1,2,\cdots,n_T)$, such that

$$\frac{1}{\Delta t^{\gamma}}\sum_{k=0}^{n}\omega_{k}^{(\gamma)}(u_{i,j}^k - u_{i,j}^0) + \frac{\Delta t^{1+\gamma}}{b_0^{(\gamma)}}L_{\Delta x,q}^{(\alpha,n)}L_{\Delta y,q}^{(\alpha,n)}\delta_t u_{i,j}^{n-\frac{1}{2}} = (L_{\Delta x,q}^{(\alpha,n)} + L_{\Delta y,q}^{(\alpha,n)})u_{i,j}^n + g_{i,j}^n. \tag{4.325}$$

In such a case, $\alpha_{n,n}$ and $\theta_{n,n}$ in (4.320) are chosen as: $\alpha_{n,n} = \omega_0^{(\gamma)} = 1$, $\theta_{n,n} = 1$.

- **Non-ADI method (2)**: The time derivative is discretized as same as that in (4.284), the non-ADI method for (4.318) is given by: Find $u_{i,j}^n$ $(i = 1,2,\cdots,N_x-1, j = 1,2,\cdots,N_y, n = 1,2,\cdots,n_T)$, such that

$$\frac{1}{\Delta t^{\gamma}}\sum_{k=0}^{n-1}b_{n-k}^{(\gamma)}(u_{i,j}^{k+1} - u_{i,j}^k) = (L_{\Delta x,q}^{(\alpha,n)} + L_{\Delta y,q}^{(\alpha,n)})u_{i,j}^n + g_{i,j}^n, \tag{4.326}$$

where $b_k^{(\gamma)} = \frac{1}{\Gamma(2-\gamma)}\left[(k+1)^{1-\gamma} - k^{1-\gamma}\right]$.
ADI method (2): From (4.320) and (4.326), the corresponding ADI method for (4.318) is given by: Find $u_{i,j}^n$ $(i = 1,2,\cdots,N_x-1, j = 1,2,\cdots,N_y, n = 1,2,\cdots,n_T)$, such that

$$\frac{1}{\Delta t^{\gamma}}\sum_{k=0}^{n-1}b_{n-k}^{(\gamma)}(u_{i,j}^{k+1} - u_{i,j}^k) + \frac{\Delta t^{1+\gamma}}{b_0^{(\gamma)}}L_{\Delta x,q}^{(\alpha,n)}L_{\Delta y,q}^{(\alpha,n)}\delta_t u_{i,j}^{n-\frac{1}{2}} = (L_{\Delta x,q}^{(\alpha,k)} + L_{\Delta y,q}^{(\alpha,k)})u_{i,j}^n + g_{i,j}^n. \tag{4.327}$$

In such a case, $\alpha_{n,n}$ and $\theta_{n,n}$ in (4.320) are chosen as: $\alpha_{n,n} = b_0^{(\gamma)}$, $\theta_{n,n} = 1$.

- **Non-ADI method (3)**: The time is discretized as in (4.286), the non-ADI method for (4.318) is given by: Find $u_{i,j}^n$ $(i = 1,2,\cdots,N_x-1, j = 1,2,\cdots,N_y, n = 1,2,\cdots,n_T)$, such that

$$\begin{aligned}\frac{1}{\Delta t^{\gamma}}\sum_{k=0}^{n}\omega_{k}^{(\gamma)}(u_{i,j}^{n-k} - u_{i,j}^0) &= \frac{1}{2^{\gamma}}\sum_{k=0}^{n}(-1)^k\omega_k^{(\gamma)}(L_{\Delta x,q}^{(\alpha,n-k)} + L_{\Delta y,q}^{(\alpha,n-k)})u_{i,j}^{n-k} \\ &\quad + B_n^{(1)}(L_{\Delta x,q}^{(\alpha,0)} + L_{\Delta y,q}^{(\alpha,0)})u_{i,j}^0 + \sum_{k=0}^{n}\omega_k^{(\gamma)}G_{i,j}^{n-k},\end{aligned} \tag{4.328}$$

where $\omega_k^{(\gamma)} = (-1)^k\binom{\gamma}{k}$, $G_{i,j}^n = \frac{1}{\Delta t^{\gamma}}\left[\mathrm{D}_{0,t}^{-\gamma}g(x_i,y_i,t)\right]_{t=t_n}$, and $B_n^{(1)}$ is defined by

(4.114) with $m = 1$.

ADI method (3): From (4.320) and (4.328), the corresponding ADI method for (4.318) is given by: Find $u_{i,j}^n$ ($i = 1,2,\cdots,N_x-1, j = 1,2,\cdots,N_y, n = 1,2,\cdots,n_T$), such that

$$\frac{1}{\Delta t^\gamma}\sum_{k=0}^{n}\omega_k^{(\gamma)}(u_{i,j}^{n-k}-u_{i,j}^0)+2^{-2\gamma}\Delta t^{1+\gamma}L_{\Delta x,q}^{(\alpha,n)}L_{\Delta y,q}^{(\alpha,n)}\delta_t u_{i,j}^{n-\frac{1}{2}}$$
$$=\frac{1}{2^\gamma}\sum_{k=0}^{n}(-1)^k\omega_k^{(\gamma)}(L_{\Delta x,q}^{(\alpha,n-k)}+L_{\Delta y,q}^{(\alpha,n-k)})u_{i,j}^{n-k}+B_n^{(1)}(L_{\Delta x,q}^{(\alpha,0)}+L_{\Delta y,q}^{(\alpha,0)})u_{i,j}^0+\sum_{k=0}^{n}\omega_k^{(\gamma)}G_{i,j}^{n-k}, \tag{4.329}$$

$\alpha_{n,n}$ and $\theta_{n,n}$ in (4.320) are chosen as: $\alpha_{n,n} = 1$, $\theta_{n,n} = 2^{-\gamma}$.

- **Non-ADI method (4)**: The time derivative is discretized as in (4.288), the non-ADI method for (4.318) is given by: Find $u_{i,j}^n$ ($i = 1,2,\cdots,N_x-1, j = 1,2,\cdots,N_y, n = 1,2,\cdots,n_T$), such that

$$\frac{1}{\Delta t^\gamma}\sum_{k=0}^{n}\omega_k^{(\gamma)}(u_{i,j}^{n-k}-u_{i,j}^0)=(1-\frac{\gamma}{2})(L_{\Delta x,q}^{(\alpha,n)}+L_{\Delta y,q}^{(\alpha,n)})u_{i,j}^n+\frac{\gamma}{2}(L_{\Delta x,q}^{(\alpha,n-1)}+L_{\Delta y,q}^{(\alpha,n-1)})u_{i,j}^{n-1}$$
$$+B_n^{(2)}(L_{\Delta x,q}^{(\alpha,0)}+L_{\Delta y,q}^{(\alpha,0)})u_{i,j}^0+(1-\frac{\gamma}{2})G_{i,j}^n+\frac{\gamma}{2}G_{i,j}^{n-1}, \tag{4.330}$$

where $\omega_k^{(\gamma)} = (-1)^k\binom{\gamma}{k}$, $G_{i,j}^n = \frac{1}{\Delta t^\gamma}\left[\mathrm{D}_{0,t}^{-\gamma}g(x_i,y_i,t)\right]_{t=t_n}$, and $B_n^{(1)}$ is defined by (4.114) with $m = 2$.

ADI method (4): From (4.320) and (4.330), the corresponding ADI method for (4.318) is given by: Find $u_{i,j}^n$ ($i = 1,2,\cdots,N_x-1, j = 1,2,\cdots,N_y, n = 1,2,\cdots,n_T$), such that

$$\frac{1}{\Delta t^\gamma}\sum_{k=0}^{n}\omega_k^{(\gamma)}(u_{i,j}^{n-k}-u_{i,j}^0)+\left(1-\frac{\gamma}{2}\right)^2\Delta t^{1+\gamma}(L_{\Delta x,q}^{(\alpha,n)}L_{\Delta y,q}^{(\alpha,n)})\delta_t u_{i,j}^{n-\frac{1}{2}}$$
$$=(1-\frac{\gamma}{2})(L_{\Delta x,q}^{(\alpha,n)}+L_{\Delta y,q}^{(\alpha,n)})u_{i,j}^n+\frac{\gamma}{2}(L_{\Delta x,q}^{(\alpha,n-1)}+L_{\Delta y,q}^{(\alpha,n-1)})u_{i,j}^{n-1} \tag{4.331}$$
$$+B_n^{(2)}(L_{\Delta x,q}^{(\alpha,n)}L_{\Delta y,q}^{(\alpha,n)})u_{i,j}^0+(1-\frac{\gamma}{2})G_{i,j}^n+\frac{\gamma}{2}G_{i,j}^{n-1}.$$

In this case, $\alpha_{n,n}$ and $\theta_{n,n}$ in (4.320) are chosen as: $\alpha_{n,n} = 1$, $\theta_{n,n} = 1-\frac{\gamma}{2}$.

- **Non-ADI method (5)**: The time derivative is discretized as in (4.290), the non-ADI method for (4.318) is given by: Find $u_{i,j}^n$ ($i = 1,2,\cdots,N_x-1, j = 1,2,\cdots,N_y, n = 1,2,\cdots,n_T$), such that

$$\frac{1}{\Delta t^\gamma}\sum_{k=0}^{n}\omega_k^{(\gamma)}(u_{i,j}^{n-k}-u_{i,j}^0)=\frac{1}{2^\gamma}\sum_{k=0}^{n}(-1)^k\omega_k^{(\gamma)}(L_{\Delta x,q}^{(\alpha,n-k)}+L_{\Delta y,q}^{(\alpha,n-k)})u_{i,j}^{n-k}$$
$$+B_n^{(1)}(L_{\Delta x,q}^{(\alpha,0)}L_{\Delta y,q}^{(\alpha,0)})u_{i,j}^0+C_n^{(1)}\left(L_{\Delta x,q}^{(\alpha,1)}L_{\Delta y,q}^{(\alpha,1)}u_{i,j}^1-L_{\Delta x,q}^{(\alpha,0)}L_{\Delta y,q}^{(\alpha,0)}u_{i,j}^0\right) \tag{4.332}$$
$$+\sum_{k=0}^{n}\omega_k^{(\gamma)}G_{i,j}^{n-k},$$

where $\omega_k^{(\gamma)} = (-1)^k\binom{\gamma}{k}$, $G_{i,j}^n = \frac{1}{\Delta t^\gamma}\left[D_{0,t}^{-\gamma}g(x_i,y_i,t)\right]_{t=t_n}$, $B_n^{(1)}$ is defined by (4.114) with $m=1$, and $C_n^{(1)}$ is defined by (4.119) with $m=1$.

ADI method (5): From (4.320) and (4.332), the corresponding ADI method for (4.318) is given by: Find $u_{i,j}^n$ $(i=1,2,\cdots,N_x-1, j=1,2,\cdots,N_y, n=1,2,\cdots,n_T)$, such that

$$\begin{aligned}&\frac{1}{\Delta t^\gamma}\sum_{k=0}^{n}\omega_k^{(\gamma)}(u_{i,j}^{n-k}-u_{i,j}^0)+2^{-2\gamma}\Delta t^{1+\gamma}(L_{\Delta x,q}^{(\alpha,n)}L_{\Delta y,q}^{(\alpha,n)})\delta_t u_{i,j}^{n-\frac{1}{2}}\\ =&\frac{1}{2^\gamma}\sum_{k=0}^{n}(-1)^k\omega_k^{(\gamma)}(L_{\Delta x,q}^{(\alpha,n-k)}+L_{\Delta y,q}^{(\alpha,n-k)})u_{i,j}^{n-k}+B_n^{(1)}(L_{\Delta x,q}^{(\alpha,0)}L_{\Delta y,q}^{(\alpha,0)})u_{i,j}^0\\ &+C_n^{(1)}\left(L_{\Delta x,q}^{(\alpha,1)}L_{\Delta y,q}^{(\alpha,1)}u_{i,j}^1-L_{\Delta x,q}^{(\alpha,0)}L_{\Delta y,q}^{(\alpha,0)}u_{i,j}^0\right)+\sum_{k=0}^{n}\omega_k^{(\gamma)}G_{i,j}^{n-k}.\end{aligned}\tag{4.333}$$

In this situation, $\alpha_{n,n}$ and $\theta_{n,n}$ in (4.320) are chosen as: $\alpha_{n,n}=1$, $\theta_{n,n}=2^{-\gamma}$ for $n>1$. If $n=1$, then $\theta_{n,n}=2^{-\gamma}+C_1^{(1)}$.

- **Non-ADI method (6)**: The time derivative is discretized as in (4.292), the non-ADI method for (4.318) is given by: Find $u_{i,j}^n$ $(i=1,2,\cdots,N_x-1, j=1,2,\cdots,N_y, n=1,2,\cdots,n_T)$, such that

$$\begin{aligned}\frac{1}{\Delta t^\gamma}\sum_{k=0}^{n}\omega_k^{(\gamma)}(u_{i,j}^{n-k}-u_{i,j}^0)=&(1-\frac{\gamma}{2})(L_{\Delta x,q}^{(\alpha,n)}+L_{\Delta y,q}^{(\alpha,n)})u_{i,j}^n+\frac{\gamma}{2}(L_{\Delta x,q}^{(\alpha,n-1)}+L_{\Delta y,q}^{(\alpha,n-1)})u_{i,j}^{n-1}\\ &+B_n^{(2)}(L_{\Delta x,q}^{(\alpha,0)}+L_{\Delta y,q}^{(\alpha,0)})u_{i,j}^0+C_n^{(2)}\left(L_{\Delta x,q}^{(\alpha,1)}L_{\Delta y,q}^{(\alpha,1)}u_{i,j}^1-L_{\Delta x,q}^{(\alpha,0)}L_{\Delta y,q}^{(\alpha,0)}u_{i,j}^0\right)\\ &+(1-\frac{\gamma}{2})G_{i,j}^n+\frac{\gamma}{2}G_{i,j}^{n-1},\end{aligned}\tag{4.334}$$

where $\omega_k^{(\gamma)} = (-1)^k\binom{\gamma}{k}$, $G_{i,j}^n = \frac{1}{\Delta t^\gamma}\left[D_{0,t}^{-\gamma}g(x_i,y_i,t)\right]_{t=t_n}$, and $B_n^{(2)}$ is defined by (4.114) with $m=2$, $C_n^{(2)}$ is defined by (4.119) with $m=2$.

ADI method (6): From (4.320) and (4.334), the corresponding ADI method for (4.318) is given by: Find $u_{i,j}^n$ $(i=1,2,\cdots,N_x-1, j=1,2,\cdots,N_y, n=1,2,\cdots,n_T)$, such that

$$\begin{aligned}&\frac{1}{\Delta t^\gamma}\sum_{k=0}^{n}\omega_k^{(\gamma)}(u_{i,j}^{n-k}-u_{i,j}^0)+\left(1-\frac{\gamma}{2}\right)^2\Delta t^{1+\gamma}L_{\Delta x,q}^{(\alpha,n)}L_{\Delta y,q}^{(\alpha,n)}\delta_t u_{i,j}^{n-\frac{1}{2}}\\ =&(1-\frac{\gamma}{2})(L_{\Delta x,q}^{(\alpha,n)}+L_{\Delta y,q}^{(\alpha,n)})u_{i,j}^n+\frac{\gamma}{2}(L_{\Delta x,q}^{(\alpha,n-1)}+L_{\Delta y,q}^{(\alpha,n-1)})u_{i,j}^{n-1}+B_n^{(2)}(L_{\Delta x,q}^{(\alpha,0)}+L_{\Delta y,q}^{(\alpha,0)})u_{i,j}^0\\ &+C_n^{(2)}\left(L_{\Delta x,q}^{(\alpha,1)}L_{\Delta y,q}^{(\alpha,1)}u_{i,j}^1-L_{\Delta x,q}^{(\alpha,0)}L_{\Delta y,q}^{(\alpha,0)}u_{i,j}^0\right)+(1-\frac{\gamma}{2})G_{i,j}^n+\frac{\gamma}{2}G_{i,j}^{n-1}.\end{aligned}\tag{4.335}$$

In such a case, $\alpha_{n,n}$ and $\theta_{n,n}$ in (4.320) are chosen as: $\alpha_{n,n}=1$, $\theta_{n,n}=1-\frac{\gamma}{2}$. If $n=1$, then $\theta_{n,n}=1-\frac{\gamma}{2}+C_1^{(2)}$.

If $d_+(x,y,t)=d_-(x,y,t)=K_1$ and $c_+(x,y,t)=c_-(x,y,t)=K_2$, K_1 and K_2 are positive constants, then the non-ADI methods (4.324), (4.326), (4.328), (4.330), (4.332), and (4.334) are unconditionally stable, which are the same for the corresponding ADI methods (4.325), (4.327), (4.329), (4.331), (4.333), and (4.335).

The convergence of the above methods (4.324)–(4.335) in space are of order $O(\Delta x^p + \Delta y^p)$. The convergence for the non-ADI methods (4.324), (4.326), (4.328), (4.330), (4.332), and (4.334) are of order $O(\Delta t)$, $O(\Delta t^{2-\gamma})$, $O(\Delta t)$, $O(\Delta t)$, $O(\Delta t^2)$, and $O(\Delta t^2)$, respectively. For the ADI methods (4.325), (4.327), (4.329), (4.331), (4.333), and (4.335), the convergence orders in time are $O(\Delta t)$, $O(\Delta t^{2-\gamma} + \Delta t^{1+\gamma})$, $O(\Delta t)$, $O(\Delta t)$, $O(\Delta t^{1+\gamma})$, and $O(\Delta t^{1+\gamma})$, respectively.

4.5.5 Time-Space Fractional Diffusion Equation with Riemann–Liouville Derivative in Time

In this subsection, we study the following two-dimensional diffusion equation

$$\begin{cases} \partial_t U = {}_{RL}\mathrm{D}_{0,t}^{1-\gamma}\left(L_x^{(\alpha)} + L_y^{(\alpha)}\right)U + g(x,y,t), & (x,y,t) \in (a,b)\times(c,d)\times(0,T], \\ U(x,y,0) = \phi_0(x,y), & (x,y) \in (a,b)\times(c,d), \\ U(a,y) = U(b,y) = 0, & (y,t) \in (c,d)\times(0,T], \\ U(x,c) = U(x,d) = 0, & (x,t) \in (a,b)\times(0,T], \end{cases} \tag{4.336}$$

where $L_x^{(\alpha)} = d_+(x,y)_{RL}\mathrm{D}_{a,x}^{\alpha} + d_-(x,y)_{RL}\mathrm{D}_{x,b}^{\alpha}$, $L_y^{(\alpha)} = c_+(x,y)_{RL}\mathrm{D}_{c,y}^{\alpha} + c_-(x,y)_{RL}\mathrm{D}_{y,d}^{\alpha}$, $0 < \gamma \le 1$, and $1 < \alpha < 2$,

We can apply the time discretization techniques for (4.224) to the time discretization of (4.336), while the space derivatives can be approximated as those in (4.318).

Next, we introduce the Crank–Nicolson type method used in (4.265) to the time discretization of (4.336) with the space discretized as in (4.318), which yields the Crank–Nicolson type non-ADI method for (4.336) as: Find $u_{i,j}^n$ ($i = 1,2,\cdots,N_x - 1, j = 1,2,\cdots,N_y, n = 1,2,\cdots,n_T$), such that

$$\begin{cases} \delta_t u_{i,j}^{n-\frac{1}{2}} = \delta_t^{(1-\gamma)}(L_{\Delta x,q}^{(\alpha,0)} + L_{\Delta y,q}^{(\alpha,0)})u_{i,j}^{n-\frac{1}{2}} + f(x_i,y_j,t_{n-\frac{1}{2}}), & n = 2,3,\cdots,n_T, \\ u_{i,j}^0 = \phi_0(x_i,y_j), \quad i = 0,1,2,\cdots,N_x, \ j = 0,1,2,\cdots,N_y, \\ u_{0,j}^n = u_{N_x,j}^n = 0, \quad j = 0,1,2,\cdots,N_y, \\ u_{i,0}^n = u_{i,N_y}^n = 0, \quad i = 0,1,2,\cdots,N_x, \end{cases} \tag{4.337}$$

where $L_{\Delta x,q}^{(\alpha,0)}$ and $L_{\Delta y,q}^{(\alpha,0)}$ are defined as in (4.303) with $d_\pm(x,y,t) = d_\pm(x,y)$ and $c_\pm(x,y,t) = c_\pm(x,y)$, and $\delta_t^{(1-\gamma)}$ is defined by

$$\delta_t^{(1-\gamma)} u_{i,j}^{n-\frac{1}{2}} = \frac{1}{\Delta t^{1-\gamma}}\left[b_0 u_{i,j}^{n-\frac{1}{2}} - \sum_{k=1}^{n-1}(b_{n-1-k} - b_{n-k})u_{i,j}^{k-\frac{1}{2}} - (b_{n-1} - B_{n-1})u_{i,j}^{\frac{1}{2}} - A_{n-1}u_{i,j}^0\right],$$

in which $A_n = B_n - \frac{\gamma(n+1/2)^{\gamma-1}}{\Gamma(1+\gamma)}$, b_n and B_n are defined by

$$b_n = \frac{1}{\Gamma(1+\gamma)}[(n+1)^\gamma - n^\gamma], \quad B_n = \frac{2}{\Gamma(1+\gamma)}[(n+1/2)^\gamma - n^\gamma].$$

From (4.259), (4.266) and (4.337), we can obtain the corresponding ADI method for (4.336) as: Find $u_{i,j}^n$ $(i=1,2,\cdots,N_x-1, j=1,2,\cdots,N_y, n=1,2,\cdots,n_T)$, such that

$$\begin{cases}\delta_t u_{i,j}^{n-\frac{1}{2}}+(a_{n,n}\Delta t^{\gamma})^2 L_{\Delta x,q}^{(\alpha,0)}L_{\Delta y,q}^{(\alpha,0)}\delta_t u_{i,j}^{n-\frac{1}{2}}=\delta_t^{(1-\gamma)}(L_{\Delta x,q}^{(\alpha,0)}+L_{\Delta y,q}^{(\alpha,0)})u_{i,j}^{n-\frac{1}{2}}+f(x_i,y_j,t_{n-\frac{1}{2}}),\\ u_{i,j}^0=\phi_0(x_i,y_j),\quad i=0,1,2,\cdots,N_x,\ j=0,1,2,\cdots,N_y,\\ u_{0,j}^n=u_{N_x,j}^n=0,\quad j=0,1,2,\cdots,N_y,\\ u_{i,0}^n=u_{i,N_y}^n=0,\quad i=0,1,2,\cdots,N_x,\end{cases} \tag{4.338}$$

where $a_{n,n}=\frac{b_0}{2}=\frac{1}{2\Gamma(1+\gamma)}$ for $n>1$ and $a_{n,n}=\frac{1}{2}B_0=\frac{2^{1-\gamma}}{\Gamma(1+\gamma)}$ for $n=1$, see also (4.266).

We give the matrix representation of the ADI method (4.338) in the case of $d_{\pm}(x,y,t)=K_1$ and $c_{\pm}(x,y,t)=K_2$ and $q=1$ as follows

$$\begin{aligned}&\underline{\mathbf{u}}^n-\underline{\mathbf{u}}^{n-1}+a_{n,n}^2\mu_1\mu_2 S_{N_x-1}^{(\alpha)}(\underline{\mathbf{u}}^n-\underline{\mathbf{u}}^{n-1})S_{N_y-1}^{(\alpha)}\\ =&-b_0\left(\mu_1 S_{N_x-1}^{(\alpha)}\underline{\mathbf{u}}^{n-\frac{1}{2}}+\mu_2\underline{\mathbf{u}}^{n-\frac{1}{2}}S_{N_y-1}^{(\alpha)}\right)\\ &+\sum_{k=1}^{n-1}(b_{n-1-k}-b_{n-k})\left(\mu_1 S_{N_x-1}^{(\alpha)}\underline{\mathbf{u}}^{k-\frac{1}{2}}+\mu_2\underline{\mathbf{u}}^{k-\frac{1}{2}}S_{N_y-1}^{(\alpha)}\right)\\ &+(b_{n-1}-B_{n-1})\left(\mu_1 S_{N_x-1}^{(\alpha)}\underline{\mathbf{u}}^{\frac{1}{2}}+\mu_2\underline{\mathbf{u}}^{\frac{1}{2}}S_{N_y-1}^{(\alpha)}\right)+A_{n-1}\left(\mu_1 S_{N_x-1}^{(\alpha)}\underline{\mathbf{u}}^{0}+\mu_2\underline{\mathbf{u}}^{0}S_{N_y-1}^{(\alpha)}\right),\end{aligned} \tag{4.339}$$

where $S_N^{(\alpha)}$ is defined as in (4.312), E_N is an identity matrix, $\mu_1=\frac{K_1\Delta t^{\gamma}}{\Delta x^{\alpha}}$, and $\mu_2=\frac{K_2\Delta t^{\gamma}}{\Delta y^{\alpha}}$.

The above equation is equivalent to the following form

$$\begin{aligned}&\left(E_{N_x-1}+\frac{\mu_1 b_0}{2}S_{N_x-1}^{(\alpha)}\right)\underline{\mathbf{u}}^n\left(E_{N_y-1}+\frac{\mu_1 b_0}{2}S_{N_y-1}^{(\alpha)}\right)\\ =&\left(E_{N_x-1}-\frac{\mu_1 b_0}{2}S_{N_x-1}^{(\alpha)}\right)\underline{\mathbf{u}}^{n-1}\left(E_{N_y-1}-\frac{\mu_1 b_0}{2}S_{N_y-1}^{(\alpha)}\right)+RHS^n,\quad n>1,\end{aligned} \tag{4.340}$$

where

$$\begin{aligned}RHS^n=&\sum_{k=1}^{n-1}(b_{n-1-k}-b_{n-k})\left(\mu_1 S_{N_x-1}^{(\alpha)}\underline{\mathbf{u}}^{k-\frac{1}{2}}+\mu_2\underline{\mathbf{u}}^{k-\frac{1}{2}}S_{N_y-1}^{(\alpha)}\right)\\ &+(b_{n-1}-B_{n-1})\left(\mu_1 S_{N_x-1}^{(\alpha)}\underline{\mathbf{u}}^{\frac{1}{2}}+\mu_2\underline{\mathbf{u}}^{\frac{1}{2}}S_{N_y-1}^{(\alpha)}\right)+A_{n-1}\left(\mu_1 S_{N_x-1}^{(\alpha)}\underline{\mathbf{u}}^{0}+\mu_2\underline{\mathbf{u}}^{0}S_{N_y-1}^{(\alpha)}\right).\end{aligned}$$

For $n=1$, one has from (4.339) with $a_{n,n}=B_0$

$$\begin{aligned}&\left(E_{N_x-1}+\frac{\mu_1 B_0}{2}S_{N_x-1}^{(\alpha)}\right)\underline{\mathbf{u}}^1\left(E_{N_y-1}+\frac{\mu_1 B_0}{2}S_{N_y-1}^{(\alpha)}\right)\\ =&\left(E_{N_x-1}-\frac{\mu_1 B_0}{2}S_{N_x-1}^{(\alpha)}\right)\underline{\mathbf{u}}^{0}\left(E_{N_y-1}-\frac{\mu_1 B_0}{2}S_{N_y-1}^{(\alpha)}\right)+A_0\left(\mu_1 S_{N_x-1}^{(\alpha)}\underline{\mathbf{u}}^{0}+\mu_2\underline{\mathbf{u}}^{0}S_{N_y-1}^{(\alpha)}\right).\end{aligned} \tag{4.341}$$

Clearly, the linear system (4.340) can be solved by the following two steps:

- Solve $\left(E_{N_x-1}+\frac{\mu_1 b_0}{2}S^{(\alpha)}_{N_x-1}\right)\mathbf{u}^*=\left(E_{N_x-1}-\frac{\mu_1 b_0}{2}S^{(\alpha)}_{N_x-1}\right)\underline{\mathbf{u}}^{n-1}\left(E_{N_y-1}-\frac{\mu_1 b_0}{2}S^{(\alpha)}_{N_y-1}\right)+RHS^n$ to obtain $\mathbf{u}^*$;
- Solve $\left(E_{N_y-1}+\frac{\mu_1 B_0}{2}S^{(\alpha)}_{N_y-1}\right)(\underline{\mathbf{u}}^n)^T=(\mathbf{u}^*)^T$ to obtain $\underline{\mathbf{u}}^n$.

It is easy to prove that the non-ADI method (4.337) and ADI method (4.338) are unconditionally stable when $d_\pm(x,y)=K_1$ and $c_\pm(x,y)=K_2$. From (4.301) and (4.56) we can derive that the truncation errors of the two methods (4.337) and (4.338) are $O(\Delta t^{1+\gamma}+\Delta x^p+\Delta y^p)$ and $O(\Delta t^{1+\gamma}+\Delta t^{2\gamma}+\Delta x^p+\Delta y^p)$, respectively.

If the time derivative is discretized as that of (4.30), (4.49), (4.67), or (4.83), we can also get the corresponding non-ADI and ADI methods, which are unconditionally stable too.

4.5.6 Numerical Examples

Example 9 *Consider the following two-dimensional time-fractional equation*

$$\begin{cases} {}_C\mathrm{D}^{\gamma}_{0,t}U=\Delta U+g(x,y,t), \quad (x,y,t)\in(0,1)\times(0,1)\times(0,1], \quad 0<\gamma<1,\\ U(x,y,0)=\sin(x+y), \quad (x,y)\in[0,1]\times[0,1],\\ U(0,y,t)=(t^2+t+1)\sin(y),\ U(1,y,t)=(t^2+t+1)\sin(1+y), \quad (y,t)\in(0,1)\times(0,1],\\ U(x,0,t)=(t^2+t+1)\sin(x),\ U(x,1,t)=(t^2+t+1)\sin(1+x), \quad (x,t)\in(0,1)\times(0,1]. \end{cases} \tag{4.342}$$

Choose the suitable right-hand side function $g(x,y,t)$ such that Eq. (4.342) *has the analytical solution $U(x,y,t)=(t^2+t+1)\sin(x+y)$.*

Here, we just test the Non-ADI methods (4.290) and (4.292), and ADI methods (4.291) and (4.293) for solving (4.342); the L^2 errors at $t=1$ are shown in Table 4.24. It is found that the numerical results are in line with the theoretical analysis.

Example 10 *Consider the following two-dimensional time-space fractional subdiffusion equation*

$$\begin{cases} {}_C\mathrm{D}^{\gamma}_{0,t}U=({}_{RL}\mathrm{D}^{\alpha}_{0,x}+{}_{RL}D^{\alpha}_{x,1})U+({}_{RL}\mathrm{D}^{\alpha}_{0,y}+{}_{RL}\mathrm{D}^{\alpha}_{y,1})U+g(x,y,t), \quad 0<\gamma<1,\\ \qquad (x,y,t)\in(0,1)\times(0,1)\times(0,1],\\ U(x,y,0)=x^4(1-x)^4y^4(1-y)^4, \quad (x,y)\in[0,1]\times[0,1],\\ U(0,y,t)=U(1,y,t)=0, \quad (y,t)\in(0,1)\times(0,1],\\ U(x,0,t)=U(x,1,t)=0, \quad (x,t)\in(0,1)\times(0,1]. \end{cases} \tag{4.343}$$

Choose the suitable right-hand side function $g(x,y,t)$ such that Eq. (4.343) *has the analytical solution $U(x,y,t)=(t^{2.5}+t+1)x^4(1-x)^4y^4(1-y)^4$.*

In this example, we simply use the ADI methods (4.333) and (4.335) to solve (4.343), the L^2 errors are shown in Tables 4.25 and 4.26, where the convergence rates in space are also displayed, which is in line with the theoretical analysis.

TABLE 4.24: The L^2 error at $t = 1$ for Example 9, $N_x = N_y = 400$.

Method	$1/\Delta t$	$\gamma = 0.2$	order	$\gamma = 0.5$	order	$\gamma = 0.8$	order
	8	6.7379e−5		1.0877e−4		6.2100e−5	
Non-ADI	16	1.7622e−5	1.934	2.6686e−5	2.027	1.5295e−5	2.021
(4.290)	32	4.5357e−6	1.958	6.6719e−6	1.999	3.9165e−6	1.965
	64	1.2120e−6	1.904	1.7409e−6	1.938	1.0570e−6	1.889
	8	4.3721e−3		1.7653e−3		6.4651e−4	
ADI	16	2.2882e−3	0.934	6.8562e−4	1.364	1.9513e−4	1.728
(4.291)	32	1.0662e−3	1.101	2.5169e−4	1.445	5.7258e−5	1.768
	64	4.7767e−4	1.158	9.0521e−5	1.475	1.6593e−5	1.786
	8	2.0417e−5		3.2626e−5		1.7588e−5	
Non-ADI	16	6.6979e−6	1.608	9.6653e−6	1.755	4.9231e−6	1.836
(4.292)	32	1.9191e−6	1.803	2.6386e−6	1.873	1.3474e−6	1.869
	64	5.7330e−7	1.743	7.4970e−7	1.815	4.1547e−7	1.697
	8	4.6235e−3		2.0422e−3		7.5212e−4	
ADI	16	2.4396e−3	0.922	7.8811e−4	1.373	2.2467e−4	1.743
(4.293)	32	1.1392e−3	1.098	2.8766e−4	1.454	6.5421e−5	1.780
	64	5.1063e−4	1.157	1.0301e−4	1.481	1.8848e−5	1.795

TABLE 4.25: The L^2 error at $t = 1$ for method (4.333), $\gamma = 0.8$, $N = N_x = N_y$, $\Delta t = 10^{-3}$.

q	N	$\alpha = 1.2$	order	$\alpha = 1.5$	order	$\alpha = 1.8$	order
	8	4.7596e−6		1.2860e−6		4.9546e−7	
	16	3.1131e−6	0.6125	8.2419e−7	0.6419	7.9589e−8	2.6381
1	32	1.8470e−6	0.7532	4.6789e−7	0.8168	3.9719e−8	1.0027
	64	1.0222e−6	0.8536	2.4942e−7	0.9076	2.7368e−8	0.5373
	8	3.7728e−7		5.3728e−7		6.8465e−7	
	16	9.0740e−8	2.0558	1.2739e−7	2.0765	1.5982e−7	2.0989
2	32	2.2493e−8	2.0123	3.1436e−8	2.0187	3.9251e−8	2.0256
	64	5.5972e−9	2.0067	7.7777e−9	2.0150	9.6470e−9	2.0246
	8	1.1109e−6		1.2108e−6		1.0371e−6	
	16	2.8707e−7	1.9523	3.0289e−7	1.9991	2.5046e−7	2.0499
3	32	7.3656e−8	1.9625	7.6649e−8	1.9825	6.2291e−8	2.0075
	64	1.8704e−8	1.9775	1.9296e−8	1.9899	1.5465e−8	2.0100
	8	6.4331e−7		1.1712e−6		1.1338e−6	
	16	2.7603e−7	1.2207	4.3286e−7	1.4360	3.2506e−7	1.8024
4	32	1.1582e−7	1.2530	1.2524e−7	1.7892	8.5447e−8	1.9276
	64	3.5216e−8	1.7175	3.3269e−8	1.9124	2.1678e−8	1.9788

TABLE 4.26: The L^2 error at $t = 1$ for method (4.335), $\gamma = 0.8$, $N = N_x = N_y$, $\Delta t = 10^{-3}$.

q	N	$\alpha = 1.2$	order	$\alpha = 1.5$	order	$\alpha = 1.8$	order
	8	4.7596e−6		1.2860e−6		4.9544e−7	
	16	3.1131e−6	0.6125	8.2420e−7	0.6418	7.9587e−8	2.6381
1	32	1.8470e−6	0.7532	4.6790e−7	0.8168	3.9735e−8	1.0021
	64	1.0222e−6	0.8536	2.4943e−7	0.9076	2.7386e−8	0.5370
	8	3.7728e−7		5.3727e−7		6.8463e−7	
	16	9.0738e−8	2.0558	1.2738e−7	2.0765	1.5980e−7	2.0991
2	32	2.2491e−8	2.0124	3.1429e−8	2.0190	3.9235e−8	2.0261
	64	5.5951e−9	2.0071	7.7700e−9	2.0161	9.6307e−9	2.0264
	8	1.1109e−6		1.2108e−6		1.0371e−6	
	16	2.8707e−7	1.9523	3.0289e−7	1.9992	2.5044e−7	2.0500
3	32	7.3654e−8	1.9626	7.6641e−8	1.9826	6.2275e−8	2.0077
	64	1.8702e−8	1.9776	1.9289e−8	1.9904	1.5449e−8	2.0112
	8	6.4331e−7		1.1712e−6		1.1338e−6	
	16	2.7603e−7	1.2207	4.3285e−7	1.4360	3.2504e−7	1.8025
4	32	1.1582e−7	1.2530	1.2523e−7	1.7893	8.5430e−8	1.9278
	64	3.5214e−8	1.7176	3.3261e−8	1.9127	2.1661e−8	1.9796

Example 11 *Consider the following two-dimensional time-space fractional subdiffusion equation*

$$\begin{cases} \partial_t U = ({}_{RL}\mathrm{D}^{\alpha}_{0,x} + {}_{RL}\mathrm{D}^{\alpha}_{x,1})U + ({}_{RL}\mathrm{D}^{\alpha}_{0,y} + {}_{RL}\mathrm{D}^{\alpha}_{y,1})U + g(x,y,t), \quad 0 < \gamma < 1, \\ \qquad (x,y,t) \in (0,1)\times(0,1)\times(0,1], \\ U(x,y,0) = x^2(1-x)^2y^2(1-y)^2, \quad (x,y) \in [0,1]\times[0,1], \\ U(0,y,t) = U(1,y,t) = 0, \quad (y,t) \in (0,1)\times(0,1], \\ U(x,0,t) = U(x,1,t) = 0, \quad (x,t) \in (0,1)\times(0,1]. \end{cases} \tag{4.344}$$

Choose the suitable right-hand side function $g(x,y,t)$ such that Eq. (4.344) *has the analytical solution $U(x,y,t) = \cos(t)x^2(1-x)^2y^2(1-y)^2$.*

In this example, we apply the Crank–Nicolson ADI method (4.303) for solving (4.344); the L^2 errors at $t = 1$ are shown in Table 4.27. We find that the numerical results fit well with the theoretical analysis.

TABLE 4.27: The L^2 error at $t = 1$ for the CN ADI method (4.303), $N = N_x = N_y, \Delta t = 10^{-3}$.

q	N	$\alpha = 1.2$	order	$\alpha = 1.5$	order	$\alpha = 1.8$	order
1	8	3.7686e−4		7.8117e−5		3.3292e−5	
	16	2.4793e−4	0.6041	5.2218e−5	0.5811	4.1518e−6	3.0033
	32	1.4452e−4	0.7787	2.9435e−5	0.8270	1.7310e−6	1.2621
	64	7.8217e−5	0.8857	1.5516e−5	0.9238	1.5218e−6	0.1859
2	8	2.4268e−5		3.3437e−5		4.8765e−5	
	16	5.3571e−6	2.1795	7.5303e−6	2.1507	1.1589e−5	2.0731
	32	1.2359e−6	2.1159	1.7218e−6	2.1288	2.7583e−6	2.0709
	64	2.9686e−7	2.0577	4.0167e−7	2.0999	6.5887e−7	2.0657
3	8	6.3347e−5		6.3789e−5		6.5025e−5	
	16	1.5670e−5	2.0153	1.5406e−5	2.0498	1.5767e−5	2.0441
	32	3.9095e−6	2.0029	3.7367e−6	2.0436	3.8143e−6	2.0474
	64	9.8287e−7	1.9919	9.1273e−7	2.0335	9.2291e−7	2.0471
4	8	2.9385e−5		7.6134e−5		7.4103e−5	
	16	2.0318e−5	0.5323	2.2857e−5	1.7359	1.9482e−5	1.9274
	32	6.7945e−6	1.5803	6.1342e−6	1.8976	4.9245e−6	1.9841
	64	1.9107e−6	1.8303	1.5728e−6	1.9635	1.2212e−6	2.0116

Chapter 5

Galerkin Finite Element Methods for Fractional Partial Differential Equations

Generally speaking, the finite difference methods for FDEs may have less accuracy. Even if the higher order difference schemes can be constructed, the strong smooth conditions must be assumed. In order to weaken the smooth conditions, the Galerkin finite element methods are established for fractional partial differential equations.

5.1 Mathematical Preliminaries

We first introduce some notations. Let $(\cdot,\cdot)_{L^2(O)}$ be the *inner product* defined on the domain O, i.e.,

$$(u,v)_{L^2(O)} = \int_O uv\,\mathrm{d}O, \qquad \forall u,v\in L^2(O),$$

where O may stand for the finite domain Ω or infinite domain $\mathbb{R}^d$, d is a positive integer. The L^2 *norm* $||\cdot||_{L^2(O)}$ is defined as

$$||u||_{L^2(O)} = \sqrt{(u,u)_{L^2(O)}}, \qquad \forall u\in L^2(O).$$

The *Sobolev space* $H^r(O), r\geq 0$ is defined as a vector space of functions $u\in L^2(O)$ such that all the distributional derivatives of u of order up to r belong to $L^2(O)$. In short,

$$H^r(O) = \{u\in L^2(O) : D^\alpha u\in L^2(O),\ |\alpha|\leq k,\ k=0,1,\cdots,r\},$$

where O is a subset of $\mathbb{R}^d$, $D^\alpha = \partial_{x_1}^{\alpha_1}\partial_{x_2}^{\alpha_2}\cdots\partial_{x_d}^{\alpha_d}$, $\alpha=(\alpha_1,\alpha_2,\cdots,\alpha_d)$ is a multi-index, and α_j is a non-negative integer with $|\alpha| = \alpha_1+\alpha_2+\cdots+\alpha_d$.

The *semi-norm* and norm associated with the Sobolev space $H^r(O)$ are

$$|u|_{H^k(O)} = \left(\sum_{|\alpha|=k} ||D^\alpha u||_{L^2(O)}\right)^{1/2}, \qquad ||u||_{H^k(O)} = \left(\sum_{j=0}^{k} |u|^2_{H^j(O)}\right)^{1/2}.$$

Next, we introduce several other spaces.

Definition 14 ([131]) *Let $\mu > 0$ and $\Omega = (a,b)$. Define the semi-norm*

$$|u|_{J_L^\mu(\Omega)} = \|{}_{RL}D_{a,x}^\mu u(x)\|_{L^2(\Omega)}$$

and the norm

$$\|u\|_{J_L^\mu(\Omega)} = \left(\|u\|_{L^2(\Omega)}^2 + |u|_{J_L^\mu(\Omega)}^2\right)^{1/2},$$

and denote $J_L^\mu(\Omega)$ (or $J_{L,0}^\mu(\Omega)$) as the closure of $C^\infty(\Omega)$ (or $C_0^\infty(\Omega)$) with respect to $\|\cdot\|_{J_L^\mu(\Omega)}$, where $C_0^\infty(\Omega)$ is the space of smooth functions with compact support in Ω.

Definition 15 ([49]) *Let $\mu > 0$. Define the semi-norm*

$$|u|_{J_R^\mu(\Omega)} = \|{}_{RL}D_{x,b}^\mu u(x,y)\|_{L^2(\Omega)}$$

and the norm

$$\|u\|_{J_R^\mu(\Omega)} = \left(\|u\|_{L^2(\Omega)}^2 + |u|_{J_R^\mu(\Omega)}^2\right)^{1/2},$$

and denote $J_R^\mu(\Omega)$ (or $J_{R,0}^\mu(\Omega)$) as the closure of $C^\infty(\Omega)$ (or $C_0^\infty(\Omega)$) with respect to $\|\cdot\|_{J_R^\mu(\Omega)}$.

Definition 16 ([49]) *Let $\mu > 0, \mu \neq n - 1/2, n \in \mathbb{N}$. Define the semi-norm*

$$|u|_{J_S^\mu(\Omega)} = |({}_{RL}D_{a,x}^\mu u(x), {}_{RL}D_{x,b}^\mu u(x))|^{\frac{1}{2}}$$

and the norm

$$\|u\|_{J_S^\mu(\Omega)} = \left(\|u\|_{L^2(\Omega)}^2 + |u|_{J_S^\mu(\Omega)}^2\right)^{1/2},$$

and let $J_S^\mu(\Omega)$ (or $J_{S,0}^\mu(\Omega)$) denote the closure of $C^\infty(\Omega)$ (or $C_0^\infty(\Omega)$) with respect to $\|\cdot\|_{J_S^\mu(\Omega)}$.

The *fractional Sobolev space $H^\mu(\Omega)$* can be defined via the Fourier transform approach [88, 116].

Definition 17 *The Fourier transform of $u \in L^2(\mathbb{R})$ is defined as*

$$\hat{u}(\omega) = \mathcal{F}(u(x)) = \int_{\mathbb{R}} e^{-i\omega x} u(x)\,\mathrm{d}x.$$

Definition 18 ([116, 131]) *Let $\mu > 0$. Define the semi-norm*

$$|u|_{H^\mu(\mathbb{R})} = \| |\omega|^\mu \hat{u} \|_{L^2(\mathbb{R})}$$

and the norm

$$\|u\|_{H^\mu(\mathbb{R})} = \left(\|u\|_{L^2(\mathbb{R})}^2 + |u|_{H^\mu(\mathbb{R})}^2\right)^{1/2},$$

where $\hat{u}(\mathbb{R})$ is the Fourier transform of function u. And let $H^\mu(\mathbb{R})$ (or $H_0^\mu(\mathbb{R})$) be the closure of $C^\infty(\mathbb{R})$ (or $C_0^\infty(\mathbb{R})$) with respect to $\|\cdot\|_{H^\mu(\mathbb{R})}$.

The following lemma presents the properties of the Fourier transform that will be used later on.

Lemma 5.1.1 ([49, 50]) *Let $\mu > 0$, $u \in L^p(\mathbb{R}), p \geq 1$. The Fourier transform of the left and right Riemann–Liouville fractional integral and derivatives satisfy:*

$$\mathcal{F}({}_{RL}\mathrm{D}^{-\mu}_{-\infty,x}u(x)) = (i\omega)^{-\mu}\hat{u}, \qquad \mathcal{F}({}_{RL}\mathrm{D}^{-\mu}_{x,\infty}u(x)) = (-i\omega)^{-\mu}\hat{u}, \tag{5.1}$$

and

$$\mathcal{F}({}_{RL}\mathrm{D}^{\mu}_{-\infty,x}u(x)) = (i\omega)^{\mu}\hat{u}, \qquad \mathcal{F}({}_{RL}\mathrm{D}^{\mu}_{x,\infty}u(x)) = (-i\omega)^{\mu}\hat{u}, \tag{5.2}$$

where

$$\hat{u}(\omega) = \mathcal{F}(u(x)) = \int_{\mathbb{R}} e^{-i\omega x}u(x)\,\mathrm{d}x.$$

Lemma 5.1.2 ([49]) *Let $\mu > 0$ be given. Then for a real valued function $u(x)$*

$$(\mathbf{D}^{\mu}u, \mathbf{D}^{\mu *}u)_{L^2(\mathbb{R})} = \cos(\mu\pi)||\mathbf{D}^{\mu}u||^2_{L^2(\mathbb{R})}, \tag{5.3}$$

in which $\mathbf{D}^{\mu}u := {}_{RL}\mathrm{D}^{\mu}_{-\infty,x}u(x)$, and $\mathbf{D}^{\mu *}u := {}_{RL}\mathrm{D}^{\mu}_{x,\infty}u(x)$.

Proof. The following Fourier transform property (the overbar denotes complex conjugate) is helpful in establishing this result:

$$\int_{\mathbb{R}} u\bar{v}\,\mathrm{d}x = \int_{\mathbb{R}} \hat{u}\bar{\hat{v}}\,\mathrm{d}\omega. \tag{5.4}$$

One can observe that

$$\overline{(i\omega)^{\mu}} = \begin{cases} \exp(-i\pi\mu)\overline{(i\omega)^{\mu}}, & \omega \geq 0, \\ \exp(i\pi\mu)\overline{(i\omega)^{\mu}}, & \omega < 0. \end{cases} \tag{5.5}$$

Thus,

$$\begin{aligned}(\mathbf{D}^{\mu}u, \mathbf{D}^{\mu *}u)_{L^2(\mathbb{R})} &= \int_{-\infty}^{\infty} (i\omega)^{\mu}\hat{u}(\omega)\,\overline{(i\omega)^{\mu}\hat{u}(\omega)}\,\mathrm{d}\omega \\ &= \int_{-\infty}^{0} (i\omega)^{\mu}\hat{u}(\omega)\,\overline{(-i\omega)^{\mu}\hat{u}(\omega)}\,\mathrm{d}\omega + \int_{0}^{\infty} (i\omega)^{\mu}\hat{u}(\omega)\,\overline{(-i\omega)^{\mu}\hat{u}(\omega)}\,\mathrm{d}\omega.\end{aligned} \tag{5.6}$$

Using (5.5) yields

$$\begin{aligned}&(\mathbf{D}^{\mu}u, \mathbf{D}^{\mu *}u)_{L^2(\mathbb{R})} \\ &= \int_{-\infty}^{0} (i\omega)^{\mu}\hat{u}(\omega)\exp(-i\pi\mu)\overline{(i\omega)^{\mu}\hat{u}(\omega)}\,\mathrm{d}\omega \\ &\quad + \int_{0}^{\infty} (i\omega)^{\mu}\hat{u}(\omega)\exp(i\pi\mu)\overline{(i\omega)^{\mu}\hat{u}(\omega)}\,\mathrm{d}\omega \\ &= \cos(\mu\pi)\int_{-\infty}^{\infty} (i\omega)^{\mu}\hat{u}(\omega)\overline{(i\omega)^{\mu}\hat{u}(\omega)}\,\mathrm{d}\omega \\ &\quad + i\sin(\mu\pi)\left(\int_{0}^{\infty} (i\omega)^{\mu}\hat{u}(\omega)\overline{(i\omega)^{\mu}\hat{u}(\omega)}\,\mathrm{d}\omega - \int_{-\infty}^{0} (i\omega)^{\mu}\hat{u}(\omega)\overline{(i\omega)^{\mu}\hat{u}(\omega)}\,\mathrm{d}\omega\right).\end{aligned} \tag{5.7}$$

For real $u(x)$ we have $\overline{\hat{u}(-\omega)} = \hat{u}(\omega)$. Hence

$$\int_0^{\infty} (i\omega)^{\mu}\hat{u}(\omega)\overline{(i\omega)^{\mu}\hat{u}(\omega)}\,\mathrm{d}\omega = \int_{-\infty}^{0} (i\omega)^{\mu}\hat{u}(\omega)\overline{(i\omega)^{\mu}\hat{u}(\omega)}\,\mathrm{d}\omega. \tag{5.8}$$

Combining (5.7) and (5.8) we obtain

$$(\mathbf{D}^{\mu}u, \mathbf{D}^{\mu*}u)_{L^2(\mathbb{R})} = \cos(\mu\pi)(\mathbf{D}^{\mu}u, \mathbf{D}^{\mu}u)_{L^2(\mathbb{R})} = \cos(\mu\pi)\|\mathbf{D}^{\mu}u\|^2_{L^2(\mathbb{R})}. \tag{5.9}$$

Thus the proof is completed. □

Lemma 5.1.3 ([49]) *Let $\mu > 0$ be given. The spaces $J^{\mu}_L(\mathbb{R})$, $J^{\mu}_R(\mathbb{R})$, and $H^{\mu}(\mathbb{R})$ are equal with equivalent semi-norms and norms.*

Proof. We first prove that a function $u\in L^2(\mathbb{R})$ belongs to $J^{\mu}_L(\mathbb{R})$ if and only if

$$|\omega|^{\mu}\hat{u}\in L^2(\mathbb{R}). \tag{5.10}$$

Let $u\in L^2(\mathbb{R})$ be given. Then $\mathbf{D}^{\mu}u\in L^2(\mathbb{R})$, and from (5.2)

$$\mathcal{F}(\mathbf{D}^{\mu}u) = \mathcal{F}({}_{RL}\mathrm{D}^{\mu}_{-\infty,x}u(x)) = (i\omega)^{\mu}\hat{u}.$$

Using Plancherel's theorem, we have

$$\int_{\mathbb{R}} |\omega|^{2\mu}|\hat{u}|\,\mathrm{d}\omega = \int_{\mathbb{R}} |\mathbf{D}^{\mu}u|^2\,\mathrm{d}x. \tag{5.11}$$

Hence,

$$|u|_{H^{\mu}(\mathbb{R})} = \|\,|\omega|^{\mu}\hat{u}\|_{L^2(\mathbb{R})} = |u|_{J^{\mu}_L(\mathbb{R})}. \tag{5.12}$$

Therefore, the spaces $J^{\mu}_L(\mathbb{R})$, and $H^{\mu}(\mathbb{R})$ are equal, with equivalent semi-norms and norms. We similarly have that the spaces $J^{\mu}_R(\mathbb{R})$, and $H^{\mu}(\mathbb{R})$ are equal with equivalent semi-norms and norms. The proof is completed. □

By almost the same reasoning, one has the following results.

Lemma 5.1.4 *For $\mu > 0, \mu\neq n-1/2, n\in\mathbb{N}$, the spaces $J^{\mu}_L(\mathbb{R})$ and $J^{\mu}_S(\mathbb{R})$ are equal, with equivalent semi-norms and norms.*

Lemma 5.1.5 *Let $\mu > 0$, $\mu\neq n-1/2, n\in\mathbb{N}$. Then the spaces $J^{\mu}_{S,0}(\Omega)$ and $H^{\mu}_0(\Omega)$ are equal, with equivalent semi-norms and norms.*

Lemma 5.1.6 *Let $\mu > 0$. Then the spaces $J^{\mu}_{L,0}(\Omega)$, $J^{\mu}_{R,0}(\Omega)$, and $H^{\mu}_0(\Omega)$ are equal. Also, if $\mu\neq n-1/2, n\in\mathbb{N}$, the spaces $J^{\mu}_{L,0}(\Omega)$, $J^{\mu}_{R,0}(\Omega)$, $J^{\mu}_{S,0}(\Omega)$, and $H^{\mu}_0(\Omega)$ have equivalent semi-norms and norms.*

Lemma 5.1.7 (*Fractional Poincaré-Friedrichs* [49]) *For $u\in J^{\mu}_{L,0}(\Omega)$, $0\leq s\leq\mu$, one has*

$$\|u\|_{J^s_L(\Omega)} \leq C|u|_{J^{\mu}_L L(\Omega)}, \tag{5.13}$$

and for $u\in J^{\mu}_{R,0}(\Omega), \mu > 0$, we have

$$\|u\|_{J^s_R(\Omega)} \leq C|u|_{J^{\mu}_R L(\Omega)}. \tag{5.14}$$

Lemma 5.1.8 *The left and right Riemann–Liouville fractional integral operators are adjoints w.r.t. the inner product in* $L^2(\Omega), \Omega = (a,b)$*, i.e.,*

$$(\mathrm{D}_{a,x}^{-\mu}u, v)_{L^2(\Omega)} = (u, \mathrm{D}_{x,b}^{-\mu}v)_{L^2(\Omega)}, \quad u,v \in L^2(\Omega), \quad \mu > 0.$$

Proof. Interchanging the order of integration yields

$$\begin{aligned}(\mathrm{D}_{a,x}^{-\mu}u, v)_{L^2(\Omega)} =& \frac{1}{\Gamma(\mu)}\int_a^b\int_a^x (x-\xi)^{\mu-1}u(\xi)v(x)\,\mathrm{d}\xi\,\mathrm{d}x \\ =& \frac{1}{\Gamma(\mu)}\int_a^b u(\xi)\int_\xi^b (x-\xi)^{\mu-1}v(x)\,\mathrm{d}x\,\mathrm{d}\xi \\ =& (u, \mathrm{D}_{x,b}^{-\mu}v)_{L^2(\Omega)}.\end{aligned} \tag{5.15}$$

The proof is completed. □

Lemma 5.1.9 ([64]) *For* $0 < \beta, \gamma < 1$*, if* $u(x) \in H^1(\Omega), \Omega = (a,b)$*, then*

$${}_{RL}\mathrm{D}_{a,x}^{\beta}\,{}_{RL}\mathrm{D}_{a,x}^{\gamma}u(x) = {}_{RL}\mathrm{D}_{a,x}^{\beta+\gamma}u(x). \tag{5.16}$$

Proof. By the definition of the Riemann–Liouville derivative,

$${}_{RL}\mathrm{D}_{a,x}^{\beta}\,{}_{RL}\mathrm{D}_{a,x}^{\gamma}u(x) = \frac{1}{\Gamma(1-\beta)}\frac{\mathrm{d}}{\mathrm{d}x}\int_a^x (x-s)^{-\beta}\frac{1}{\Gamma(1-\gamma)}\frac{\mathrm{d}}{\mathrm{d}s}\int_a^s (s-\tau)^{-\gamma}u(\tau)\,\mathrm{d}\tau\,\mathrm{d}s. \tag{5.17}$$

Interchanging the order of integration yields

$$\begin{aligned}&{}_{RL}\mathrm{D}_{a,x}^{\beta}\,{}_{RL}\mathrm{D}_{a,x}^{\gamma}u(x) \\ =& \frac{1}{\Gamma(1-\beta)}\frac{1}{\Gamma(1-\gamma)}\frac{\mathrm{d}}{\mathrm{d}x}\int_a^x (x-s)^{-\beta}\left[(s-a)^{-\gamma}u(a) + \int_a^s (s-\tau)^{-\gamma}u'(\tau)\,\mathrm{d}\tau\right]\mathrm{d}s \\ =& \begin{cases}\dfrac{1}{\Gamma(1-\beta-\gamma)}\dfrac{\mathrm{d}}{\mathrm{d}x}\displaystyle\int_a^x (x-\tau)^{-\beta-\gamma}u(\tau)\,\mathrm{d}\tau, & 0 < \beta+\gamma < 1, \\ \dfrac{1}{\Gamma(2-\beta-\gamma)}\dfrac{\mathrm{d}^2}{\mathrm{d}x^2}\displaystyle\int_a^x (x-\tau)^{1-\beta-\gamma}u(\tau)\,\mathrm{d}\tau, & 1 \le \beta+\gamma < 2\end{cases} \\ =& {}_{RL}\mathrm{D}_{a,x}^{\beta+\gamma}u(x),\end{aligned} \tag{5.18}$$

which completes the proof. □

Lemma 5.1.10 ([94, 171, 175]) *Let* $0 < \beta < 2$*,* $\Omega = (a,b)$*. Then for any* $u \in H_0^{\beta}(\Omega)$*,* $v \in H_0^{\beta/2}(\Omega)$*, we have*

$$({}_{RL}\mathrm{D}_{a,x}^{\beta}u, v)_{L^2(\Omega)} = ({}_{RL}\mathrm{D}_{a,x}^{\beta/2}u, {}_{RL}\mathrm{D}_{x,b}^{\beta/2}v)_{L^2(\Omega)},$$

$$({}_{RL}\mathrm{D}_{x,b}^{\beta}u, v)_{L^2(\Omega)} = ({}_{RL}\mathrm{D}_{x,b}^{\beta/2}u, {}_{RL}\mathrm{D}_{a,x}^{\beta/2}v)_{L^2(\Omega)}.$$

Lemma 5.1.11 ([94, 175]) *Let* $0 < \beta < 1$, $\Omega = (a,b)$. *Then for any* $u \in H^{\beta}(\Omega)$, $v \in H^{\beta/2}(\Omega)$, $u(a) = u(b) = 0$, *we have*

$$({}_{RL}\mathrm{D}_{a,x}^{\beta}u, v)_{L^2(\Omega)} = ({}_{RL}\mathrm{D}_{a,x}^{\beta/2}u, {}_{RL}\mathrm{D}_{x,b}^{\beta/2}v)_{L^2(\Omega)}.$$

Lemma 5.1.12 (*Gronwall's inequality*) *Let* $a(t), q(t) \in L[t_0,t_1]$, $u(t), b(t)$, $t \in [t_0,t_1]$ *be real valued continuous functions;* $b(t)$ *and* $q(t)$ *are nonnegative functions satisfying*

$$u(t) \leq a(t) + q(t)\int_{t_0}^{t} b(s)u(s)\,\mathrm{d}s, \quad \forall t \in [t_0,t_1].$$

Then we have

$$u(t) \leq a(t) + q(t)\int_{t_0}^{t} a(s)b(s)\exp\left(\int_{s}^{t} q(r)b(r)\,\mathrm{d}r\right)\mathrm{d}s, \quad \forall t \in [t_0,t_1].$$

Lemma 5.1.13 (*Discrete Gronwall's inequality*) *Let* x_n *be real positive numbers,* $H, C, \Delta t > 0$, $x_0 \leq H$. x_n *satisfies*

$$x_n \leq C\Delta t\sum_{k=0}^{n-1} x_k + H.$$

Then we have

$$x_n \leq H\exp(Cn\Delta t).$$

In the following sections, we introduce the Galerkin FEM for the fractional differential equations. We mainly focus on stationary fractional advection dispersion equations [49, 152], space-fractional diffusion equations [178, 181, 182], time-fractional differential equations [52, 167, 168], time-space fractional differential equations [92, 179]. Other numerical methods such as *Discontinuous Galerkin methods* [23, 109, 180, 181] are not going to be presented in this book.

5.2 Galerkin FEM for Stationary Fractional Advection Dispersion Equation

This section deals with the following *steady state fractional advection dispersion equation* [49]

$$Lu = -\mathrm{D}a(p\mathrm{D}_{0,x}^{-\beta} + q\mathrm{D}_{x,1}^{-\beta})\mathrm{D}u + b(x)\mathrm{D}u + c(x)u = f, \quad x \in \Omega = (x_L, x_R) = (0,1), \tag{5.19}$$

with the boundary conditions

$$u = 0, \quad x \in \partial\Omega, \tag{5.20}$$

where D represents a single spatial derivative, $0 \leq \beta < 1$, $a > 0$, $b(x) \in C^1(\bar{\Omega})$, $c(x) \in C(\bar{\Omega})$ with $c - \mathrm{D}b/2 \geq 0$, and $p + q = 1$, $0 \leq p, q \leq 1$. The main results in this section come from [49].

5.2.1 Notations and Polynomial Approximation

In this subsection, we introduce some notations and lemmas that are needed in the following sections.

Let $\Omega = (x_L, x_R)$ be a general finite domain, and denote by $(\cdot,\cdot)$ the inner product on the space $L^2(\Omega)$ with the L^2 norm $\|\cdot\|$ and the maximum norm $\|\cdot\|_\infty$. Denote $H^r(\Omega)$ and $H_0^r(\Omega)$ as the commonly used Sobolev spaces with the norm $\|\cdot\|_r$ and semi-norm $|\cdot|_r$, respectively. Define $\mathbb{P}_r(\Omega)$ as the space of polynomials defined on Ω with the degree no greater than r, $r\in Z^+$. Let S_h be a uniform partition of Ω, which is given by

$$x_L = x_0 < x_1 < \cdots < x_{N-1} < x_N = x_R, \quad N \in Z^+.$$

Denote by $h = (x_R - x_L)/N = x_i - x_{i-1}$ and $\Omega_i = [x_{i-1}, x_i]$ for $i = 1, 2, \cdots, N$. We define the *finite element space* X_h^r as the set of piecewise polynomials with degree at most r $(r \geq 1)$ on the mesh S_h, which can be expressed by

$$X_h^r = \{v : v|_{\Omega_i} \in \mathbb{P}_r(\Omega_i), v \in C(\Omega)\}.$$

Introduce the *piecewise interpolation operator* $I_h : C(\bar{\Omega}) \to X_h^r$ as

$$I_h u\Big|_{\Omega_i} = \sum_{k=0}^{r} u(x_k^i) F_k^i(x), \quad u \in C(\bar{\Omega}),$$

where $F_k^i(x)$ are *Lagrangian basis functions* defined by

$$F_k^i(x) = \prod_{l=0, l\neq k}^{r} \frac{x - x_l^i}{x_k^i - x_l^i}, \quad i = 1, 2, \cdots, N,$$

and $\{x_k^i, k = 0, 1, \cdots, r\}$ are the interpolation nodes on the interval Ω_i with $x_0^i = x_{i-1}$ and $x_r^i = x_i$.

Define φ^i $(i = 0, 1, \cdots, N)$ and φ_k^i $(k = 1, 2, \cdots, r-1; i = 1, 2, \cdots, N)$ as

$$\varphi_k^i(x) = \begin{cases} F_k^i(x), & x \in [x_{i-1}, x_i], \quad k = 1, 2, \cdots, r-1, i = 1, \cdots, N, \\ 0, & \text{others}, \end{cases}$$

$$\varphi^i(x) = \begin{cases} F_r^i(x), & x \in [x_{i-1}, x_i], \quad i = 1, \cdots, N-1, \\ F_0^{i+1}(x), & x \in [x_i, x_{i+1}], \quad i = 1, \cdots, N-1, \\ 0, & \text{others}, \end{cases}$$

$$\varphi^0(x) = \begin{cases} F_0^1(x), & x \in [x_0, x_1], \\ 0, & \text{others}, \end{cases}$$

$$\varphi^N(x) = \begin{cases} F_r^N(x), & x \in [x_{N-1}, x_N], \\ 0, & \text{others}. \end{cases}$$

Let $X_{h0}^r = X_h^r \cap H_0^1(\Omega)$. Then the spaces X_{h0}^r and X_h^r can be expressed as

$$X_{h0}^r = \text{span}\left\{\varphi_k^i, k = 1, 2, \cdots, r-1, i = 1, 2, \cdots, N\right\} \cup \left\{\varphi^i, i = 1, 2, \cdots, N-1\right\},$$
$$X_h^r = \text{span}\left\{\varphi_k^i, k = 1, 2, \cdots, r-1, i = 1, 2, \cdots, N\right\} \cup \left\{\varphi^i, i = 0, 1, \cdots, N\right\}.$$

Denote by

$$\phi_j(x) = \begin{cases} \varphi_k^i(x), & j=(i-1)r+k,\ k=1,2,\cdots,r-1, i=1,2,\cdots,N, \\ \varphi^i(x), & j=ir,\ i=0,1,\cdots,N. \end{cases} \tag{5.21}$$

Then

$$X_{h0}^r = \text{span}\left\{\phi_j, j=1,2,\cdots,Nr-1\right\},$$
$$X_h^r = \text{span}\left\{\phi_j, j=0,1,2,\cdots,Nr\right\}.$$

The *orthogonal projection operator* $\Pi_h^{1,0}: H_0^1(\Omega) \to X_{h0}^r$ is defined as

$$(\partial_x(u-\Pi_h^{1,0}u), \partial_x v) = 0, \quad u \in H_0^1(\Omega), \forall v \in X_{h0}^r. \tag{5.22}$$

Next, we introduce the properties of the projector $\Pi_h^{1,0}$ and interpolation operator I_h that will be used later on.

Lemma 5.2.1 ([9]) *Let $m, r \in Z^+$, $r \geq 1$, and $u \in H^m(\Omega) \cap H_0^1(\Omega)$. If $1 \leq m \leq r+1$, then there exists a positive constant C independent of h, such that*

$$\|u-\Pi_h^{1,0}u\|_{H^l(\Omega)} \leq Ch^{m-l}\|u\|_{H^m(\Omega)}, \quad 0 \leq l \leq 1.$$

Lemma 5.2.2 ([3]) *Let m, l be nonnegative numbers, $r \in Z^+$, $r \geq 1$, and $u \in H^m(\Omega)$. If $0 \leq l \leq m \leq r+1$, then there exists a positive constant C independent of h, such that*

$$\|u-I_h u\|_{H^l(\Omega)} \leq Ch^{m-l}\|u\|_{H^m(\Omega)}, \quad 0 \leq l \leq 1.$$

5.2.2 Variational Formulation

In order to derive a variational form of the problem (5.19)-(5.20), we assume that u is a sufficiently smooth solution of (5.19)-(5.20). We multiply by an arbitrary $v \in C_0^\infty(\Omega)$ to obtain

$$\int_\Omega \left(-\mathrm{D}(ap\mathrm{D}_{0,x}^{-\beta} + aq\mathrm{D}_{x,1}^{-\beta})\mathrm{D}u + b(x)\mathrm{D}u + c(x)u\right)v\,\mathrm{d}x = \int_\Omega fv\,\mathrm{d}x.$$

Integrating by parts and noting that $v=0$ on $\partial\Omega$ gives

$$\int_\Omega \left[a(p\mathrm{D}_{0,x}^{-\beta} + q\mathrm{D}_{x,1}^{-\beta})\mathrm{D}u\mathrm{D}v + b(x)\mathrm{D}u\,v + c(x)uv\right]\mathrm{d}x = \int_\Omega fv\,\mathrm{d}x.$$

For convenience, when we are working on a fixed domain Ω, we often omit the set in the notations and write simply $(\cdot,\cdot) = (\cdot,\cdot)_{L^2(\Omega)}$, $\|\cdot\| = \|\cdot\|_{L^2(\Omega)}$, $|\cdot|_r = |\cdot|_{H^r(\Omega)}$, and $\|\cdot\|_r = \|\cdot\|_{H^r(\Omega)}$.

For $0 \leq \beta < 1$ and $u=0$ on $\partial\Omega$, we have

$$\mathrm{D}_{0,x}^{-\beta}\mathrm{D}u = {}_{RL}\mathrm{D}_{0,x}^{1-\beta}u, \quad \mathrm{D}\mathrm{D}_{0,x}^{-\beta}\mathrm{D}u = {}_{RL}\mathrm{D}_{0,x}^{2-\beta}u,$$

$$\mathrm{D}_{x,1}^{-\beta}\mathrm{D}u = {}_{RL}\mathrm{D}_{x,1}^{1-\beta}u, \quad \mathrm{D}\mathrm{D}_{x,1}^{-\beta}\mathrm{D}u = {}_{RL}\mathrm{D}_{x,1}^{2-\beta}u.$$

Let $\alpha = 1-\beta/2$. Then $1/2 < \alpha \leq 1$. Define the bilinear form $A : H_0^\alpha(\Omega)\times H_0^\alpha(\Omega) \to \mathbb{R}$ as

$$A(u,v) = ap(\mathrm{D}_{0,x}^{-\beta}u, \mathrm{D}v) + aq(\mathrm{D}_{x,1}^{-\beta}\mathrm{D}u, \mathrm{D}v) + (b\mathrm{D}u, v) + (cu, v). \tag{5.23}$$

For a given function $f \in H^{-\alpha}(\Omega)$, we define the associated linear functional $F : H_0^\alpha(\Omega) \to \mathbb{R}$ as

$$F(v) = (f, v). \tag{5.24}$$

Obviously, the duality pairing in (5.24) is well defined for $u, v \in H_0^\alpha(\Omega)$. Therefore, we have

$$A(u,v) = F(v), \qquad v \in H_0^\alpha(\Omega). \tag{5.25}$$

Next, we prove that the variational form (5.25) has a unique solution in $H_0^\alpha(\Omega)$. We need to introduce some concepts and conclusions.

Definition 19 ([3]) *A linear space V together with an inner product $(\cdot,\cdot)$ is called an inner-product space and is denoted by $(V, (\cdot,\cdot))$.*

Definition 20 ([3]) *Let $(V, (\cdot,\cdot))$ be an inner-product space. If the associated normed linear space $(V, \|\cdot\|_V)$ is complete, then $(V, (\cdot,\cdot))$ is called a Hilbert space.*

Definition 21 ([3]) *A bilinear form $A(\cdot,\cdot)$ on a normed linear space H is said to be **bounded** (or **continuous**) if there exists a positive constant C such that*

$$A(v,w) \leq C\|v\|_H\|w\|_H, \qquad v, w \in H,$$

*and **coercive** on $V \subset H$ if there exists a positive constant c_0 such that*

$$A(v,v) \geq c_0\|v\|_H^2, \qquad v \in V.$$

Theorem 34 (*Lax-Milgram Theorem* [3]) *Given a Hilbert space $(V, (\cdot,\cdot))$, a continuous, coercive bilinear form $A(\cdot,\cdot)$ and a continuous linear functional $F \in V'$, V' is the dual space of V, there exists a unique $u \in V$ such that*

$$A(u,v) = F(v), \qquad v \in V. \tag{5.26}$$

From Theorem 34, we can obtain that there exists a unique solution to (5.25).

Theorem 35 *There exists a unique solution $u \in H_0^\alpha(\Omega)$ to (5.25) satisfying*

$$\|u\|_{H^\alpha(\Omega)} \leq C\|f\|_{H^{-\alpha}(\Omega)}. \tag{5.27}$$

Proof. In order to prove the uniqueness of the solution u to (5.25), we need to prove that the bilinear form $A(u,v)$ is continuous and coercive. We first prove the coercivity. It is easy to obtain

$$\begin{aligned}
A(u,u) =& ap(\mathrm{D}_{0,x}^{-\beta}\mathrm{D}u, \mathrm{D}u) + aq(\mathrm{D}_{x,1}^{-\beta}\mathrm{D}u, \mathrm{D}u) + (b\mathrm{D}u, u) + (cu, u)\\
=& ap(\mathrm{D}_{0,x}^{-\beta}\mathrm{D}u, \mathrm{D}u) + aq(\mathrm{D}_{x,1}^{-\beta}\mathrm{D}u, \mathrm{D}u) + ((c-\mathrm{D}b/2)u, u)\\
\geq& ap(\mathrm{D}_{0,x}^{-\beta}u, \mathrm{D}u) + aq(\mathrm{D}_{x,1}^{-\beta}\mathrm{D}u, \mathrm{D}u),
\end{aligned}$$

where we have used $(c-\mathrm{D}b/2)\geq 0$.

Noting that $u=0$ on $\partial\Omega$, and using Lemmas 5.1.8, 5.1.9, and 5.1.10, we have

$$(\mathrm{D}_{0,x}^{-\beta}\mathrm{D}u,\mathrm{D}u)=(\mathrm{D}_{0,x}^{-\beta/2}\mathrm{D}_{0,x}^{\alpha}u,\mathrm{D}u)=-(\mathrm{D}_{0,x}^{\alpha}u,\mathrm{D}_{x,1}^{\alpha}u),$$
$$(\mathrm{D}_{x,1}^{-\beta}\mathrm{D}u,\mathrm{D}u)=-(\mathrm{D}_{x,1}^{\alpha}u,\mathrm{D}_{0,x}^{\alpha}u).$$

Therefore,

$$A(u,u)\geq -a(\mathrm{D}_{0,x}^{\alpha}u,\mathrm{D}_{x,1}^{\alpha}u)=a|u|^2_{J_S^{\alpha}(\Omega)}.$$

The semi-norm equivalence of $J_{S,0}^{\alpha}(\Omega)$ and $H_0^{\alpha}(\Omega)$ (see Lemma 5.1.5) implies that

$$A(u,u)\geq a|u|^2_{J_S^{\alpha}(\Omega)}\geq C|u|^2_{H^{\alpha}(\Omega)}.$$

Since $u=0$ on $\partial\Omega$, from fractional Poincaré–Friedrichs inequality (see Lemma 5.1.7) and Lemma 5.1.6, we have

$$||u||^2_{H^{\alpha}(\Omega)}\leq C|u|^2_{H^{\alpha}(\Omega)}.$$

Therefore,

$$A(u,u)\geq C_0||u||^2_{H^{\alpha}(\Omega)}. \tag{5.28}$$

Next, we prove the continuity of $A(u,v)$. From the definition of $A(u,v)$ (see Eqs. (5.24)) and (5.2.2) we have

$$\begin{aligned}
|A(u,v)|\leq& ap|(\mathrm{D}_{0,x}^{-\beta}\mathrm{D}u,\mathrm{D}v)|+aq|(\mathrm{D}_{x,1}^{-\beta}\mathrm{D}u,\mathrm{D}v)|+|(b\mathrm{D}u,v)|+|(cu,v)|\\
=&ap|(\mathrm{D}_{0,x}^{\alpha}u,\mathrm{D}_{x,1}^{\alpha}v)|+aq|(\mathrm{D}_{x,1}^{\alpha}u,\mathrm{D}_{0,x}^{\alpha}v)|+|(b\mathrm{D}u,v)|+|(cu,v)|\\
\leq& ap||u||_{J_L^{\alpha}(\Omega)}||v||_{J_R^{\alpha}(\Omega)}+aq||u||_{J_R^{\alpha}(\Omega)}||v||_{J_L^{\alpha}(\Omega)}\\
&+C||u||_{H^{\alpha}(\Omega)}||v||_{H^{\alpha}(\Omega)}+||c||_{L^{\infty}(\Omega)}||u||_{L^2(\Omega)}||v||_{L^2(\Omega)}\\
\leq& C||u||_{H^{\alpha}(\Omega)}||v||_{H^{\alpha}(\Omega)},
\end{aligned} \tag{5.29}$$

where we have used $||bu||_{H^{\alpha}(\Omega)}\leq C||u||_{H^{\alpha}(\Omega)}$ for $b\in C^1(\bar{\Omega})$ (see Lemma 3.2 in [49]).

For the linear functional $F(v)$, we have

$$|F(v)|=|(f,v)|\leq ||f||_{H^{-\alpha}(\Omega)}||v||_{H^{\alpha}(\Omega)}.$$

Therefore,

$$C||u||^2_{H^{\alpha}(\Omega)}\leq |A(u,u)|\leq C_1||f||_{H^{-\alpha}(\Omega)}||u||_{H^{\alpha}(\Omega)},$$

which yields (5.27).

From Theorem 34, we know that there exists a unique solution $u\in H_0^{\alpha}(\Omega)$ of (5.25) satisfying (5.27). All this completes the proof. □

5.2.3 Finite Element Solution and Error Estimates

From (5.25), we have the finite element solution to (5.19)-(5.20) as: Find $u_h \in X_{h0}^r$, such that

$$A(u_h, v) = F(v), \qquad v \in X_{h0}^r. \tag{5.30}$$

Theorem 36 (***Cea's lemma***) *Let u be the solution to (5.25). Then the finite element solution u_h to (5.30) satisfies*

$$\|u - u_h\|_{H^\alpha(\Omega)} \le C\|u - v\|_{H^\alpha(\Omega)}, \quad \forall v \in X_{h0}^r. \tag{5.31}$$

Proof. From (5.25) and (5.30), we have

$$A(u - u_h, v) = 0, \qquad \forall v \in X_{h0}^r.$$

From (5.28) and (5.29) we have for $u_h, v_h \in X_{h0}^r$

$$\begin{aligned}\|u - u_h\|_{H^\alpha(\Omega)}^2 &\le \frac{1}{C_0} A(u - u_h, u - u_h) = \frac{1}{C_0} A(u - u_h, u - v_h + v_h - u_h) \\ &= \frac{1}{C_0} A(u - u_h, u - v_h) \le \frac{C}{C_0} \|u - u_h\|_{H^\alpha(\Omega)} \|u - v_h\|_{H^\alpha(\Omega)},\end{aligned}$$

which yields (5.26). The proof is completed. □

Next, we discuss the error estimate.

Theorem 37 *Let $u \in H_0^\alpha(\Omega) \cap H^r(\Omega), \alpha \le r$ and u_h be the solution to (5.25) and (5.30), respectively. Then there exists a positive constant C independent of h such that*

$$\|u - u_h\|_{H^\alpha(\Omega)} \le Ch^{r-\alpha} \|u\|_{H^r(\Omega)}. \tag{5.32}$$

Proof. From Theorem 36 we have

$$\|u - u_h\|_{H^\alpha(\Omega)} \le C\|u - I_h u\|_{H^\alpha(\Omega)} \le Ch^{r-\alpha} \|u\|_{H^r(\Omega)}, \tag{5.33}$$

where we have used Lemma 5.1.12. All this ends the proof. □

Next, we apply the Aubin–Nitsche trick to obtain the error estimate in L^2 norm. Consider the following problem

$$\begin{cases} -\mathrm{D}a(p\mathrm{D}_{0,x}^{-\beta} + q\mathrm{D}_{x,1}^{-\beta})\mathrm{D}w + b(x)\mathrm{D}w + c(x)w = g, \quad x \in \Omega, \\ w = 0, \quad x \in \partial\Omega. \end{cases} \tag{5.34}$$

Suppose that w is the solution to (5.34) with $g = e = u - u_h$. Then w satisfies the following variational form

$$A(w, v) = (e, v), \qquad v \in H_0^\alpha(\Omega). \tag{5.35}$$

According to [131], the solution w to (5.35) satisfies the following regularity

$$\|w\|_{H^{2\alpha}(\Omega)} \le C\|e\|_{L^2(\Omega)}, \text{ for } \alpha \ne \frac{3}{4}. \tag{5.36}$$

Now, we give the following convergence estimate in the L^2 norm.

Theorem 38 *Let $u \in H_0^\alpha(\Omega) \cap H^r(\Omega), 3/4 < \alpha \leq r$ and u_h be the solutions to (5.25) and (5.30), respectively. Then there exists a positive constant C independent of h such that*

$$\|u - u_h\|_{L^2(\Omega)} \leq Ch^r \|u\|_{H^r(\Omega)}. \tag{5.37}$$

Proof. Substituting $v = e = u - u_h$ in (5.35), and applying Galerkin orthogonality $A(e,v) = 0, \forall v \in X_{h0}^r$, we have

$$\begin{aligned}
\|e\|_{L^2(\Omega)}^2 &= A(e,w) = A(e, w - I_h w + I_h w) \\
&= A(e, w - I_h w) \leq C \|e\|_{H^\alpha(\Omega)} \|w - I_h w\|_{H^\alpha(\Omega)} \\
&\leq Ch^\alpha \|e\|_{H^\alpha(\Omega)} \|w\|_{H^{2\alpha}(\Omega)} \\
&\leq Ch^\alpha \|e\|_{H^\alpha(\Omega)} \|e\|_{L^2(\Omega)}.
\end{aligned}$$

It follows that

$$\|u - u_h\|_{L^2(\Omega)} \leq Ch^\alpha \|u - u_h\|_{H^\alpha(\Omega)} \leq Ch^r \|u\|_{H^r(\Omega)},$$

where we have used (5.32), which completes the proof. □

5.3 Galerkin FEM for Space-Fractional Diffusion Equation

In this section, we introduce the Galerkin FEM for the space-fractional partial differential equations in one space dimension. For the case of two space dimensions, see [6, 7, 8, 170].

Consider the following model of the *space-fractional diffusion equation*

$$\begin{cases}
\partial_t u = {}_{RZ}\mathrm{D}_x^{2\alpha} u + f(x,t), & (x,t) \in \Omega \times (0,T], \\
u(x,0) = \phi_0(x), & x \in \Omega, \\
u = 0, & (x,t) \in \partial\Omega \times (0,T],
\end{cases} \tag{5.38}$$

where $1/2 < \alpha \leq 1, \Omega = (a,b)$, and ${}_{RZ}\mathrm{D}_x^{2\alpha}$ is the Riesz space fractional derivative operator of order 2α defined as

$${}_{RZ}\mathrm{D}_x^{2\alpha} u = -c_{2\alpha}({}_{RL}\mathrm{D}_{a,x}^{2\alpha} + {}_{RL}\mathrm{D}_{x,b}^{2\alpha})u, \qquad c_{2\alpha} = \frac{1}{2\cos(\alpha\pi)}.$$

5.3.1 Semi-Discrete Approximation

We first write the variational formulation for (5.38). Multiplying $v \in H_0^\alpha(\Omega)$ on both sides of (5.38) and integrating in space yield

$$\begin{aligned}
(\partial_t u, v) &= ({}_{RZ}\mathrm{D}_x^{2\alpha} u, v) + (f, v) \\
&= -c_{2\alpha}({}_{RL}\mathrm{D}_{a,x}^{\alpha} u, {}_{RL}\mathrm{D}_{x,b}^{\alpha} v) + ({}_{RL}\mathrm{D}_{x,b}^{\alpha} u, {}_{RL}\mathrm{D}_{a,x}^{\alpha} v) + (f, v),
\end{aligned}$$

where Lemma 5.1.10 is used. Let

$$A(u,v) = c_{2\alpha}\left[({}_{RL}\mathrm{D}^{\alpha}_{a,x}u, {}_{RL}\mathrm{D}^{\alpha}_{x,b}v) + ({}_{RL}\mathrm{D}^{\alpha}_{x,b}u, {}_{RL}\mathrm{D}^{\alpha}_{a,x}v)\right]. \tag{5.39}$$

Then we have

$$(\partial_t u, v) + A(u,v) = (f,v), \quad \forall v \in H^{\alpha}_{0}(\Omega). \tag{5.40}$$

Therefore, the weak formulation of (5.38) reads as: Find $U(t) = U(\cdot,t) \in H^{\alpha}_{0}(\Omega)$, $U(0) = u(0)$ such that

$$(\partial_t U, v) + A(U,v) = (f,v), \quad \forall v \in H^{\alpha}_{0}(\Omega). \tag{5.41}$$

Now we give the following theorem.

Theorem 39 *Let* $1/2 < \alpha < 1$ *and* $t \in (0,T]$. *Suppose that* $u(t) \in H^{\alpha}_{0}(\Omega)$ *is a solution to* (5.41). *Then* u *is the unique solution to* (5.41) *satisfying*

$$\|U(t)\|^2 + C\int_0^t \|U(s)\|^2_{H^{\alpha}(\Omega)}\,\mathrm{d}s \le \|u(0)\|^2 + \frac{1}{C}\int_0^t \|f(s)\|^2\,\mathrm{d}s,$$

where C *is a positive constant.*

Proof. We show that $A(u,v)$ defined by (5.39) is continuous and coercive. It is easy to verify that

$$\begin{aligned} |A(u,v)| &\le -c_{2\alpha}\Big(|({}_{RL}\mathrm{D}^{\alpha}_{a,x}u, {}_{RL}\mathrm{D}^{\alpha}_{x,b}v)| + |({}_{RL}\mathrm{D}^{\alpha}_{x,b}u, {}_{RL}\mathrm{D}^{\alpha}_{a,x}v)|\Big) \\ &\le -c_{2\alpha}\Big(|u|_{J^{\alpha}_{L}(\Omega)}|v|_{J^{\alpha}_{R}(\Omega)} + |u|_{J^{\alpha}_{R}(\Omega)}|v|_{J^{\alpha}_{L}(\Omega)}\Big) \\ &\le C\|u\|_{H^{\alpha}(\Omega)}\|v\|_{H^{\alpha}(\Omega)}, \end{aligned} \tag{5.42}$$

where Lemma 5.1.6 is used when $u, v \in H^{\alpha}_{0}(\Omega)$. Inequality (5.42) means that $A(u,v)$ is continuous. For coercivity, we have

$$\begin{aligned} A(u,u) &= 2c_{2\alpha}({}_{RL}\mathrm{D}^{\alpha}_{a,x}u, {}_{RL}\mathrm{D}^{\alpha}_{x,b}u) = 2c_{2\alpha}\cos(\alpha\pi)\|{}_{RL}\mathrm{D}^{\alpha}_{a,x}\tilde{u}\|^2_{L^2(\mathbb{R})} \\ &= \|{}_{RL}\mathrm{D}^{\alpha}_{a,x}\tilde{u}\|^2_{L^2(\mathbb{R})} = |u|^2_{H^{\alpha}(\Omega)} \\ &\ge C\|u\|^2_{H^{\alpha}(\Omega)}, \end{aligned} \tag{5.43}$$

where $\tilde{u}$ is the zero extension of u outside Ω.

Letting $v = U$ in (5.41) yields

$$(\partial_t U, U) + A(U,U) = \frac{1}{2}\frac{\mathrm{d}}{\mathrm{d}t}\|U(t)\|^2 + A(U,U) = (f,U). \tag{5.44}$$

From (5.43) and (5.44), we have

$$\begin{aligned} \frac{1}{2}\frac{\mathrm{d}}{\mathrm{d}t}\|U(t)\|^2 + C_0\|U(t)\|^2_{H^{\alpha}(\Omega)} &\le \frac{1}{4\epsilon}\|f(t)\|^2 + \epsilon\|U(t)\|^2 \\ &\le \frac{1}{4\epsilon}\|f(t)\|^2 + \epsilon\|U(t)\|^2_{H^{\alpha}(\Omega)}, \end{aligned} \tag{5.45}$$

where ϵ is a suitable positive constant. Choosing $\epsilon = \frac{C_0}{2}$ and integrating over $(0,t)$, we have

$$\|U(t)\|^2 + C_0 \int_0^t \|U(s)\|^2_{H^\alpha(\Omega)} \mathrm{d}s \le \|u(0)\|^2 + \frac{1}{C_0} \int_0^t \|f(s)\|^2 \mathrm{d}s. \tag{5.46}$$

Next, we prove the uniqueness. Suppose that u, w are two solutions to (5.41). Let $e = u - w$ with $e(0) = 0$. Then we have

$$(\partial_t e, v) + A(e, v) = 0. \tag{5.47}$$

From (5.46), we have

$$\|e(t)\|^2 + C_0 \int_0^t \|e(s)\|^2_{H^\alpha(\Omega)} \mathrm{d}s \le 0,$$

which yields $e(t) = 0$, i.e., $u = w$. The proof is completed. □

Next, we give the semi-discrete approximation for (5.38), which reads as: Find $u_h = u_h(\cdot, t) \in X^\alpha_{h0}$ such that

$$(\partial_t u_h, v) + A(u_h, v) = (I_h f, v), \quad \forall v \in X^\alpha_{h0} \tag{5.48}$$

with initial condition $u_h(0) = I_h u(0)$.

Now, we give the matrix representation of (5.48). Suppose that $u_h(t) \in X^\alpha_{h0}$ has the following representation

$$u_h(t) = u_h(x,t) = \sum_{j=1}^{N_r - 1} c_j(t) \phi_j(x), \tag{5.49}$$

where ϕ_j is defined by (5.21). Inserting $u_h(t)$ into (5.48) and letting $v = \phi_j, j = 1, 2, \cdots, Nr - 1$, we can obtain

$$M \frac{\mathrm{d}\mathbf{c}(t)}{\mathrm{d}t} + S\,\mathbf{c}(t) = \mathbf{F}(t), \tag{5.50}$$

in which $\mathbf{c}(t) = (c_1(t), c_2(t), \cdots, c_{Nr-1}(t))^T$, $(\mathbf{F}(t))_j = (I_h f(t), \phi_j)$, and

$$(M)_{i,j} = (\phi_i, \phi_j), \qquad (S)_{i,j} = A(\phi_i, \phi_j). \tag{5.51}$$

Eq. (5.50) is a linear ordinary differential equation, which can be solved by using the Euler method, the trapezoidal rule, or high order methods. The initial value $\mathbf{c}(0)$ can be obtained from the initial condition $\phi_0(x)$ in (5.38).

Similar to Theorem 39, we can prove that the semi-discrete approximation (5.48) has a unique solution, which has the similar bound as that of (5.46).

Theorem 40 *Let $1/2 < \alpha < 1$ and $t \in (0, T]$. Suppose that $u_h(t) \in X^r_{h0}$ is a solution to (5.48). Then u_h is the unique solution to (5.48) satisfying*

$$\|u_h(t)\|^2 + C \int_0^t \|u_h(s)\|^2_{H^\alpha(\Omega)} \mathrm{d}s \le \|u_h(0)\|^2 + \frac{1}{C} \int_0^t \|I_h f(s)\|^2 \mathrm{d}s,$$

where C is a positive constant.

Next, we discuss the error estimate for the semi-discrete scheme (5.48). We first introduce a projector $\Pi_h^{\alpha,0}$: $H_0^\alpha(\Omega) \to X_{h0}^r$ as

$$A(\Pi_h^{\alpha,0}u - u, v) = 0, \qquad u \in H_0^\alpha(\Omega), \quad \forall v \in X_{h0}^r, \tag{5.52}$$

in which $A(u, v)$ is defined by (5.39).

Lemma 5.3.1 *Let* $u \in H^m(\Omega) \cap H_0^\alpha(\Omega), \alpha \le m \le r+1, 1/2 < \alpha < 1$. *Then there exists a positive constant C independent of h such that*

$$\|\Pi_h^{\alpha,0}u - u\|_{H^\alpha(\Omega)} \le Ch^{m-\alpha}\|u\|_{H^m(\Omega)}. \tag{5.53}$$

Proof. We first prove (5.53). From (5.52), we have

$$\begin{aligned} A(\Pi_h^{\alpha,0}u - u, \Pi_h^{\alpha,0}u - u) =& A(\Pi_h^{\alpha,0}u - u, \Pi_h^{\alpha,0}u - I_h u + I_h u - u) \\ =& A(\Pi_h^{\alpha,0}u - u, I_h u - u) \\ \le& C\|\Pi_h^{\alpha,0}u - u\|_{H^\alpha(\Omega)}\|I_h u - u\|_{H^\alpha(\Omega)}. \end{aligned}$$

From the coercivity of $A(u, v)$, we have

$$\begin{aligned} \|\Pi_h^{\alpha,0}u - u\|_{H^\alpha(\Omega)}^2 \le& C_0 A(\Pi_h^{\alpha,0}u - u, \Pi_h^{\alpha,0}u - u) \\ \le& C\|\Pi_h^{\alpha,0}u - u\|_{H^\alpha(\Omega)}\|I_h u - u\|_{H^\alpha(\Omega)}. \end{aligned}$$

Canceling the factor $\|\Pi_h^{\alpha,0}u - u\|_{H^\alpha(\Omega)}$ in the above equation and using Lemma 5.1.12 yields

$$\|\Pi_h^{\alpha,0}u - u\|_{H^\alpha(\Omega)} \le C\|I_h u - u\|_{H^\alpha(\Omega)} \le Ch^{m-\alpha}\|u\|_{H^m(\Omega)}. \tag{5.54}$$

Hence, inequality (5.53) holds. □

Let $u_*(t) = \Pi_h^{\alpha,0}u(t)$, $\eta(t) = u(t) - u_*(t)$ and $e(t) = u_*(t) - u_h(t)$. Then we can obtain the error equation of the semi-discrete approximation (5.48) as

$$(\partial_t e, v) + A(e, v) = -(\partial_t \eta, v) + (f - I_h f, v), \quad \forall v \in X_{h0}^\alpha. \tag{5.55}$$

Now we can get the following result.

Theorem 41 *Let* $1/2 < \alpha < 1$ *and* $t \in (0, T]$. *Suppose that* $u(t) \in H_0^\alpha(\Omega) \cap H^{r+1}(\Omega)$ *is a solution to* (5.41), $u_h(t)$ *is the solution to* (5.48), *and* $f(t) \in H^{r+1}(\Omega)$. *Then*

$$\left(\int_0^t \|u(s) - u_h(s)\|_{H^\alpha(\Omega)}^2 \mathrm{d}s\right)^{1/2} \le Ch^{r+1-\alpha},$$

where C is a positive constant.

Proof. From (5.55) and Theorem 39, we have

$$\begin{aligned}
&\|e(t)\|^2 + C\int_0^t \|e(s)\|^2_{H^\alpha(\Omega)}\,\mathrm{d}s \\
\leq &\|e(0)\|^2 + \frac{1}{C}\int_0^t \left(\|\partial_s\eta(s)\|^2 + \|f(s) - I_h f(s)\|^2\right)\mathrm{d}s \\
\leq &Ch^{2r+2}\|u(0)\|^2_{H^{r+1}(\Omega)} + Ch^{2r+2-2\alpha}\int_0^t \|\partial_s u(s)\|^2_{H^{r+1}(\Omega)}\,\mathrm{d}s \\
&+ Ch^{2r+2}\int_0^t \|f(s)\|^2_{H^{r+1}(\Omega)}\,\mathrm{d}s \\
\leq &Ch^{2r+2-2\alpha}.
\end{aligned} \tag{5.56}$$

Hence,

$$\begin{aligned}
\int_0^t \|u(s) - u_h(s)\|^2_{H^\alpha(\Omega)}\,\mathrm{d}s \leq &\int_0^t \|e(s)\|^2_{H^\alpha(\Omega)}\,\mathrm{d}s + \int_0^t \|\eta(s)\|^2_{H^\alpha(\Omega)}\,\mathrm{d}s \\
\leq &Ch^{2r+2-2\alpha}.
\end{aligned} \tag{5.57}$$

The proof is completed. □

5.3.2 Fully Discrete Approximation

In the previous subsection, we investigate the semi-discrete approximation for (5.41), where the space is approximated by the finite element method. In application, the semi-discrete scheme is not suitable for real computations. In this subsection, we discuss the fully discrete approximation for (5.41). The time discretization can be accomplished in several possible ways, such as the Euler method, the trapezoidal method, etc.

Next, we present the first fully discrete algorithm. The time direction is discretized by the backward Euler method, the space is discretized by the finite element method, the fully discretized approximation for (5.41) reads as: Find $u_h^n \in X_{h0}^r, n = 1, 2, \cdots, n_T$, such that

$$\begin{cases}
(\delta_t u_h^{n-\frac{1}{2}}, v) + A(u_h^n, v) = (I_h f^n, v), \quad \forall v \in X_{h0}^\alpha, \\
u_h^0 = I_h \phi_0,
\end{cases} \tag{5.58}$$

where

$$\delta_t u_h^{n-\frac{1}{2}} = \frac{u_h^n - u_h^{n-1}}{\Delta t}. \tag{5.59}$$

If the time direction is discretized by the Crank–Nicolson method, we can obtain the fully discrete approximation for (5.41) as: Find $u_h^n \in X_{h0}^r, n = 1, 2, \cdots, n_T$, such that

$$\begin{cases}
(\delta_t u_h^{n-\frac{1}{2}}, v) + A(u_h^{n-\frac{1}{2}}, v) = (I_h f(t_{n-\frac{1}{2}}), v), \quad \forall v \in X_{h0}^\alpha, \\
u_h^0 = I_h \phi_0,
\end{cases} \tag{5.60}$$

where

$$u_h^{n-\frac{1}{2}} = \frac{u_h^n + u_h^{n-1}}{2}. \tag{5.61}$$

We investigate the stability and convergence for (5.58).

Theorem 42 *Let $1/2 < \alpha < 1$. Suppose that $u_h^k, k = 1, 2, \cdots, n_T$ is a solution to* (5.58) *and $f \in C([0,T], L^2(\Omega))$. Then*

$$||u_h^k||^2_{H^\alpha(\Omega)} \le C||u_h^0||^2_{H^\alpha(\Omega)} + C\Delta t \sum_{n=1}^{k} ||f^n||^2,$$

where C is a positive constant independent of n and h.

Proof. Letting $v = \delta_t u_h^{n-\frac{1}{2}}$ in (5.58) yields

$$(\delta_t u_h^{n-\frac{1}{2}}, \delta_t u_h^{n-\frac{1}{2}}) + A(u_h^n, \delta_t u_h^{n-\frac{1}{2}}) = (I_h f^n, \delta_t u_h^{n-\frac{1}{2}}). \tag{5.62}$$

Using the Cauchy–Schwarz inequality and the coercivity of $A(u,v)$ yields

$$A(u_h^n, u_h^n) - A(u_h^{n-1}, u_h^{n-1}) \le \frac{\Delta t}{4} ||I_h f^n||^2. \tag{5.63}$$

Hence

$$A(u_h^n, u_h^n) \le A(u_h^0, u_h^0) + C\Delta t \sum_{k=1}^{n} ||I_h f^k||^2.$$

Summing n from 1 to k and using $||I_h f^n|| \le C||f^n||$ lead to

$$A(u_h^n, u_h^n) \le A(u_h^0, u_h^0) + C\Delta t \sum_{k=1}^{n} ||I_h f^k||^2.$$

Applying (5.42) and (5.43) gives the desired result.
The proof is thus completed. □

Let $u_*^n = \Pi_h^{\alpha,0} u^n$, $\eta^n = u^n - u_*^n$ and $e^n = u_*^n - u_h^n$. From (5.40) we have

$$(\partial_t u(t_n), v) + A(u(t_n), v) = (f^n, v), \quad \forall v \in H_0^\alpha(\Omega). \tag{5.64}$$

Replacing u_h in (5.60) with u_* leads to

$$(\delta_t u_*^{n-\frac{1}{2}}, v) + A(u_*^n, v) = (I_h f(t_n), v) + R. \tag{5.65}$$

Eliminating R from (5.60), (5.64) and (5.65) yields

$$(\delta_t e^{n-\frac{1}{2}}, v) + A(e^n, v) = (\delta_t u_*^n - \partial_t u(t_n), v) + (f^n - I_h f^n, v), \quad \forall v \in X_{h0}^r. \tag{5.66}$$

Next we can obtain the following convergence result.

Theorem 43 *Let $1/2<\alpha<1$, $m\geq r+1$. Suppose that u_h^k, $k=1,2,\cdots,n_T$ is a solution to (5.58), u is the solution to (5.38). If $u\in H^2([0,T],H^m(\Omega)\cap H_0^1(\Omega))$, $\phi_0\in H^m(\Omega)$, and $f\in C([0,T],H^m(\Omega))$, then*

$$\|u(t_k)-u_h^k\|_{H^\alpha(\Omega)}\leq C(\Delta t+h^{r+1-\alpha}),$$

where C is a positive constant independent of $k,\Delta t$ and h.

Proof. From Theorem 42, we have

$$\|e^k\|_{H^\alpha(\Omega)}^2\leq\|e^0\|_{H^\alpha(\Omega)}^2+C\Delta t\sum_{n=1}^{k}(\|\delta_t u_*^{n-\frac{1}{2}}-\partial_t u(t_n)\|^2+\|f^n-I_hf^n\|^2).$$

Noting that

$$\|e^0\|=\|I_h\phi_0-\phi_0+\phi_0-\Pi_h^{\alpha,0}\phi_0\|\leq Ch^{r+1-\alpha}\|\phi_0\|_{H^{r+1}(\Omega)},$$
$$\|f^n-I_hf^n\|\leq Ch^{r+1-\alpha}\|f^n\|_{H^{r+1}(\Omega)},$$

and

$$\|\delta_t u_*^{n-\frac{1}{2}}-\partial_t u(t_n)\|^2=\|-\delta_t\eta^{n-\frac{1}{2}}+\delta_t u^{n-\frac{1}{2}}-\partial_t u(t_n)\|^2$$
$$\leq C\left(h^{2r+2}\Delta t^{-1}\int_{t_{n-1}}^{t_n}\|\partial_t u\|_{H^{r+1-\alpha}(\Omega)}^2\,\mathrm{d}t+\int_{t_{n-1}}^{t_n}\|\partial_t^2 u\|\,\mathrm{d}t\right),$$

we have $\|e^k\|_{H^\alpha(\Omega)}\leq Ch^{r+1-\alpha}$. Using $\|u(t_k)-u_h^k\|_{H^\alpha(\Omega)}\leq\|\eta^k\|_{H^\alpha(\Omega)}+\|e^k\|_{H^\alpha(\Omega)}$ yields the desired result. The proof is completed. □

Next, we analyze the method (5.60).

Theorem 44 *Let $1/2<\alpha<1$. Suppose that $u_h^k, k=1,2,\cdots,n_T$ is a solution to (5.60) and $f\in C([0,T],L^2(\Omega))$. Then one has*

$$\|u_h^k\|_{H^\alpha(\Omega)}^2\leq C\|u_h^0\|_{H^\alpha(\Omega)}^2+C\Delta t\sum_{n=1}^{k}\|f(t_{n-\frac{1}{2}})\|^2, \tag{5.67}$$

where C is a positive constant independent of $n,\Delta t$ and h.

Proof. Letting $v=\delta_t u_h^{n-\frac{1}{2}}$ in (5.60) yields

$$(\delta_t u_h^{n-\frac{1}{2}},\delta_t u_h^{n-\frac{1}{2}})+A(u_h^{n-\frac{1}{2}},\delta_t u_h^{n-\frac{1}{2}})=(I_hf^{n-\frac{1}{2}},\delta_t u_h^{n-\frac{1}{2}}). \tag{5.68}$$

Using the Cauchy–Schwarz inequality yields

$$\|\delta_t u_h^{n-\frac{1}{2}}\|^2+A(u_h^{n-\frac{1}{2}},\delta_t u_h^{n-\frac{1}{2}})\leq\|\delta_t u_h^{n-\frac{1}{2}}\|^2+C\|I_hf(t_{n-\frac{1}{2}})\|^2. \tag{5.69}$$

Rearranging the above inequality and using the property $A(u,v)=A(v,u)$ gives

$$A(u_h^n,u_h^n)\leq A(u_h^{n-1},u_h^{n-1})+C\Delta t\|I_hf(t_{n-\frac{1}{2}})\|^2. \tag{5.70}$$

Summing n from 1 to k and using $||I_hf^n|| \leq C||f^n||$ lead to

$$A(u_h^k, u_h^k) \leq A(u_h^0, u_h^0) + C\Delta t \sum_{n=1}^{k} ||f(t_{n-\frac{1}{2}})||^2. \quad (5.71)$$

Using the coercivity and continuity of $A(u, v)$ yields (5.67).

The proof is completed. □

Next, we show the convergence of (5.60).

Theorem 45 *Let $1/2 < \alpha < 1, m \geq r+1$. Suppose that $u_h^k, k = 1, 2, \cdots, n_T$ is solution of (5.60), u is the solution to (5.38). If $u \in H^3([0,T], H^m(\Omega) \cap H_0^1(\Omega))$, $\phi_0 \in H^m(\Omega)$, and $f \in C([0,T], H^m(\Omega))$, then*

$$||u(t_k) - u_h^k||_{H^\alpha(\Omega)} \leq C(\Delta t^2 + h^{r+1-\alpha}), \quad (5.72)$$

where C is a positive constant independent of $k, \Delta t$ and h.

Proof. The error equation of (5.66) can be written as

$$(\delta_t e^{n-\frac{1}{2}}, v) + A(e^{n-\frac{1}{2}}, v) = (\delta_t u_*^{n-\frac{1}{2}} - \partial_t u(t_{n-\frac{1}{2}}), v) + (f(t_{n-\frac{1}{2}}) - I_h f(t_{n-\frac{1}{2}}), v). \quad (5.73)$$

From Theorem 44, we have

$$||e^k||^2_{H^\alpha(\Omega)} \leq ||e^0||^2_{H^\alpha(\Omega)} + C\Delta t \sum_{n=1}^{k} (||\delta_t u_*^{n-\frac{1}{2}} - \partial_t u(t_{n-\frac{1}{2}})||^2 + ||f(t_{n-\frac{1}{2}}) - I_h f(t_{n-\frac{1}{2}})||^2). \quad (5.74)$$

For e^0, we have

$$\begin{aligned} ||e^0||_{H^\alpha(\Omega)} &= ||I_h\phi_0 - \Pi_h^{\alpha,0} u^0||_{H^\alpha(\Omega)} \\ &\leq ||I_h\phi_0 - \phi_0||_{H^\alpha(\Omega)} + ||\phi_0 - \Pi_h^{\alpha,0} u^0||_{H^\alpha(\Omega)} \\ &\leq Ch^{r+1-\alpha} ||\phi_0||_{H^{r+1}(\Omega)}. \end{aligned}$$

For $||\delta_t u_*^{n-\frac{1}{2}} - \partial_t u(t_{n-\frac{1}{2}})||$ and $||f(t_{n-\frac{1}{2}}) - I_h f(t_{n-\frac{1}{2}})||$, we have

$$\begin{aligned} ||\delta_t u_*^{n-\frac{1}{2}} - \partial_t u(t_{n-\frac{1}{2}})||^2 &\leq ||\delta_t u_*^{n-\frac{1}{2}} - \delta_t u^{n-\frac{1}{2}}||^2 + ||\delta_t u^{n-\frac{1}{2}} - \partial_t u(t_{n-\frac{1}{2}})||^2 \\ &\leq C\Delta t^{-1} h^{2r+2} \int_{t_{n-1}}^{t_n} ||\partial_t u(t)||_{H^{r+1}(\Omega)} \, dt + C\Delta t^3 \int_{t_{n-1}}^{t_n} ||\partial_t^3 u(t)|| \, dt, \end{aligned} \quad (5.75)$$

and

$$||f(t_{n-\frac{1}{2}}) - I_h f(t_{n-\frac{1}{2}})|| \leq Ch^{r+1} ||f(t)||_{H^{r+1}(\Omega)}. \quad (5.76)$$

Hence,

$$||e^k||_{H^\alpha(\Omega)} \leq C(\Delta t^2 + h^{r+1}).$$

Using $||u(t_k) - u_h^k||_{H^\alpha(\Omega)} \leq ||u(t_k) - \Pi_h^{\alpha,0} u(t_k)||_{H^\alpha(\Omega)} + ||\Pi_h^{\alpha,0} u(t_k) - u_h^k||_{H^\alpha(\Omega)} = ||e^k||_{H^\alpha(\Omega)} + ||\Pi_h^{\alpha,0} u(t_k) - u_h^k||_{H^\alpha(\Omega)}$ and Lemma 5.3.1 give the desired result. The proof is thus finished. □

5.4 Galerkin FEM for Time-Fractional Differential Equations

In this section, we introduce the finite element method for the time-fractional differential equations. These equations include time-fractional diffusion equations, the time fractional cable equation, and the time fractional Fokker–Planck equation, etc. In order to illustrate how to use the Galerkin finite element method to solve time-fractional equations, we mainly investigate two kinds of model problems.

The *Riemann–Liouville type time-fractional diffusion equation* is given below:

$$\begin{cases} \partial_t u = {}_{RL}\mathrm{D}_{0,t}^{1-\gamma}\left(K_\gamma \partial_x^2 u\right) + f(x,t), & (x,t) \in \Omega \times (0,T], \\ u(x,0) = \phi_0(x), \quad x \in \Omega, \\ u(x,t) = 0, \quad (x,t) \in \partial\Omega \times (0,T], \end{cases} \tag{5.77}$$

where $\Omega = (a,b)$, $K_\gamma > 0$ and $0 < \gamma < 1$.

The *Caputo type time-fractional diffusion equation* reads as:

$$\begin{cases} {}_C\mathrm{D}_{0,t}^{\gamma} u = K_\gamma \partial_x^2 u + g(x,t), & (x,t) \in \Omega \times (0,T], \\ u(x,0) = \phi_0(x), \quad x \in \Omega, \\ u(x,t) = 0, \quad (x,t) \in \partial\Omega \times (0,T], \end{cases} \tag{5.78}$$

where $\Omega = (a,b)$, $K_\gamma > 0$ and $0 < \gamma < 1$.

5.4.1 Semi-Discrete Schemes

We first consider the semi-discrete approximations for (5.77) and (5.78). Multiplying by $v \in H_0^1(\Omega)$ on both sides of (5.77) and integrating by parts, we obtain

$$(\partial_t u, v) + K_\gamma({}_{RL}\mathrm{D}_{0,t}^{1-\gamma}\partial_x u, \partial_x v) = (f, v). \tag{5.79}$$

From the above equation, we can derive the semi-discrete scheme for (5.77) as: Find $u_h(t) \in X_{h0}^r$, such that

$$\begin{cases} (\partial_t u_h, v) + K_\gamma({}_{RL}\mathrm{D}_{0,t}^{1-\gamma}\partial_x u_h, \partial_x v) = (I_h f, v), & v \in X_{h0}^r, \\ u_h(0) = \Pi_h^{1,0}\phi_0(x). \end{cases} \tag{5.80}$$

We can similarly give the semi-discrete scheme for (5.78) as: Find $u_h(t) \in X_{h0}^r$, such that

$$\begin{cases} ({}_C\mathrm{D}_{0,t}^{\gamma} u_h, v) + K_\gamma(\partial_x u_h, \partial_x v) = (I_h g, v), & v \in X_{h0}^r, \\ u_h(0) = \Pi_h^{1,0}\phi_0(x). \end{cases} \tag{5.81}$$

Next, we present the matrix representations for (5.80) and (5.81). Suppose that the solution to (5.80) or (5.81) has the expression as in (5.49). Denote by $\mathbf{c}(t) = (c_1(t), c_2(t), \cdots, c_{Nr-1}(t))^T$, $(\mathbf{F}(t))_j = (f(t), \phi_j)$, $(\mathbf{G}(t))_j = (g(t), \phi_j)$, and

$$(M)_{i,j} = (\phi_i, \phi_j), \qquad (S)_{i,j} = (\partial_x \phi_i, \partial_x \phi_j). \tag{5.82}$$

Then the matrix representation of (5.80) or (5.81) can be expressed as

$$M\frac{\mathrm{d}\mathbf{c}(t)}{\mathrm{d}t}+K_\gamma S\,{}_{RL}D_{0,t}^{1-\gamma}\mathbf{c}(t)=\mathbf{F}(t), \tag{5.83}$$

or

$$M_C\mathrm{D}_{0,t}^{\gamma}\mathbf{c}(t)+K_\gamma S\,\mathbf{c}(t)=\mathbf{G}(t). \tag{5.84}$$

Eqs. (5.83) and (5.84) are fractional ordinary differential equations, so they can be solved by the methods developed in the previous chapters. For example, Eq. (5.83) can be solved by time discretization techniques as presented in Subsection 4.2.1 for (4.10), and Eq. (5.84) can be solved by time discretization as derived in Subsection 4.2.2 for (4.86), or see the numerical methods used for FODE (3.1).

Next, we present the error estimate for (5.80) and (5.81). Let $u_*(t)=\Pi_h^{1,0}u(t)$, $\eta(t)=u(t)-u_*(t)$ and $e(t)=u_*(t)-u_h(t)$. We can obtain the error equation for (5.80) as

$$\begin{cases}(\partial_t e,v)+K_\gamma({}_{RL}\mathrm{D}_{0,t}^{1-\gamma}\partial_x e,\partial_x v)=-(\partial_t\eta,v)+(f-I_hf,v), & v\in X_{h0}^r,\\ e(0)=0.\end{cases} \tag{5.85}$$

Theorem 46 *Let $0<\gamma<1$ and $t\in(0,T]$. Suppose that $u\in C^1(0,T;H^{r+1}(\Omega)\cap H_0^1(\Omega))$ is a solution to (5.77), and $u_h(t)$ is the solution to (5.80), $f\in L^2(0,T;H^{r+1}(\Omega))$. Then*

$$\|u_h(t)-u(t)\|^2\le Ch^{2r+2}\|u(t)\|_{H^{r+1}(\Omega)}^2+Ch^{2r+2}\int_0^t\left(\|\partial_s u(s)\|_{H^{r+1}(\Omega)}^2+\|f(s)\|_{H^{r+1}(\Omega)}^2\right)\mathrm{d}s,$$

where C is a positive constant.

Proof. Letting $v=e$ in (5.85) and using the Cauchy–Schwarz inequality yield

$$\begin{aligned}\frac{1}{2}\frac{\mathrm{d}}{\mathrm{d}t}\|e\|^2+K_\gamma({}_{RL}\mathrm{D}_{0,t}^{1-\gamma}\partial_x e,\partial_x e)&=(\partial_t e,e)+K_\gamma({}_{RL}\mathrm{D}_{0,t}^{1-\gamma}\partial_x e,\partial_x e)\\&=(-\partial_t\eta+f-I_hf,e)\\&\le\frac{1}{2}\left(\|e\|^2+\|-\partial_t\eta+f-I_hf\|^2\right).\end{aligned} \tag{5.86}$$

Integrating on the interval $(0,t]$ gives

$$\begin{aligned}\|e(t)\|^2\le&\|e(t)\|^2+2K_\gamma\int_0^t({}_{RL}\mathrm{D}_{0,s}^{1-\gamma}\partial_x e(s),\,\partial_x e(s))\,\mathrm{d}s\\\le&\int_0^t\|e(s)\|^2\,\mathrm{d}s+\int_0^t\|-\partial_s\eta(s)+f(s)-I_hf(s)\|^2\,\mathrm{d}s,\end{aligned} \tag{5.87}$$

where $\int_0^t({}_{RL}D_{0,t}^{1-\gamma}\partial_x e(s),\,\partial_x e(s))\,\mathrm{d}s\ge0$ because of Lemmas 5.1.2 and 5.1.11. Applying Gronwall's inequality (see Lemma 5.1.12) yields

$$\begin{aligned}\|e(t)\|^2\le&C\int_0^t\|-\partial_s\eta(s)+f(s)-I_hf(s)\|^2\,\mathrm{d}s\\\le&C\int_0^t\|\partial_s\eta(s)\|^2\,\mathrm{d}s+C\int_0^t\|f(s)-I_hf(s)\|^2\,\mathrm{d}s\\\le&Ch^{2r+2}\int_0^t\left(\|\partial_s u(s)\|_{H^{r+1}(\Omega)}^2+\|f(s)\|_{H^{r+1}(\Omega)}^2\right)\mathrm{d}s.\end{aligned} \tag{5.88}$$

Hence,

$$\begin{aligned} &\|u(t)-u_h(t)\| \le \|u(t)-u_*(t)\| + \|u_*(t)-u_h(t)\| = \|e(t)\| + \|\eta(t)\| \\ &\le Ch^{r+1}\|u(t)\|_{H^{r+1}(\Omega)} + Ch^{r+1}\left[\int_0^t \left(\|\partial_s u(s)\|^2_{H^{r+1}(\Omega)} + \|f(s)\|^2_{H^{r+1}(\Omega)}\right) \mathrm{d}s\right]^{1/2}. \end{aligned} \tag{5.89}$$

The proof is completed. □

We can similarly write the error equation for (5.81) as

$$\begin{cases} ({}_C\mathrm{D}^{\gamma}_{0,t}e(t), v) + K_\gamma(\partial_x e, \partial_x v) = -({}_C\mathrm{D}^{\gamma}_{0,t}\eta(t), v) + (g - I_h g, v), \quad v \in X^r_{h0}, \\ e(0) = 0. \end{cases} \tag{5.90}$$

Theorem 47 *Let $0 < \gamma < 1$ and $t \in (0,T]$. Suppose that $u(t) \in C^1(0,T; H^{r+1}(\Omega) \cap H^1_0(\Omega))$ is a solution to (5.78), and $u_h(t)$ is the solution to (5.81), $g(t) \in L^2(0,T; H^{r+1}(\Omega))$. Then*

$$\int_0^t \|{}_C\mathrm{D}^{\gamma/2}_{0,s}(u_h(s) - u(s))\|^2 \mathrm{d}s \le Ch^{2r+2},$$

where C is a positive constant independent of h.

Proof. Letting $v = e(t)$ in (5.90) yields

$$\begin{aligned} ({}_C\mathrm{D}^{\gamma}_{0,t}e(t), e(t)) + K_\gamma(\partial_x e, \partial_x e) &= -({}_C\mathrm{D}^{\gamma}_{0,t}\eta(t), e) + (g - I_h g, e) \\ &\le \epsilon\|e(t)\|^2 + C(\|{}_C\mathrm{D}^{\gamma}_{0,t}\eta(t)\|^2 + \|g - I_h g\|^2), \end{aligned} \tag{5.91}$$

where ϵ is a suitable constant such that $\epsilon\|e(t)\|^2 \le K_\gamma\|\partial_x e\|^2$. So

$$({}_C\mathrm{D}^{\gamma}_{0,t}e(t), e(t)) \le C(\|{}_C\mathrm{D}^{\gamma}_{0,t}\eta(t)\|^2 + \|g - I_h g\|^2). \tag{5.92}$$

Integrating on $[0,t]$ yields

$$\begin{aligned} \int_0^t ({}_C\mathrm{D}^{\gamma}_{0,s}e(s), e(s))\mathrm{d}s &\le C\int_0^t (\|{}_C\mathrm{D}^{\gamma}_{0,s}\eta(s)\|^2 + \|g(s) - I_h g(s)\|^2)\mathrm{d}s \\ &\le Ch^{2(r+1)}\int_0^t (\|{}_C D^{\gamma}_{0,s}u(s)\|^2_{H^{r+1}(\Omega)} + \|g(s)\|^2_{H^{r+1}(\Omega)})\mathrm{d}s. \end{aligned} \tag{5.93}$$

Since $e(0) = 0$, ${}_C\mathrm{D}^{\gamma}_{0,t}e(t) = {}_{RL}\mathrm{D}^{\gamma}_{0,t}e(t)$. Using Lemmas 5.1.11, 5.1.2, and 5.1.6 yields

$$\int_0^t \|{}_C\mathrm{D}^{\gamma/2}_{0,s}e(s)\|^2 \mathrm{d}s \le Ch^{2(r+1)}. \tag{5.94}$$

Applying $\|{}_C\mathrm{D}^{\gamma/2}_{0,s}(u_h(s) - u(s))\|^2 \le \|{}_C\mathrm{D}^{\gamma/2}_{0,s}e(s)\|^2 + \|{}_C\mathrm{D}^{\gamma/2}_{0,s}\eta(s)\|^2$ gives the desired result. The proof is finished. □

5.4.2 Fully Discrete Schemes

In the present subsection, we present the fully discrete algorithms for the time-fractional differential equations in forms of (5.77) and (5.78). As is known, we introduce several finite difference schemes for Eqs. (5.77) and (5.78) in Chapter 4, which can be directly extended to solve Eqs. (5.77) and (5.78), except that the space is approximated by the finite element in this chapter. Here, we just present several schemes to illustrate how to construct the fully discrete finite element schemes for equations in forms of (5.77) and (5.78), and how to analyze the stability and convergence.

- **The fully discrete finite element methods for** (5.77)

We first consider the fully discrete schemes for (5.77). For the first fully discrete scheme, the integer time derivative and the Riemann–Liouville derivative are discretized by the backward Euler formula and the Grünwald–Letnikov formula (see the time discretization for (4.30)), respectively. Therefore, the fully discrete finite element method for (5.77) is given by: Find $u_h^n \in X_{h0}^r$ for $n = 0, 1, \cdots, n_T - 1$, such that

$$\begin{cases} (\delta_t u_h^{n-\frac{1}{2}}, v) + K_\gamma({}^{GL}\delta_t^{(1-\gamma)}\partial_x u_h^n, \partial_x v) = (I_h f^n, v), \quad v \in X_{h0}^r, \\ u_h^0 = \Pi_h^{1,0}\phi_0, \end{cases} \tag{5.95}$$

where ${}^{GL}\delta_t^{(\gamma)}, 0 < \gamma < 1$ is defined by

$${}^{GL}\delta_t^{(\gamma)} u_h^n = \frac{1}{\Delta t^\gamma}\sum_{k=0}^{n}\omega_{n-k}^{(\gamma)} u_h^k, \qquad \omega_k^{(\gamma)} = (-1)^k\binom{\gamma}{k}.$$

We can also construct the Crank–Nicolson finite element method for (5.77), in which the time discretization of (5.77) is approximated as that of Eq. (4.10), see the Crank–Nicolson finite difference method (4.59). The fully discrete Crank–Nicolson finite element method for (5.77) is given by: Find $u_h^n \in X_{h0}^r$ for $n = 0, 1, \cdots, n_T - 1$, such that

$$\begin{cases} (\delta_t u_h^{n+\frac{1}{2}}, v) = \dfrac{\mu}{\tau^{1-\beta}}\Big[-b_0(\partial_x u_h^{n+\frac{1}{2}}, \partial_x v) + \displaystyle\sum_{k=1}^{n}(b_{n-k} - b_{n-k+1})(\partial_x u_h^{k-\frac{1}{2}}, \partial_x v) \\ \qquad + (b_n - B_n)(\partial_x u_h^{\frac{1}{2}}, \partial_x v) + A_n(\partial_x u_h^0, \partial_x v)\Big] + (I_h f(t_{n+\frac{1}{2}}), v), \quad \forall v \in X_{h0}^r, \\ u_h^0 = \Pi_h^{1,0}\phi_0, \end{cases} \tag{5.96}$$

where $A_n = B_n - \frac{\gamma(n+1/2)^{\gamma-1}}{\Gamma(1+\gamma)}$, b_n and B_n are defined by

$$b_n = \frac{1}{\Gamma(1+\gamma)}\Big[(n+1)^\gamma - n^\gamma\Big], \quad B_n = \frac{2}{\Gamma(1+\gamma)}\Big[(n+1/2)^\gamma - n^\gamma\Big].$$

In the following we analyze the stability and convergence for (5.95). We first give the following theorem.

Theorem 48 *Let u_h^n be the solution to (5.95), and $f \in C(0,T;L^2(\Omega))$. Then there exists a positive constant C independent of n, Δt and h, such that*

$$||u_h^n||^2 \leq ||u_h^0||^2 + \Delta t^\gamma K_\gamma ||\partial_x u_h^0||^2 + C \max_{0\leq k\leq n_T} ||f^k||^2.$$

Proof. The proof is similar to that of Theorem 23. Letting $v = u_h^n$ yields

$$(u_h^n, u_h^n) = (u_h^{n-1}, u_h^n) - \Delta t^\gamma K_\gamma \sum_{k=0}^{n} \omega_{n-k}^{(1-\gamma)} (\partial_x u_h^k, \partial_x u_h^n) + \Delta t (I_h f^n, u_h^n). \tag{5.97}$$

Denote by

$$b_n = \sum_{k=0}^{n} \omega_k^{(1-\gamma)} = \frac{\Gamma(n+\gamma)}{\Gamma(\gamma)\Gamma(n+1)} = \frac{(n+1)^{\gamma-1}}{\Gamma(\gamma)} + O((n+1)^{-2+\gamma}).$$

Then one has $b_n - b_{n-1} = \omega_n^{(1-\gamma)}$ and b_n satisfies $C_0 b_n \Delta t^\gamma \leq \Delta t \leq C_1 b_n \Delta t^\gamma$, C_0, C_1 are positive constants independent of n.

Using the Cauchy–Schwarz inequality, one obtains

$$\begin{aligned} &||u_h^n||^2 + \Delta t^\gamma K_\gamma ||\partial_x u_h^n||^2 \\ \leq &\frac{1}{2}(||u_h^{n-1}||^2 + ||u_h^n||^2) + \frac{\Delta t^\gamma K_\gamma}{2} \sum_{k=0}^{n-1} (b_{n-k-1} - b_{n-k})(||\partial_x u_h^k||^2 + ||\partial_x u_h^n||^2) \\ &+ \Delta t(\epsilon ||u_h^n||^2 + \frac{1}{4\epsilon}||I_h f^n||^2), \end{aligned} \tag{5.98}$$

where ϵ is a suitable positive constant. Denote by

$$E^n = ||u_h^n||^2 + \Delta t^\gamma K_\gamma \sum_{k=0}^{n} b_{n-k} ||\partial_x u_h^k||^2.$$

Then one has

$$\begin{aligned} E^n + \Delta t^\gamma K_\gamma b_n ||\partial_x u_h^n|| &\leq E^{n-1} + \Delta t \left(\frac{1}{2\epsilon} ||I_h f^n||^2 + 2\epsilon ||u_h^n||^2 \right) \\ &\leq E^{n-1} + \Delta t \left(\frac{1}{2\epsilon} ||I_h f^n||^2 + 2C_2 \epsilon ||u_h^n||^2 \right), \end{aligned} \tag{5.99}$$

where $||u_h^n|| \leq C ||\partial_x u_h^n||$ has been used. Choose suitable $\epsilon = \frac{K_\gamma}{2C_1 C_2}$ satisfying

$$2C_2 \epsilon \Delta t \leq 2C_2 \epsilon C_1 b_n \Delta t^\gamma \leq K_\gamma b_n \Delta t^\gamma.$$

Therefore, one obtains

$$\begin{aligned} E^n &\leq E^{n-1} + C\Delta t ||I_h f^n||^2 \leq E^0 + C\Delta t \sum_{k=1}^{n} ||f^k||^2 \\ &= ||u_h^0||^2 + \Delta t^\gamma K_\gamma ||\partial_x u_h^0||^2 + C\Delta t \sum_{k=1}^{n} ||f^k||^2. \end{aligned} \tag{5.100}$$

By the definition of E^n, one has

$$\|u_h^n\|^2 \le E^n \le \|u_h^0\|^2 + \Delta t^\gamma K_\gamma \|\partial_x u_h^0\|^2 + C \max_{0\le k\le n_T} \|f^k\|^2.$$

The proof is completed. □

The error estimate for the scheme (5.95) is given in the following theorem.

Theorem 49 *Let $0<\gamma<1$ and $t\in(0,T]$. Suppose that $u(t)\in C^3(0,T;H^{r+1}(\Omega)\cap H_0^1(\Omega))$ is a solution to (5.78), and u_h^n is the solution to (5.95), $f\in C(0,T;H^{r+1}(\Omega))$. Then there exists a positive constant C such that*

$$\|u_h^n - u(t_n)\| \le C(\Delta t + h^{r+1}).$$

Proof. We first write the error equation. Let $u_* = \Pi_h^{1,0}u$, $\eta^n = u^n - u_*^n$ and $e^n = u_*^n - u_h^n$. From (4.26), (4.27) and (4.29), we have

$$\delta_t u^{n-\frac{1}{2}} = K_\gamma {}^{GL}\delta_t^{(1-\gamma)} \partial_x^2 u^n + f^n + O(\Delta t). \tag{5.101}$$

Replacing u_h^n in (5.95) with u_*^n, we have

$$(\delta_t u_*^{n-\frac{1}{2}}, v) + K_\gamma({}^{GL}\delta_t^{(1-\gamma)} \partial_x u_*^n, \partial_x v) = (I_h f^n, v) + r^n. \tag{5.102}$$

Removing r^n from (5.95), (5.101) and (5.102) yields

$$(\delta_t e^{n-\frac{1}{2}}, v) + K_\gamma({}^{GL}\delta_t^{(1-\gamma)} \partial_x e^n, \partial_x v) = (R^n, v), \tag{5.103}$$

where $(\partial_x(u^n - \Pi_h^{1,0}u), \partial_x v) = 0$ for $v\in X_{h0}^r$, and $R^n = R_1^n + R_2^n + R_3^n$ satisfies

$$R_1^n = O(\Delta t), \quad R_2^n = f^n - \Pi_h^{1,0} f^n, \quad R_3^n = -\delta_t \eta^{n-1/2}. \tag{5.104}$$

From (5.103) and Theorem 48, we only need to estimate

$$\|e^0\|^2 + \Delta t^\gamma K_\gamma \|\partial_x e^0\|^2 + C \max_{0\le k\le n_T} \|R^k\|^2$$

to derive the error bound. Obviously, $e^0 = 0$ and

$$\|R^k\| \le \|R_1^n\| + \|R_2^n\| + \|R_3^n\| \le C(\Delta t + h^{r+1}\|f\|_{C(0,T;H^{r+1}(\Omega))} + h^{r+1}\|u\|_{C^1(0,T;H^{r+1}(\Omega))}).$$

Hence, $\|e^n\| \le C(\Delta t + h^{r+1})$. Using $\|u_h^n - u(t_n)\| \le \|e^n\| + \|\eta^n\|$ ends the proof. □

For the Crank–Nicolson finite element method (5.96), we have the similar result as that of Theorem 48, and the convergence rate of (5.96) is $O(\Delta t^{1+\gamma} + h^{r+1})$.

Of course, all the finite difference methods (see for example, (4.49), (4.67), (4.79), (4.83), and (4.84)) developed in Subsection 4.2.1 for (4.10) can be directly extended to (5.77), except that the finite difference discretization in space is replaced by the finite element discretization, and the stability and convergence are almost similar, so we do not list all these methods here.

- **The fully discrete finite element methods for** (5.78)

Next, we consider fully discrete approximations for (5.78). As is known in Subsection 4.2.2, the Caputo derivative in (5.78) (see also (4.86)) can be directly discretized by the Grünwald–Letnikov (see (4.96)) formula or the L1 method (see (4.97)). The fractional linear multi-step methods can be used to discretize the time direction of (5.78) (see (4.116) and (4.117)).

The first fully discrete finite element method for (5.78) with the time direction approximated by the L1 method (see also (4.97)) is given by: Find u_h^n $(n = 1, 2, \cdots, n_T)$, such that

$$\begin{cases} (\delta_t^{(\gamma)} u_h^n, v) + K_\gamma(\partial_x u_h^n, \partial_x v) = (I_h g^n, v), \\ u_h^0 = \Pi_h^{1,0}\phi_0, \end{cases} \tag{5.105}$$

where $\delta_t^{(\gamma)}$ is defined by

$$\delta_t^{(\gamma)} u_h^n = \frac{1}{\Delta t^\gamma}\sum_{k=0}^{n-1} b_{n-k}^{(\gamma)}(u_h^{k+1} - u_h^k), \quad b_k^{(\gamma)} = \frac{1}{\Gamma(2-\gamma)}\left[(k+1)^{1-\gamma} - k^{1-\gamma}\right].$$

The following theorem indicates that the scheme (5.105) is unconditionally stable.

Theorem 50 *Let u_h^n be the solution to* (5.105)*, and $f \in C(0,T; L^2(\Omega))$. Then there exists a positive constant C independent of n, Δt and h, such that*

$$\|u_h^n\|^2 \leq 2\|u_h^0\|^2 + 2C \max_{0\leq k\leq n_T} \|f^k\|^2.$$

Proof. Denote by $\mu = \Delta t^\gamma / b_0^{(\gamma)}$ and $c_k^{(\gamma)} = b_k^{(\gamma)}/b_0^{(\gamma)} = (k+1)^{1-\gamma} - k^{1-\gamma}$, so $c_0^{(\gamma)} = 1$. Letting $v = u_h^n$ in (5.105) yields

$$\begin{aligned} &(u_h^n, u_h^n) + \mu K_\gamma(\delta_x u_h^n, \delta_x u_h^n) \\ =& \sum_{k=1}^{n-1}(c_{n-k-1}^{(\gamma)} - c_{n-k}^{(\gamma)})(u_h^k, u_h^n) + c_{n-1}^{(\gamma)}(u_h^0, u_h^n) + \mu(I_h g^n, u_h^n). \end{aligned} \tag{5.106}$$

Using the Cauchy inequality, one has

$$\begin{aligned} \|u_h^n\|^2 + K_\gamma \mu \|\partial_x u_h^n\|^2 \leq & \frac{1}{2}\sum_{k=1}^{n-1}(c_{n-k-1}^{(\gamma)} - c_{n-k}^{(\gamma)})(\|u_h^k\|^2 + \|u_h^n\|^2) \\ & + \frac{c_{n-1}^{(\gamma)}}{4}\|u_h^n\|^2 + c_{n-1}^{(\gamma)}\|u_h^0\|^2 + \frac{c_{n-1}^{(\gamma)}}{4}\|u_h^n\|_N^2 + \frac{\mu^2}{c_{n-1}^{(\gamma)}}\|I_h g^n\|^2. \end{aligned} \tag{5.107}$$

One immediately gets from (5.107) that

$$\begin{aligned} \|u_h^n\|^2 \leq & \sum_{k=1}^{n-1}(c_{n-k-1}^{(\gamma)} - c_{n-k}^{(\gamma)})\|u_h^k\|^2 + 2c_{n-1}^{(\gamma)}\|u_h^0\|^2 + \frac{2\mu^2}{c_{n-1}^{(\gamma)}}\|g^n\|^2 \\ \leq & \sum_{k=1}^{n-1}(c_{n-k-1}^{(\gamma)} - c_{n-k}^{(\gamma)})\|u_h^k\|^2 + c_{n-1}^{(\gamma)}\left(2\|u_h^0\|^2 + 2C\|g^n\|^2\right), \end{aligned} \tag{5.108}$$

where the positive constant C is independent of $n, \Delta t$ and Δx, but satisfies $\frac{\mu^2}{(c_{n-1}^{(\gamma)})^2} \le C$. Next, we prove that

$$\|u_h^n\|^2 \le 2\|u_h^0\|^2 + 2C \max_{0\le k\le n_T} \|g^k\|^2 = E. \tag{5.109}$$

We use the mathematical induction method in the proof of (5.109). For $n = 1$, one has from (5.108) that

$$\|u_h^1\|_N^2 + K_\gamma \mu\|\partial_x u_h^1\|^2 = (u_h^0, u_h^1) + \mu(I_h g^1, u_h^1) \le \frac{1}{2}\|u_h^1\|^2 + \|u_h^0\|^2 + \mu^2\|I_h g^1\|^2, \tag{5.110}$$

which leads to

$$\|u_h^1\|^2 \le 2\|u_h^0\|^2 + 2\mu^2\|I_h g^1\|^2 \le 2\|u_h^0\|^2 + 2C\|g^1\|^2 \le E.$$

Hence, (5.109) holds for $n = 1$. Suppose that (5.109) holds for $n = 1, 2, \cdots, m-1$. For $n = m$, one has from (5.108)

$$\begin{aligned}\|u_h^m\|^2 &\le \sum_{k=1}^{m-1}(c_{m-k-1}^{(\gamma)} - c_{m-k}^{(\gamma)})\|u_h^k\|^2 + c_{m-1}^{(\gamma)} E \\ &\le \sum_{k=1}^{m-1}(c_{m-k-1}^{(\gamma)} - c_{m-k}^{(\gamma)})E + c_{m-1}^{(\gamma)} E = E.\end{aligned} \tag{5.111}$$

Therefore, $\|u_h^n\|^2 \le 2\|u_h^0\|^2 + 2C \max_{0\le k\le n_T} \|g^k\|^2$ holds for all n. The proof is thus completed. □

Next, we discuss the convergence for (5.105).

Theorem 51 *Let $0 < \gamma < 1$ and $t \in (0, T]$. Suppose that $u(t) \in C^3(0, T; H^{r+1}(\Omega) \cap H_0^1(\Omega))$ is a solution to (5.78), and u_h^n is the solution to (5.105), $f \in C(0, T; H^{r+1}(\Omega))$. Then there exists a positive constant C such that*

$$\|u_h^n - u(t_n)\| \le C(\Delta t^{2-\gamma} + h^{r+1}).$$

Proof. We first write the error equation. Let $u_* = \Pi_h^{1,0} u$, $\eta^n = u^n - u_*^n$ and $e^n = u_*^n - u_h^n$. From (4.97), we have

$$\delta_t^{(\gamma)} u^n = K_\gamma \partial_x^2 u^n + g^n + O(\Delta t^{2-\gamma}). \tag{5.112}$$

Replacing u_h^n in (5.95) with u_*^n, we have

$$(\delta_t^{(\gamma)} u_*^n,\ v) + K_\gamma(\partial_x u_*^n,\ \partial_x v) = (I_h g^n,\ v) + r^n. \tag{5.113}$$

Removing r^n from (5.105), one has

$$(\delta_t^{(\gamma)} e^n, v) + K_\gamma(\partial_x e^n, \partial_x v) = (R^n, v), \tag{5.114}$$

where $(\partial_x(u^n - \Pi_h^{1,0}u), \partial_x v) = 0$ for $v \in X_{h0}^r$, and $R^n = R_1^n + R_2^n + R_3^n$ satisfies

$$R_1^n = O(\Delta t^{2-\gamma}), \quad R_2^n = g^n - \Pi_h^{1,0} g^n, \quad R_3^n = -\delta_t^{(\gamma)} \eta^n. \tag{5.115}$$

From (5.114) and Theorem 51, we need only to estimate

$$\|e^0\|^2 + C \max_{0 \le k \le n_T} \|R^k\|^2$$

to derive the error bound. Obviously, $e^0 = 0$ and

$$\|R^k\| \le \|R_1^n\| + \|R_2^n\| + \|R_3^n\| \le C(\Delta t^{2-\gamma} + h^{r+1}\|f\|_{C(0,T;H^{r+1}(\Omega))} + h^{r+1}\|u\|_{C^1(0,T;H^{r+1}(\Omega))}),$$

where $\|R_3^n\| = \|\delta_t^{(\gamma)} \eta^n\| \le \|[{}_C D_{0,t}^{\gamma} \eta(t)]_{t=t_n}\| + O(\Delta t^{2-\gamma})$.

Hence, $\|e^n\| \le C(\Delta t^{2-\gamma} + h^{r+1})$. Using $\|u_h^n - u(t_n)\| \le \|e^n\| + \|\eta^n\|$ yields the desired result. The proof is completed. □

If the time derivative is approximated by the Grünwald formula as that of method (4.96), we can derive the following finite element scheme:

$$\begin{cases} (\delta_t^{(\gamma)}(u_h^n - u_h^0), v) = K_\gamma(\partial_x u_h^n, \partial_x v) + (I_h g^n, v), \quad v \in X_{h0}^r, \\ u_h^0 = \Pi_h^{1,0} \phi_0, \end{cases} \tag{5.116}$$

where $\delta_t^{(\gamma)}(u_h^n - u_h^0)$ is defined by

$$\delta_t^{(\gamma)}(u_h^n - u_h^0) = \frac{1}{\Delta t^\gamma} \sum_{k=0}^{n} \omega_{n-k}^{(\gamma)}(u_h^n - u_h^0), \quad \omega_k^{(\gamma)} = (-1)^k \binom{\gamma}{k}.$$

The stability of (5.116) can be proved as that of method (5.78). The convergence in the sense of L^2 norm is $O(\Delta t + h^{r+1})$.

If the time direction is discretized by the FLMMs as those in (4.116), (4.117), (4.120), or (4.121), then we can obtain the corresponding finite element methods with much better convergence rates in time.

- FLMM-FEM-I: Find $u_h^n \in X_{h0}^r$ $(n = 1, 2, \cdots, n_T)$, such that

$$\begin{cases} \dfrac{1}{\Delta t^\gamma} \displaystyle\sum_{k=0}^{n} \omega_k^{(\gamma)}(u_h^{n-k} - u_h^0, v) = -\dfrac{K_\gamma}{2^\beta} \displaystyle\sum_{k=0}^{n} (-1)^k \omega_k^{(\gamma)}(\partial_x u_h^{n-k}, \partial_x v) - K_\gamma B_n^{(1)}(\partial_x u_h^0, \partial_x v) \\ \qquad + \dfrac{1}{\Delta t^\gamma} \displaystyle\sum_{k=0}^{n} \omega_{n-k}^{(\gamma)}(G^{n-k}, v), \quad v \in X_{h0}^r, \\ u_h^0 = \Pi_h^{1,0} \phi_0, \end{cases} \tag{5.117}$$

where $\omega_k^{(\gamma)} = (-1)^k \binom{\gamma}{k}$, $G^n = \left[\mathrm{D}_{0,t}^{-\gamma} I_h g(t) \right]_{t=t_n}$, and $B_n^{(1)}$ is defined by (4.114) with $m = 1$.

- FLMM-FEM-II: Find $u_h^n \in X_{h0}^r$ $(n = 1, 2, \cdots, n_T)$, such that

$$\begin{cases} \dfrac{1}{\Delta t^\gamma}\displaystyle\sum_{k=0}^{n}\omega_k^{(\gamma)}(u_h^{n-k}-u_h^0, v) = -K_\gamma(1-\frac{\gamma}{2})(\partial_x u_h^n, \partial_x v) + \frac{\gamma}{2}(\partial_x u_h^{n-1}, \partial_x v) \\ \qquad - K_\gamma B_n^{(2)}(\partial_x u_h^0, \partial_x v) + \dfrac{1}{\Delta t^\gamma}\displaystyle\sum_{k=0}^{n}\omega_{n-k}^{(\gamma)}(G^{n-k}, v), \quad v \in X_{h0}^r, \\ u_h^0 = \Pi_h^{1,0}\phi_0, \end{cases} \tag{5.118}$$

where $\omega_k^{(\gamma)} = (-1)^k\binom{\gamma}{k}$, $G^n = \left[\mathrm{D}_{0,t}^{-\gamma} I_h g(t)\right]_{t=t_n}$, and $B_n^{(2)}$ is defined by (4.114) with $m = 2$.

- FLMM-FEM-III: Find $u_h^n \in X_{h0}^r$ $(n = 1, 2, \cdots, n_T)$, such that

$$\begin{cases} \dfrac{1}{\Delta t^\gamma}\displaystyle\sum_{k=0}^{n}\omega_k^{(\gamma)}(u_h^{n-k}-u_h^0, v) = -\frac{K_\gamma}{2^\beta}\displaystyle\sum_{k=0}^{n}(-1)^k\omega_k^{(\gamma)}(\partial_x u_h^{n-k}, \partial_x v) - K_\gamma B_n^{(1)}(\partial_x u_h^0, \partial_x v) \\ \qquad - K_\gamma C_n^{(1)}(\partial_x(u_h^1 - u_h^0), \partial_x v) + \dfrac{1}{\Delta t^\gamma}\displaystyle\sum_{k=0}^{n}\omega_{n-k}^{(\gamma)}(G^{n-k}, v), \quad v \in X_{h0}^r, \\ u_h^0 = \Pi_h^{1,0}\phi_0, \end{cases} \tag{5.119}$$

where $\omega_k^{(\gamma)} = (-1)^k\binom{\gamma}{k}$, $G^n = \left[\mathrm{D}_{0,t}^{-\gamma} I_h g(t)\right]_{t=t_n}$, $B_n^{(1)}$ is defined by (4.114) with $m = 1$, and $C_n^{(1)}$ is defined by (4.119) with $m = 1$.

- FLMM-FEM-IV: Find $u_h^n \in X_{h0}^r$ $(n = 1, 2, \cdots, n_T)$, such that

$$\begin{cases} \dfrac{1}{\Delta t^\gamma}\displaystyle\sum_{k=0}^{n}\omega_k^{(\gamma)}(u_h^{n-k}-u_h^0, v) = -K_\gamma(1-\frac{\gamma}{2})(\partial_x u_h^n, \partial_x v) + \frac{\gamma}{2}(\partial_x u_h^{n-1}, \partial_x v) \\ \qquad - K_\gamma B_n^{(2)}(\partial_x u_h^0, \partial_x v) - K_\gamma C_n^{(2)}(\partial_x(u_h^1 - u_h^0), \partial_x v) \\ \qquad + \dfrac{1}{\Delta t^\gamma}\displaystyle\sum_{k=0}^{n}\omega_{n-k}^{(\gamma)}(G^{n-k}, v), \quad v \in X_{h0}^r, \\ u_h^0 = \Pi_h^{1,0}\phi_0, \end{cases} \tag{5.120}$$

where $\omega_k^{(\gamma)} = (-1)^k\binom{\gamma}{k}$, $G^n = \left[\mathrm{D}_{0,t}^{-\gamma} I_h g(t)\right]_{t=t_n}$, $B_n^{(2)}$ is defined by (4.114) with $m = 2$, and $C_n^{(1)}$ is defined by (4.119) with $m = 1$.

The four methods (5.117), (5.118), (5.119), and (5.120) are all unconditionally stable [168, 169], which are reduced to the Crank–Nicolson finite element methods with second-order accuracy in time when $\gamma = 1$.

Theorem 52 *Let u_h^n be the solution to (5.117), (5.118), (5.119), or (5.120), $g \in C(0, T; L^2(\Omega))$. Then there exists a positive constant C_0 independent of n, Δt, h* *and* T, *and a positive constant* C_1 *independent of* n, Δt *and* h, *such that*

$$||u_h^n||^2 + \frac{1}{2}\Delta t^{\gamma}||\partial_x u_h^n||^2 \leq C_0\left(||u_h^0||^2 + \Delta t^{\gamma}||\partial_x u_h^0||^2\right) + C_1 \max_{0\leq k\leq n_T} ||g^k||^2. \tag{5.121}$$

The error bounds for (5.117) and (5.118) are the same, which are given as

$$||u_h^n - u(t_n)|| \leq C(\Delta t + h^{r+1}), \qquad \sqrt{\Delta t \sum_{k=0}^{n} ||u_h^k - U(t_k)||^2} \leq C(\Delta t^{1.5} + h^{r+1}).$$

The error estimates for (5.119) and (5.120) are the same, which are given as

$$||u_h^n - u(t_n)|| \leq C(\Delta t^2 + h^{r+1}).$$

Readers can refer to [168, 169] for more detailed information.

5.4.3 Numerical Examples

In this subsection, we present numerical examples to verify the theoretical results. For convenience, we use interpolation operator I_h to replace the projector $\Pi_h^{1,0}$ in schemes (5.96) and (5.117)–(5.120) in the following numerical experiments. We first numerically verify the error estimate and the corresponding convergence order of the method CNFEM (5.96).

Example 12 *Consider the following subdiffusion equation [173]*

$$\begin{cases} \partial_t u = {}_{RL}D_{0,t}^{1-\beta}\partial_x^2 u + f(x,t), & (x,t)\in(0,1)\times(0,1], \\ u(0,t) = t^2+t+1, \quad u(1,t) = \exp(1)(t^2+t+1), & t\in(0,1], \\ u(x,0) = \exp(x), \quad x\in[0,1], \end{cases} \tag{5.122}$$

where

$$f(x,t) = \left[2t+1-\left(\frac{2t^{2-\beta}}{\Gamma(3-\beta)} + \frac{t^{1-\beta}}{\Gamma(2-\beta)} + \frac{t^{-\beta}}{\Gamma(1-\beta)}\right)\right]\exp(x).$$

The exact solution of (5.122) *is*

$$u = \exp(x)(t^2+t+1).$$

Denote by $\varepsilon^n = u(x,t_n) - u_h^n$. Then the L^∞-error and L^2-error at t_n are defined as

$$||\varepsilon^n||_\infty = \max_{0\leq i\leq N} |u(x_i,t_n) - u_h^n(x_i)|, \quad ||\varepsilon^n|| = \left(h\sum_{i=0}^{N-1}(u(x_i,t_n) - u_h^n(x_i))^2\right)^{1/2}.$$

We first verify the convergence orders in time and space for CNFEM (5.96). The linear element is used in this example. Tables 5.1 and 5.2 display the maximum L^∞-error $\max_{0\leq n\leq n_T} ||\varepsilon^n||_\infty$ and the maximum L^2-error $\max_{0\leq n\leq n_T} ||\varepsilon^n||$ with the parameter values

$\beta = 0.25, 0.5, 0.75$. From Tables 5.1 and 5.2, we can find that the numerical solutions fit well with the analytical solutions, and the convergence orders in time and space also fit well with the theoretical analysis.

Next, we compare CNFEM (5.96) with the Crank–Nicolson type finite difference method (CNFDM) developed in [173], in which the space is discretized by the central difference method, and the convergence order in time is $\min\{1+\beta, 2-\beta/2\}$. So we use the linear element in the computation; the maximum L^∞-error is shown in Table 5.3. We find that when β is small, the two methods CNFEM and CNFDM have almost the same numerical results. When β increases, the method CNFEM achieves better numerical results in this example. Theoretically, one can find that if $0<\beta\leq 2/3$, then CNFEM and CNFDM have the same convergence orders in time, otherwise $(2/3<\beta\leq 1)$, CNFEM displays better convergence orders than CNFDM in time, which is also illustrated in the numerical experiments; see the numerical results shown in Table 5.3.

TABLE 5.1: The maximum L^∞ errors for Example 12 with $N = 1/h = 1000$.

β	$1/\Delta t$	L^∞-error	order	L^2-error	order
0.25	16	4.1782e−3		3.0566e−3	
	32	1.8315e−3	1.1898	1.3386e−3	1.1912
	64	7.8822e−4	1.2164	5.7592e−4	1.2168
	128	3.3579e−4	1.2310	2.4535e−4	1.2311
	256	1.4227e−4	1.2389	1.0395e−4	1.2389
0.5	16	1.2444e−3		9.1043e−4	
	32	4.7937e−4	1.3762	3.5020e−4	1.3784
	64	1.7888e−4	1.4222	1.3067e−4	1.4222
	128	6.5551e−5	1.4483	4.7884e−5	1.4483
	256	2.3765e−5	1.4638	1.7359e−5	1.4638
0.75	16	1.6258e−4		1.1881e−4	
	32	6.4940e−5	1.3239	4.7437e−5	1.3246
	64	2.3415e−5	1.4717	1.7099e−5	1.4721
	128	8.0019e−6	1.5490	5.8424e−6	1.5493
	256	2.6623e−6	1.5877	1.9436e−6	1.5878

Next, we numerically verify the error estimates and the convergence orders of the FEMs (5.117)–(5.120).

Example 13 *Consider the following subdiffusion equation [65, 96]*

$$\begin{cases} {}_C D_{0,t}^{\beta} = \partial_x^2 u + f(x,t), & (x,t)\in(0,1)\times(0,1], \\ u(x,0) = 2\sin(2\pi x), & x\in[0,1], \\ u(0,t) = u(1,t) = 0, & t\in(0,1]. \end{cases} \tag{5.123}$$

Choose a suitable right-hand side function f such that the exact solution to (5.123) *is*

$$u = (t^{2+\beta} + t + 2)\sin(2\pi x).$$

TABLE 5.2: The maximum L^∞ errors for Example 12 with $\Delta t = 1e-4$.

β	$N = 1/h$	L^∞-error	order	L^2-error	order
	4	2.9246e−3		2.1485e−3	
	8	7.2782e−4	2.0066	5.3600e−4	2.0031
0.25	16	1.8404e−4	1.9835	1.3462e−4	1.9933
	32	4.7114e−5	1.9658	3.4450e−5	1.9664
	64	1.2889e−5	1.8700	9.4175e−6	1.8711
	4	2.9765e−3		2.1870e−3	
	8	7.3959e−4	2.0088	5.4475e−4	2.0053
0.5	16	1.8598e−4	1.9915	1.3605e−4	2.0014
	32	4.6570e−5	1.9977	3.4057e−5	1.9982
	64	1.1727e−5	1.9896	8.5690e−6	1.9907
	4	2.9801e−3		2.1897e−3	
	8	7.4041e−4	2.0090	5.4536e−4	2.0055
0.75	16	1.8612e−4	1.9921	1.3615e−4	2.0020
	32	4.6532e−5	1.9999	3.4028e−5	2.0004
	64	1.1645e−5	1.9985	8.5093e−6	1.9996

TABLE 5.3: The maximum L^∞ errors for Example 12 with $h = 1/1000$.

method	$1/\Delta t$	$\beta = 0.4$	$\beta = 0.5$	$\beta = 0.8$	$\beta = 0.9$	$\beta = 1$
	16	1.9912e−3	1.9437e−3	3.6981e−3	4.3373e−3	4.9609e−3
	32	8.1666e−4	6.3689e−4	1.1981e−3	1.3732e−3	1.5280e−3
(4.67)	64	3.2417e−4	2.0120e−4	3.6055e−4	4.0218e−4	4.3742e−4
	128	1.2651e−4	6.3193e−5	1.0259e−4	1.1219e−4	1.2048e−4
	256	4.8861e−5	2.3015e−5	2.8182e−5	3.0469e−5	3.2450e−5
	16	2.0946e−3	1.2444e−3	1.4399e−4	1.3778e−4	2.0685e−4
	32	8.4560e−4	4.7937e−4	2.8191e−5	3.0204e−5	5.1681e−5
(5.96)	64	3.3272e−4	1.7888e−4	1.1328e−5	6.6085e−6	1.2893e−5
	128	1.2907e−4	6.5551e−5	4.0791e−6	1.4141e−6	3.1999e−6
	256	4.9656e−5	2.3765e−5	1.4023e−6	2.8688e−7	7.8039e−7

The cubic element ($r = 3$) is used in this example, the space and time step sizes are chosen as $h = 1/1000$ and $\Delta t = 1/32, 1/64, 1/128, 1/256, 1/512$.

We first check the global maximum L^2 error $\max_{0\le n\le n_T} \|u_h^n - u^n\|$, the average L^2 error $(\Delta t \sum_{n=0}^{n_T} \|u_h^n - u^n\|^2)^{1/2}$, and the L^2 error $\|u_h^n - u^n\|$ at $n = n_T$, which are shown in Tables 5.4–5.6. From Table 5.4, we find that Schemes I and II show the first-order accuracy in time for $\beta = 0.1, 0.5$. When $\beta = 0.9$, (5.117) and (5.118) show much better results than the theoretical analysis. Obviously, (5.119) and (5.120) show the convergence rates, even better than expected. Table 5.5 gives the average L^2 errors, which shows that the four algorithms yield the desired convergence rates even better than anticipated. Table 5.6 displays the L^2 error at $t = 1$. Obviously, (5.117) and (5.118) show second-order accuracy in time, and (5.119) and (5.120) really show second-order accuracy as expected. Briefly speaking, (5.117) and (5.118) show better numerical results than the theoretical analysis, and (5.119) and (5.120) show the second-order accuracy as expected.

Next, we compare the present FEMs (5.117)–(5.120) with the FEM in [65]. See also (5.105); where the time derivative was discretized by the L1 method, we denote it by L1FEM. The L1FEM has convergence order of $O(\tau^{2-\beta}+h^{r+1})$. We choose the same parameters in the computations; the results are shown in Table 5.7. Obviously, the present methods show better performances than the L1FEM, especially when β increases. It is easy to verify that the present four algorithms show second-order experimental accuracy and the L1FEM shows $(2-\beta)$th-order experimental accuracy, which is in line with the theoretical analysis.

TABLE 5.4: The global maximum L^2 errors $\max_{0\leq n\leq n_T}\|u_h^n-u^n\|$ for Example 13, $N=1000, r=3$.

Method	$1/\Delta t$	$\beta=0.1$	order	$\beta=0.5$	order	$\beta=0.9$	order
	32	4.7141e−4		1.0490e−3		1.3175e−4	
	64	2.4645e−4	0.935	5.2706e−4	0.993	6.7038e−5	0.974
(5.117)	128	1.2553e−4	0.973	2.4738e−4	1.091	2.3457e−5	1.515
	256	6.3191e−5	0.990	1.1150e−4	1.149	7.2003e−6	1.703
	512	3.1637e−5	0.998	4.8600e−5	1.198	2.0743e−6	1.795
	32	7.4577e−5		6.8243e−5		1.4795e−4	
	64	5.0606e−5	0.559	1.5696e−5	2.120	2.9862e−5	2.308
(5.118)	128	2.8356e−5	0.835	8.3982e−6	0.902	7.3010e−6	2.032
	256	1.4864e−5	0.931	4.7785e−6	0.813	1.9438e−6	1.909
	512	7.5776e−6	0.972	2.2498e−6	1.086	5.2847e−7	1.879
	32	5.2941e−5		1.2396e−4		9.3790e−5	
	64	1.2332e−5	2.102	2.1236e−5	2.545	1.2378e−5	2.921
(5.119)	128	2.8724e−6	2.102	5.1989e−6	2.030	3.1014e−6	1.996
	256	6.6894e−7	2.102	1.3003e−6	1.999	7.7612e−7	1.998
	512	1.5581e−7	2.102	3.2515e−7	1.999	1.9403e−7	2.000
	32	5.2941e−5		1.2396e−4		9.3790e−5	
	64	1.2332e−5	2.102	2.1236e−5	2.545	1.1039e−5	3.086
(5.120)	128	2.8726e−6	2.102	3.6080e−6	2.557	1.3108e−6	3.074
	256	6.6894e−7	2.102	6.0779e−7	2.569	1.5970e−7	3.037
	512	1.5571e−7	2.103	1.0149e−7	2.582	3.1009e−8	2.364

5.5 Galerkin FEM for Time-Space Fractional Differential Equations

In this subsection, we introduce the finite element methods for the *time-space fractional differential equations*. We first consider the following fractional diffusion

TABLE 5.5: The average L^2 errors $(\Delta t \sum_{n=0}^{nT} \|u_h^n - u^n\|^2)^{1/2}$ for Example 13, $N = 1000, r = 3$.

Methods	$1/\Delta t$	$\beta = 0.1$	order	$\beta = 0.5$	order	$\beta = 0.9$	order
(5.117)	32	1.1498e−4		3.3381e−4		4.2705e−5	
	64	4.1847e−5	1.458	1.0495e−4	1.669	1.1796e−5	1.856
	128	1.4962e−5	1.483	3.1468e−5	1.737	2.9836e−6	1.983
	256	5.3064e−6	1.495	9.2010e−6	1.774	7.3814e−7	2.015
	512	1.8747e−6	1.501	2.6420e−6	1.800	1.8227e−7	2.017
(5.118)	32	1.8257e−5		1.5479e−5		2.7191e−5	
	64	8.0767e−6	1.176	3.0432e−6	2.346	4.3535e−6	2.642
	128	3.1225e−6	1.371	1.2367e−6	1.299	9.0489e−7	2.266
	256	1.1462e−6	1.445	4.5034e−7	1.457	2.1165e−7	2.096
	512	4.1152e−7	1.477	1.5235e−7	1.563	5.1627e−8	2.035
(5.119)	32	1.7649e−5		6.3719e−5		3.5855e−5	
	64	4.2297e−6	2.061	1.5076e−5	2.079	8.0732e−6	2.150
	128	1.0439e−6	2.018	3.7252e−6	2.016	1.9916e−6	2.019
	256	2.6030e−7	2.003	9.2916e−7	2.003	4.9700e−7	2.002
	512	6.5029e−8	2.001	2.3217e−7	2.000	1.2415e−7	2.001
(5.120)	32	9.8452e−6		2.2325e−5		1.8170e−5	
	64	1.6221e−6	2.601	2.8235e−6	2.983	2.3677e−6	2.940
	128	2.6828e−7	2.596	4.0332e−7	2.807	4.9677e−7	2.252
	256	4.4794e−8	2.582	7.3597e−8	2.454	1.2155e−7	2.031
	512	7.6500e−9	2.549	1.6522e−8	2.155	3.0315e−8	2.003

equation

$$\begin{cases} {}_C\mathrm{D}_{0,t}^{\gamma} u = {}_{RZ}\mathrm{D}_x^{2\alpha} u + g(x,t), & (x,t) \in \Omega \times (0,T], \\ u(x,0) = \phi_0(x), & x \in \Omega, \\ u = 0, & (x,t) \in \partial\Omega \times (0,T], \end{cases} \tag{5.124}$$

where $\Omega = (a,b), 0 < \gamma \le 1, 1/2 < \alpha < 1$.

Then we introduce the corresponding schemes for another type of time-space fractional diffusion equation in the following form

$$\begin{cases} \partial_t u = {}_{RL}\mathrm{D}_{0,t}^{1-\gamma} ({}_{RZ}\mathrm{D}_x^{2\alpha} u) + f(x,t), & (x,t) \in \Omega \times (0,T], \\ u(x,0) = \phi_0(x), & x \in \Omega, \\ u = 0, & (x,t) \in \partial\Omega \times (0,T], \end{cases} \tag{5.125}$$

where $\Omega = (a,b), 0 < \gamma \le 1, 1/2 < \alpha < 1$.

TABLE 5.6: The L^2 errors $\|u_h^n - u^n\|$ at $n = n_T(t = 1)$ for Example 13, $N = 1000, r = 3$.

Method	$1/\Delta t$	$\beta = 0.1$	order	$\beta = 0.5$	order	$\beta = 0.9$	order
(5.117)	32	2.3997e−6		1.5308e−5		4.8896e−5	
	64	9.3699e−7	1.356	1.1924e−5	0.360	1.2355e−5	1.984
	128	3.2532e−7	1.526	4.2184e−6	1.499	3.1005e−6	1.994
	256	1.0460e−7	1.637	1.1979e−6	1.816	7.7624e−7	1.997
	512	3.2047e−8	1.706	3.1472e−7	1.928	1.9409e−7	1.999
(5.118)	32	3.1243e−6		4.6797e−6		7.9556e−6	
	64	8.1059e−7	1.946	1.1587e−6	2.013	1.9910e−6	1.998
	128	1.9968e−7	2.021	2.8364e−7	2.030	4.9787e−7	1.999
	256	4.8185e−8	2.051	6.9455e−8	2.029	1.2441e−7	2.000
	512	1.1575e−8	2.057	1.7069e−8	2.024	3.0986e−8	2.005
(5.119)	32	1.8316e−5		8.4548e−5		4.9286e−5	
	64	4.5797e−6	1.999	2.0821e−5	2.021	1.2378e−5	1.993
	128	1.1438e−6	2.001	5.1989e−6	2.001	3.1014e−6	1.996
	256	2.8575e−7	2.001	1.3003e−6	1.999	7.7612e−7	1.998
	512	7.1316e−8	2.002	3.2515e−7	1.999	1.9403e−7	2.000
(5.120)	32	1.3694e−7		3.9832e−6		7.9564e−6	
	64	1.1867e−7	0.206	1.0307e−6	1.950	1.9917e−6	1.998
	128	3.8589e−8	1.620	2.5996e−7	1.987	4.9811e−7	1.999
	256	1.0583e−8	1.866	6.5024e−8	1.999	1.2448e−7	2.000
	512	2.7949e−9	1.920	1.6227e−8	2.002	3.1005e−8	2.005

TABLE 5.7: Comparison of the L^2 errors $\|u_h^n - u^n\|$ at $n = n_T(t = 1)$ for Example 13, $N = 1000, r = 3$.

β	$1/\Delta t$	Method (5.117)	Method (5.118)	Method (5.119)	Method (5.120)	L1FEM [65]
0.4	32	2.6133e−5	4.2566e−6	7.3488e−5	2.9381e−6	4.0588e−5
	64	1.3567e−6	1.0444e−6	1.7955e−5	7.8792e−7	1.3740e−5
	128	1.2998e−6	2.5085e−7	4.4601e−6	2.0074e−7	4.6194e−6
	256	6.6074e−7	6.0246e−8	1.1137e−6	5.0390e−8	1.5450e−6
	512	2.1850e−7	1.4488e−8	2.7828e−7	1.2532e−8	5.1494e−7
0.6	32	5.9934e−5	5.3346e−6	8.9378e−5	5.0233e−6	1.5276e−4
	64	1.9695e−5	1.3308e−6	2.2267e−5	1.2773e−6	5.8654e−5
	128	5.3544e−6	3.2966e−7	5.5749e−6	3.2041e−7	2.2413e−5
	256	1.3765e−6	8.1818e−8	1.3950e−6	8.0201e−8	8.5394e−6
	512	3.4737e−7	2.0098e−8	3.4871e−7	1.9809e−8	3.2467e−6
0.8	32	7.2323e−5	7.0592e−6	7.4360e−5	7.0381e−6	5.1564e−4
	64	1.8542e−5	1.7682e−6	1.8673e−5	1.7661e−6	2.2617e−4
	128	4.6721e−6	4.4172e−7	4.6794e−6	4.4170e−7	9.8860e−5
	256	1.1713e−6	1.1048e−7	1.1712e−6	1.1054e−7	4.3132e−5
	512	2.9302e−7	2.7603e−8	2.9286e−7	2.7632e−8	1.8799e−5

5.5.1 Semi-Discrete Approximations

Let us multiply by $v \in H_0^{\alpha}(\Omega)$ on both sides of (5.124), which yields

$$\begin{aligned}({}_C\mathrm{D}_{0,t}^{\gamma}u(t), v) =& ({}_{RZ}\mathrm{D}_x^{2\alpha}u(t), v) + (g(t), v)\\ =& -c_{2\alpha}\left[({}_{RL}\mathrm{D}_{a,x}^{\alpha}u(t), {}_{RL}\mathrm{D}_{x,b}^{\alpha}v) + ({}_{RL}\mathrm{D}_{x,b}^{\alpha}u(t), {}_{RL}\mathrm{D}_{a,x}^{\alpha}v)\right] + (g(t), v).\end{aligned} \quad (5.126)$$

Denote by

$$A(u,v) = c_{2\alpha}\left[({}_{RL}\mathrm{D}_{a,x}^{\alpha}u, {}_{RL}\mathrm{D}_{x,b}^{\alpha}v) + ({}_{RL}\mathrm{D}_{x,b}^{\alpha}u, {}_{RL}\mathrm{D}_{a,x}^{\alpha}v)\right]. \tag{5.127}$$

Then

$$({}_{C}\mathrm{D}_{0,t}^{\gamma}u(t), v) = -A(u(t), v) + (g(t), v). \tag{5.128}$$

Therefore, we can obtain the semi-discrete approximation for (5.124) as: Find $u_h(t) \in X_{h0}^r$, such that

$$\begin{cases} ({}_{C}\mathrm{D}_{0,t}^{\gamma}u_h, v) + A(u_h, v) = (I_h g, v), & v \in X_{h0}^r, \\ u_h(0) = \Pi_h^{\alpha,0}\phi_0. \end{cases} \tag{5.129}$$

We can similarly obtain the semi-discrete approximation for (5.125) as: Find $u_h(t) \in X_{h0}^r$, such that

$$\begin{cases} (\partial_t u_h, v) + A({}_{RL}\mathrm{D}_{0,t}^{1-\gamma}u_h, v) = (I_h f, v), & v \in X_{h0}^r, \\ u_h(0) = \Pi_h^{\alpha,0}\phi_0. \end{cases} \tag{5.130}$$

Next, we consider the convergence.

Let $u^*(t) = \Pi_h^{\alpha,0}u(t)$, $\eta(t) = u(t) - u^*(t)$ and $e(t) = u^*(t) - u_h(t)$. Then we can obtain the error equation of the semi-discrete approximation (5.129) as

$$({}_{C}\mathrm{D}_{0,t}^{\gamma}e, v) + A(e, v) = -({}_{C}\mathrm{D}_{0,t}^{\gamma}\eta, v) + (g - I_h g, v), \quad \forall v \in X_{h0}^r. \tag{5.131}$$

Theorem 53 *Let $1/2 < \alpha < 1$ and $t \in (0,T]$. Suppose that $u(t) = u(\cdot,t) \in H_0^{\alpha}(\Omega) \cap H^{r+1}(\Omega)$ is a solution to (5.124), and $u_h(t)$ is the solution to (5.129). Then*

$$\left(\int_0^t \|{}_{C}\mathrm{D}_{0,s}^{\gamma/2}(u(s) - u_h(s))\|^2 \,\mathrm{d}s\right)^{1/2} \le Ch^{r+1-\alpha},$$

where C is a positive constant.

Proof. Letting $v = e(t)$ in (5.131) yields

$$\begin{aligned} ({}_{C}\mathrm{D}_{0,t}^{\gamma}e, e) + A(e, e) &= -({}_{C}\mathrm{D}_{0,t}^{\gamma}\eta, e) + (g - I_h g, e) \\ &\le \epsilon\|e\|^2 + \frac{1}{4\epsilon}(\|{}_{C}\mathrm{D}_{0,t}^{\gamma}\eta\|^2 + \|g - I_h g\|^2), \end{aligned} \tag{5.132}$$

where ϵ is a suitable positive constant satisfying $\epsilon\|e\|^2 \le A(e,e)$. Hence, we have

$$({}_{C}\mathrm{D}_{0,t}^{\gamma}e, e) \le C(\|{}_{C}\mathrm{D}_{0,t}^{\gamma}\eta\|^2 + \|g - I_h g\|^2). \tag{5.133}$$

Integrating in time leads to

$$\int_0^t ({}_{C}\mathrm{D}_{0,s}^{\gamma}e(s), e(s))\,\mathrm{d}s \le C\int_0^t (\|{}_{C}\mathrm{D}_{0,s}^{\gamma}\eta(s)\|^2 + \|g(s) - I_h g(s)\|^2)\,\mathrm{d}s. \tag{5.134}$$

Since $e(0) = 0$, from Lemmas 5.1.11 and 5.1.2 we have

$$\int_0^t ({}_C\mathrm{D}_{0,s}^{\gamma} e(s),\, e(s))\,\mathrm{d}s = \int_0^t ({}_{RL}\mathrm{D}_{0,s}^{\gamma} e(s),\, e(s))\,\mathrm{d}s = \int_0^t ({}_{RL}\mathrm{D}_{0,s}^{\gamma/2} e(s),\, {}_{RL}\mathrm{D}_{s,t}^{\gamma/2} e(s))\,\mathrm{d}s.$$

Using Lemma 5.1.2 yields

$$\begin{aligned}\int_0^t \|{}_C\mathrm{D}_{0,s}^{\gamma/2} e(s)\|^2\,\mathrm{d}s &\le C\int_0^t (\|{}_C\mathrm{D}_{0,s}^{\gamma}\eta(s)\|^2 + \|g(s) - I_h g(s)\|^2)\,\mathrm{d}s \\ &\le Ch^{2r+2-2\alpha}\int_0^t (\|{}_C\mathrm{D}_{0,s}^{\gamma} u(s)\|_{H^{r+1}(\Omega)}^2 + \|g(s)\|_{H^{r+1}(\Omega)}^2)\,\mathrm{d}s,\end{aligned} \tag{5.135}$$

where we have used (5.53) and Lemma 5.1.7. Applying $\|u(t) - u_h(t)\| = \|\eta(t) + e(t)\|$ finishes the proof. □

For the semi-discrete scheme (5.130), we have the following error estimate.

Theorem 54 *Let $1/2 < \alpha < 1$ and $t \in (0,T]$. Suppose that $u(t) = u(\cdot,t) \in H_0^{\alpha}(\Omega) \cap H^{r+1}(\Omega)$ is a solution to (5.125), and $u_h(t)$ is the solution to (5.130). Then*

$$\|u(t) - u_h(t)\| \le Ch^{r+1-\alpha},$$

where C is a positive constant.

Proof. Similar to (5.131), one can write the error equation for (5.130) as

$$(\partial_t e(t), v) + A({}_{RL}\mathrm{D}_{0,t}^{1-\gamma} e(t), v) = -(\partial_t \eta, v) + (f - I_h f, v), \quad \forall v \in X_{h0}^r. \tag{5.136}$$

Letting $v = e(t)$ in the above equation yields

$$(\partial_t e(t), e(t)) + A({}_{RL}\mathrm{D}_{0,t}^{1-\gamma} e(t), e(t)) = -(\partial_t \eta, e(t)) + (f - I_h f, e(t)). \tag{5.137}$$

Integrating on $(0,t]$ gives

$$\begin{aligned}&\|e(t)\|^2 - \|e(0)\|^2 + 2\int_0^t A({}_{RL}\mathrm{D}_{0,s}^{1-\gamma} e(s), e(s))\,\mathrm{d}s \\ =&2\int_0^t (\partial_s e(s), e(s))\,\mathrm{d}s + 2\int_0^t A({}_{RL}\mathrm{D}_{0,s}^{1-\gamma} e(s), e(s))\,\mathrm{d}s \\ \le&2C_0\int_0^t \|e(s)\|^2\,\mathrm{d}s + C\int_0^t (\|\partial_s\eta(s)\|^2 + \|f(s) - I_h f(s)\|^2)\,\mathrm{d}s.\end{aligned} \tag{5.138}$$

From Lemmas 5.1.11, 5.1.7, 5.1.6 and $e(0) = 0$, we have

$$\int_0^t A({}_{RL}\mathrm{D}_{0,s}^{1-\gamma} e(s), e(s))\,\mathrm{d}s \ge C_0\int_0^t \|e(s)\|^2\,\mathrm{d}s,$$

where C_0 is a positive constant independent of h. Therefore,

$$\begin{aligned}\|e(t)\|^2 &\le \|e(0)\|^2 + C\int_0^t (\|\partial_s\eta(s)\|^2 + \|f(s) - I_h f(s)\|^2)\,\mathrm{d}s \\ &\le Ch^{2r+2-2\alpha}\int_0^t (\|\partial_s u(s)\|_{H^{r+1}(\Omega)}^2 + \|f(s)\|_{H^{r+1}(\Omega)}^2)\,\mathrm{d}s,\end{aligned} \tag{5.139}$$

where (5.53) is used. Using $||u(t)-u_h(t)||=||\eta(t)+e(t)||$ yields the desired result. The proof is completed. □

5.5.2 Fully Discrete Schemes

In this subsection, we introduce the fully discrete finite element approximations for the time-space fractional partial differential equations as (5.124) and (5.125). We find that the time discretization techniques for (5.78) can be applied to (5.124).

- **The fully discrete finite element methods for** (5.124)

The time discretization is the same as (5.105); we present the first fully discrete approximations for (5.124) as: Find $u_h^n \in X_{h0}^{\alpha}, n=1,2,\cdots,n_T$, such that

$$\begin{cases} (\delta_t^{(\gamma)} u_h^n, v) + A(u_h^n, v) = (I_h g^n, v), \forall v \in X_{h0}^{\alpha}, \\ u_h^0 = \Pi_h^{\alpha,0}\phi_0, \end{cases} \tag{5.140}$$

where $\delta_t^{(\gamma)}$ and $A(u,v)$ are respectively defined by

$$\begin{aligned} &\delta_t^{(\gamma)} u_h^n = \frac{1}{\Delta t^{\gamma}} \sum_{k=0}^{n-1} b_{n-k}^{(\gamma)} (u_h^{k+1} - u_h^k), \quad b_k^{(\gamma)} = \frac{1}{\Gamma(2-\gamma)}\left[(k+1)^{1-\gamma} - k^{1-\gamma}\right], \\ &A(u,v) = c_{2\alpha}\left[({}_{RL}\mathrm{D}_{a,x}^{\alpha} u, {}_{RL}\mathrm{D}_{x,b}^{\alpha} v) + ({}_{RL}\mathrm{D}_{x,b}^{\alpha} u, {}_{RL}\mathrm{D}_{a,x}^{\alpha} v)\right]. \end{aligned} \tag{5.141}$$

Next, we consider the stability and error estimate for (5.140). We first give the following theorem.

Theorem 55 *Let u_h^n be the solution to* (5.140)*, and $g \in C(0,T;L^2(\Omega))$. Then there exists a positive constant C independent of n, Δt and h, such that*

$$||u_h^n||_{H^{\alpha}(\Omega)}^2 \le 2||u_h^0||_{H^{\alpha}(\Omega)}^2 + C \max_{0\le k\le n_T} ||g^k||^2.$$

Proof. The proof is very similar to that of Theorem 50, so we omit the details. The proof is completed. □

Theorem 56 *Let u_h^n be the solution to* (5.140)*, and $g \in C(0,T;L^2(\Omega))$. Then*

$$||u_h^n||^2 \le C_1 ||u_h^0||^2 + C_2 \max_{0\le k\le n_T} ||g^k||^2,$$

where the positive constant C_1 is independent of n, $\Delta t, h$ and T, and C_2 is independent of n, Δt and h.

Proof. Letting $v = \delta_t^{(\gamma)} u_h^n$ in (5.140) yields

$$(\delta_t^{(\gamma)} u_h^n, \delta_t^{(\gamma)} u_h^n) + A(u_h^n, \delta_t^{(\gamma)} u_h^n) = (I_h g^n, \delta_t^{(\gamma)} u_h^n) \le ||\delta_t^{(\gamma)} u_h^n||^2 + \frac{1}{4}||I_h g^n||. \tag{5.142}$$

Hence, one has

$$b_0^{(\gamma)}A(u_h^n,u_h^n) \le \sum_{k=1}^{n}(b_{k-1}^{(\gamma)}-b_k^{(\gamma)})A(u_h^n,u_h^{n-k})+b_n^{(\gamma)}A(u_h^n,u_h^0)+C\Delta t^{\gamma}\|g^n\|. \tag{5.143}$$

In fact, $A(u,v)$ defines a kind of *inner product*. Hence, one can derive from the relation $A(u,v) \le \epsilon A(u,u)+\frac{1}{\epsilon}A(v,v)$, $\epsilon > 0$ and (5.143) that

$$\begin{aligned} b_0^{(\gamma)}A(u_h^n,u_h^n) \le& \frac{1}{2}\sum_{k=1}^{n}(b_{k-1}^{(\gamma)}-b_k^{(\gamma)})(A(u_h^n,u_h^n)+A(u_h^{n-k},u_h^{n-k})) \\ &+\frac{1}{2}b_n^{(\gamma)}(A(u_h^n,u_h^n)+A(u_h^0,u_h^0))+C\Delta t^{\gamma}\|g^n\|, \end{aligned} \tag{5.144}$$

which leads to

$$\begin{aligned} b_0^{(\gamma)}A(u_h^n,u_h^n) &\le \sum_{k=1}^{n}(b_{k-1}^{(\gamma)}-b_k^{(\gamma)})A(u_h^{n-k},u_h^{n-k})+b_n^{(\gamma)}A(u_h^0,u_h^0)+C\Delta t^{\gamma}\|g^n\| \\ &\le \sum_{k=1}^{n}(b_{k-1}^{(\gamma)}-b_k^{(\gamma)})A(u_h^{n-k},u_h^{n-k})+b_n^{(\gamma)}\left[A(u_h^0,u_h^0)+C\max_{0\le k\le n_T}\|g^k\|^2\right], \end{aligned} \tag{5.145}$$

where $\Delta t^{\gamma} \le C_{\gamma}b_n^{(\gamma)}$ is used. Using the mathematical induction method, one can easily derive

$$A(u_h^n,u_h^n) \le A(u_h^0,u_h^0)+C\max_{0\le k\le n_T}\|g^k\|^2.$$

Note that $c_1\|u\|_{H^{\alpha}(\Omega)} \le A(u,u) \le c_2\|u\|_{H^{\alpha}(\Omega)}, c_1,c_2 > 0$. Therefore, the proof is completed. □

Let $u_* = \Pi_h^{\alpha,0}u$, $\eta^n = u^n-u_*^n$ and $e^n = u_*^n-u_h^n$. Similarly to (5.114), we can obtain the error equation for (5.140) as:

$$(\delta_t^{(\gamma)}e^n,v)+A(e^n,v) = (R^n,v), \tag{5.146}$$

where $A(\eta^n,v) = 0$ for $v \in X_{h0}^r$, and $R^n = R_1^n+R_2^n+R_3^n$ satisfies

$$R_1^n = O(\Delta t^{2-\gamma}), \quad R_2^n = g^n-\Pi_h^{1,0}g^n, \quad R_3^n = -\delta_t^{(\gamma)}\eta^n. \tag{5.147}$$

Theorem 57 *Let $0<\gamma<1, 1/2<\alpha<1$. Suppose that $u(t) \in C^3(0,T;H^{r+1}(\Omega)\cap H_0^1(\Omega))$ is a solution to (5.124), and u_h^n is the solution to (5.140), $g \in C(0,T;H^{r+1}(\Omega))$. Then there exists a positive constant C independent of $n,\Delta t$ and h such that*

$$\|u_h^n-u(t_n)\|_{H^{\alpha}(\Omega)} \le C(\Delta t^{2-\gamma}+h^{r+1-\alpha}).$$

Proof. From Theorem 56, Lemmas 5.1.12 and 5.3.1 we have

$$\|e^n\|_{H^{\alpha}(\Omega)} \le \left(C_1\|e^0\|_{H^{\alpha}(\Omega)}^2+C_2\max_{0\le k\le n_T}\|R^k\|^2\right)^{1/2} \le C(\Delta t^{2-\gamma}+h^{r+1-\alpha}).$$

Using $\|u_h^n - u(t_n)\|_{H^\alpha(\Omega)} \le \|e^n\|_{H^\alpha(\Omega)} + \|\eta^n\|_{H^\alpha(\Omega)}$ and Lemma 5.3.1 yield the desired result. The proof is completed. □

Similar to (5.117)–(5.120), we can also construct the corresponding FLMM finite element methods for (5.124), which have the similar forms as those of (5.117)–(5.120). We just need to replace $(\partial_x u, \partial_x v)$ in (5.117)–(5.120) by $A(u,v)$ to obtain the corresponding algorithms, which are listed below.

- FLMM-FEM-I: Find $u_h^n \in X_{h0}^r (n = 1,2,\cdots,n_T)$, such that

$$\begin{cases} \dfrac{1}{\Delta t^\gamma}\displaystyle\sum_{k=0}^{n}\omega_k^{(\gamma)}(u_h^{n-k}-u_h^0,v) = -\dfrac{1}{2^\beta}\displaystyle\sum_{k=0}^{n}(-1)^k\omega_k^{(\gamma)}A(u_h^{n-k},v) - B_n^{(1)}A(u_h^0,v) \\ \qquad + \dfrac{1}{\Delta t^\gamma}\displaystyle\sum_{k=0}^{n}\omega_{n-k}^{(\gamma)}(G^{n-k},v), \quad v\in X_{h0}^r, \\ u_h^0 = \Pi_h^{1,0}\phi_0, \end{cases} \tag{5.148}$$

where $\omega_k^{(\gamma)} = (-1)^k\binom{\gamma}{k}$, $G^n = \left[\mathrm{D}_{0,t}^{-\gamma} I_h g(t)\right]_{t=t_n}$, and $B_n^{(1)}$ is defined by (4.114) with $m = 1$.

- FLMM-FEM-II: Find $u_h^n \in X_{h0}^r (n = 1,2,\cdots,n_T)$, such that

$$\begin{cases} \dfrac{1}{\Delta t^\gamma}\displaystyle\sum_{k=0}^{n}\omega_k^{(\gamma)}(u_h^{n-k}-u_h^0,v) = -(1-\dfrac{\gamma}{2})A(u_h^n,v) - \dfrac{\gamma}{2}A(u_h^{n-1},v) \\ \qquad - B_n^{(2)}A(u_h^0,v) + \dfrac{1}{\Delta t^\gamma}\displaystyle\sum_{k=0}^{n}\omega_{n-k}^{(\gamma)}(G^{n-k},v), \quad v\in X_{h0}^r, \\ u_h^0 = \Pi_h^{1,0}\phi_0, \end{cases} \tag{5.149}$$

where $\omega_k^{(\gamma)} = (-1)^k\binom{\gamma}{k}$, $G^n = \left[\mathrm{D}_{0,t}^{-\gamma} I_h g(t)\right]_{t=t_n}$, and $B_n^{(2)}$ is defined by (4.114) with $m = 2$.

- FLMM-FEM-III: Find $u_h^n \in X_{h0}^r (n = 1,2,\cdots,n_T)$, such that

$$\begin{cases} \dfrac{1}{\Delta t^\gamma}\displaystyle\sum_{k=0}^{n}\omega_k^{(\gamma)}(u_h^{n-k}-u_h^0,v) = -\dfrac{1}{2^\beta}\displaystyle\sum_{k=0}^{n}(-1)^k\omega_k^{(\gamma)}A(u_h^{n-k},v) - B_n^{(1)}A(u_h^0,v) \\ \qquad - C_n^{(1)}A(u_h^1-u_h^0,v) + \dfrac{1}{\Delta t^\gamma}\displaystyle\sum_{k=0}^{n}\omega_{n-k}^{(\gamma)}(G^{n-k},v), \quad v\in X_{h0}^r, \\ u_h^0 = \Pi_h^{1,0}\phi_0, \end{cases} \tag{5.150}$$

where $\omega_k^{(\gamma)} = (-1)^k\binom{\gamma}{k}$, $G^n = \left[\mathrm{D}_{0,t}^{-\gamma} I_h g(t)\right]_{t=t_n}$, $B_n^{(1)}$ is defined by (4.114) with $m = 1$, and $C_n^{(1)}$ is defined by (4.119) with $m = 1$.

- FLMM-FEM-IV: Find $u_h^n \in X_{h0}^r (n = 1,2,\cdots,n_T)$, such that

$$\begin{cases} \frac{1}{\Delta t^{\gamma}}\sum_{k=0}^{n}\omega_k^{(\gamma)}(u_h^{n-k}-u_h^0,v) = -(1-\frac{\gamma}{2})A(u_h^n,v)-\frac{\gamma}{2}A(u_h^{n-1},v)-B_n^{(2)}A(u_h^0,v) \\ \qquad -C_n^{(2)}A(u_h^1-u_h^0,v)+\frac{1}{\Delta t^{\gamma}}\sum_{k=0}^{n}\omega_{n-k}^{(\gamma)}(G^{n-k},v), \quad v\in X_{h0}^r, \\ u_h^0 = \Pi_h^{1,0}\phi_0, \end{cases} \tag{5.151}$$

where $\omega_k^{(\gamma)} = (-1)^k\binom{\gamma}{k}$, $G^n = \left[D_{0,t}^{-\gamma} I_h g(t)\right]_{t=t_n}$, $B_n^{(2)}$ is defined by (4.114) with $m = 2$, and $C_n^{(2)}$ is defined by (4.119) with $m = 2$.

Next, we give the following theorem.

Theorem 58 *Let u_h^n be the solution to* (5.148), (5.149), (5.150), *or* (5.151), *$g\in C(0,T;L^2(\Omega))$. Then there exist a positive constant C_0 independent of n, Δt, h and T, and a positive constant C_1 independent of n, Δt and h, such that*

$$\|u_h^n\|^2+\frac{1}{2}\Delta t^{\gamma}A(u_h^n,u_h^n)\leq C_0\left(\|u_h^0\|^2+\Delta t^{\gamma}A(u_h^0,u_h^0)\right)+C_1\max_{0\leq k\leq n_T}\|g^k\|^2. \tag{5.152}$$

Theorem 59 *Suppose that u_h^n $(n = 1,2,\cdots,n_T)$ are the solutions of* (5.148), (5.149), (5.150), *or* (5.151), *u is the solution of* (5.124), *$u\in C^2(0,T;H^{r+1}(\Omega))$, $g\in C(0,T;H^{r+1}(\Omega))$, $\phi_0\in H^{r+1}(\Omega)$. Then there exists a positive constant C independent of n, h, and Δt, such that*

$$\|u_h^n-u(t_n)\|\leq C(\Delta t^q+h^{r+1-\alpha}),$$

where $q = 1$ if u_n^n is the solution to method (5.148) *or* (5.149), *$q = 2$ if u_n^n is the solution to method* (5.150) *or* (5.151).

Proof. We only consider method (5.148). The other three methods (5.149), (5.150), and (5.151) can be similarly considered. We first write the error equation for (5.148) as

$$\frac{1}{\Delta t^{\gamma}}\sum_{k=0}^{n}\omega_k^{(\gamma)}(e^{n-k}-e^0,v) = -\frac{1}{2^{\beta}}\sum_{k=0}^{n}(-1)^k\omega_k^{(\gamma)}A(e^{n-k},v)+(R^n,v), \tag{5.153}$$

where $e = \Pi_h^{\alpha,0}u-u_h, \eta = u-\Pi_h^{\alpha,0}u$, $R^n = R_1^n+R_2^n+R_3^n$ satisfies

$$\begin{aligned} &R_1^n = O(\Delta t), \quad R_2^n = \frac{1}{2^{\beta}}\sum_{k=0}^{n}(-1)^k\omega_k^{(\gamma)}(g^n-I_h g^{n-k}), \\ &R_3^n = \frac{1}{\Delta t^{\gamma}}\sum_{k=0}^{n}\omega_k^{(\gamma)}(\eta^{n-k}-\eta^0). \end{aligned} \tag{5.154}$$

From Theorem 58 and (5.153), we obtain

$$||e^n||^2 \leq C_0||e^0||^2 + C_0\Delta t^\gamma A(e^0,e^0) + C_1 \max_{0\leq k\leq n_T} ||R^k||^2.$$

Note that $e^0 = 0$ and R^n satisfies (5.154), so we have

$$||e^n||^2 \leq C(\Delta t + h^{r+1-\alpha}),$$

where (5.53) and the following bounds are utilized

$$||R_1^n|| \leq C\Delta t, \quad ||R_2^n|| \leq Ch^{r+1}, \quad ||R_3^n|| \leq C(\Delta t + h^{r+1-\alpha}).$$

Using (5.53) once more yields

$$||u_h^n - u(t_n)|| \leq ||e^n|| + ||\eta^n|| \leq C(\Delta t + h^{r+1-\alpha}).$$

The proof is thus completed. □

- **The fully discrete finite element methods for** (5.125)

Next, we introduce the fully discrete approximations for (5.125). Obviously, the time derivative in (5.124) can be discretized the same as that of (5.77) or see the time discretization for (4.10). In the following, we present several fully discrete approximations for (5.125),

(1) The fully discrete implicit Euler type finite element method based on the Grünwald formula of the discretization of the time fractional derivative (see the time discretization method in the finite difference method (4.30)) for (5.125) is given by: Find $u_h^n \in X_{h0}^r$ $(n = 1,2,\cdots,n_T)$, such that

$$\begin{cases} (\delta_t u_h^{n-\frac{1}{2}}, v) = -\delta_t^{(1-\gamma)} A(u_h^n, v) + (I_h f^n, v), \quad v \in X_{h0}^r, \\ u_h^0 = \Pi_h^{\alpha,0}\phi_0, \end{cases} \tag{5.155}$$

where $A(u,v)$ is defined by (5.141), and

$$\delta_t^{(\gamma)} A(u_h^n, v) = \frac{1}{\Delta t^\gamma}\sum_{k=0}^{n} \omega_{n-k}^{(\gamma)} A(u_h^k, v), \qquad \omega_k^{(\gamma)} = (-1)^k \binom{\gamma}{k}.$$

(2) The fully discrete implicit Euler type finite element method based on the L1 method for the discretization of the time fractional derivative (see the time discretization method in the finite difference method (4.49)) for (5.125) is given by: Find $u_h^n \in X_{h0}^r$ $(n = 1,2,\cdots,n_T)$, such that

$$\begin{cases} (\delta_t u_h^{n-\frac{1}{2}}, v) = -\delta_t^{(1-\gamma)} A(u_h^n, v) + (I_h f^n, v), \quad v \in X_{h0}^r, \\ u_h^0 = \Pi_h^{\alpha,0}\phi_0, \end{cases} \tag{5.156}$$

where $A(u,v)$ is defined by (5.141), and

$$\delta_t^{(\gamma)}A(u_h^n,v)=\frac{1}{\Delta t^{\gamma}}\left(\sum_{k=0}^{n-1}b_{n-k-1}^{(1-\gamma)}\left[A(u_h^{k+1},v)-A(u_h^k,v)\right]+\frac{n^{\gamma-1}}{\Gamma(1+\gamma)}A(u_h^0,v)\right),$$

$$b_k^{(1-\gamma)}=\frac{1}{\Gamma(1+\gamma)}[(k+1)^{\gamma}-k^{\gamma}].$$

(3) The fully discrete implicit Crank–Nicolson type finite element method (see the time discretization method in the finite difference method (4.59)) for (5.125) is given by: Find $u_h^n\in X_{h0}^r$ $(n=1,2,\cdots,n_T)$, such that

$$\begin{cases}(\delta_t u_h^{n-\frac{1}{2}},v)=-\delta_t^{(1-\gamma)}A(u_h^n,v)+(I_hf^n,v), & v\in X_{h0}^r,\\ u_h^0=\Pi_h^{\alpha,0}\phi_0,\end{cases}\tag{5.157}$$

where $A(u,v)$ is defined by (5.141), and

$$\delta_t^{(\gamma)}A(u_h^n,v)=\frac{1}{\Delta t^{\gamma}}\left(\sum_{k=0}^{n-1}b_{n-k-1}^{(\gamma)}\left[A(u_h^{k+1},v)-A(u_h^k,v)\right]+\frac{n^{\gamma-1}}{\Gamma(1+\gamma)}A(u_h^0,v)\right),$$

$$b_k^{(1-\gamma)}=\frac{1}{\Gamma(1+\gamma)}[(k+1)^{\gamma}-k^{\gamma}].$$

(4) The time direction is discretized the same way as that in (4.81); then we have the fully discrete approximation for (5.125) as: Find $u_h^n\in X_{h0}^r$ $(n=1,2,\cdots,n_T)$, such that

$$\begin{cases}(\delta_t u_h^{n-\frac{1}{2}},v)=\delta_t^{(1-\gamma)}A(u_h^n,v)+(I_hf^n,v), & v\in X_{h0}^r,\\ u_h^0=\Pi_h^{\alpha,0}\phi_0,\end{cases}\tag{5.158}$$

where $A(u,v)$ is defined by (5.141), and

$$\delta_t^{(1-\gamma)}A(u_h^n,v)=\frac{1}{\Delta t^{1-\gamma}}\left[\sum_{l=1}^{n}b_{n-l}^{(1-\gamma)}A(u_h^l,v)-\sum_{l=1}^{n-1}b_{n-l-1}^{(1-\gamma)}A(u_h^l,v)\right],$$

$$b_l^{(1-\gamma)}=\frac{1}{\Gamma(1+\gamma)}[(l+1)^{\gamma}-l^{\gamma}].$$

Next, we consider the stability and convergence for (5.155)–(5.158), the proof of which is similar to that of Theorems 58 and 59. Next, we analyze the algorithm (5.155).

We now give the following result.

Theorem 60 *Let* u_h^n *be the solution to* (5.155), $f\in C(0,T;L^2(\Omega))$. *Then there exists a positive constant* C *independent of* n, Δt *and* h, *such that*

$$||u_h^n||^2\le||u_h^0||^2+\Delta t^{\gamma}A(u_h^0,u_h^0)+C\Delta t\sum_{k=1}^{n}||f^k||^2.\tag{5.159}$$

Proof. From (5.155), we have

$$(u_h^n, v) + b_0\Delta t^\gamma A(u_h^n, u_h^n) = (u_h^{n-1}, v) + \Delta t^\gamma \sum_{k=1}^{n}(b_{k-1} - b_k)A(u_h^k, v) + \Delta t(I_h f^n, u_h^n), \tag{5.160}$$

where $b_n = \sum_{k=0}^{n}\omega_k^{(1-\gamma)}$. Letting $v = u_h^n$ in the above equality, and applying the Cauchy–Schwarz inequality yields

$$\begin{aligned}\|u_h^n\|^2 + \Delta t^\gamma A(u_h^n, u_h^n) \le & \frac{1}{2}(\|u_h^n\|^2 + \|u_h^{n-1}\|^2) + \Delta t(I_h f^n, u_h^n)\\ & + \frac{1}{2}\Delta t^\gamma \sum_{k=1}^{n}(b_{k-1} - b_k)\big(A(u_h^n, u_h^n) + A(u_h^{n-k}, u_h^{n-k})\big).\end{aligned} \tag{5.161}$$

Rearranging the above inequality yields

$$\begin{aligned}\|u_h^n\|^2 + \Delta t^\gamma \sum_{k=0}^{n} b_k A(u_h^{n-k}, u_h^{n-k}) \le & \|u_h^{n-1}\|^2 + \Delta t^\gamma \sum_{k=0}^{n-1} b_k A(u_h^{n-1-k}, u_h^{n-1-k})\\ & - b_n \Delta t^\gamma A(u_h^n, u_h^n) + 2\Delta t(I_h f^n, u_h^n).\end{aligned} \tag{5.162}$$

Denote by

$$E^n = \|u_h^n\|^2 + \Delta t^\gamma \sum_{k=0}^{n} b_k A(u_h^{n-k}, u_h^{n-k}).$$

Then we have

$$E^n \le E^{n-1} + 2\Delta t(I_h f^n, u_h^n) \le E^0 + 2\Delta t \sum_{k=1}^{n} \Delta t(I_h f^k, u_h^k). \tag{5.163}$$

Hence,

$$\begin{aligned}\|u_h^n\|^2 + \Delta t^\gamma \sum_{k=0}^{n} b_{n-k} A(u_h^k, u_h^k) \le & E^0 + 2\Delta t \sum_{k=1}^{n} (I_h f^k, u_h^k)\\ \le & E^0 + \sum_{k=1}^{n}\left(\frac{\Delta t^2}{\epsilon b_{n-k}\Delta t^\gamma}\|I_h f^k\|^2 + \epsilon b_{n-k}\Delta t^\gamma \|u_h^k\|^2\right),\end{aligned} \tag{5.164}$$

where ϵ is a suitable positive constant such that $\epsilon\|u_h^k\|^2 \le \|u_h^k\|^2_{H^\alpha(\Omega)} \le CA(u_h^k, u_h^k)$.

Hence

$$\|u_h^n\|^2 \le E^0 + \sum_{k=1}^{n} \frac{\Delta t^{2-\gamma}}{\epsilon b_{n-k}}\|I_h f^k\|^2 \le E^0 + C\Delta t \sum_{k=1}^{n} \|f^k\|^2, \tag{5.165}$$

where we have used $\Delta t^{1-\gamma} \le C_\gamma b_k$ and $\|I_h f^k\| \le C\|f^k\|$. Using $E^0 = \|u_h^0\|^2 + \Delta t^\gamma A(u_h^0, u_h^0)$ yields the desired result. The proof is completed. □

For the methods (5.156)–(5.158), the similar results as (5.159) can be obtained. Next, we consider the convergence for (5.155).

Theorem 61 *Suppose that* u_h^n $(n = 1,2,\cdots,n_T)$ *is the solution of* (5.155), *and that* U *is the solution of* (5.125), $U \in C^2(0,T;H^{r+1}(\Omega))$, $f \in C(0,T;H^{r+1}(\Omega))$, $\phi_0 \in H^{r+1}(\Omega)$. *Then there exists a positive constant* C *independent of* n, h, *and* Δt, *such that*

$$\|u_h^n - u(t_n)\| \leq C(\Delta t + h^{r+1-\alpha}).$$

Proof. We write the error equation for (5.155) as

$$(\delta_t e^{n-\frac{1}{2}}, v) = -\delta_t^{(1-\gamma)} A(e^n, v) + (R^n, v), \quad v \in X_{h0}^r, \tag{5.166}$$

where $R^n = R_1^n + R_2^n + R_3^n$ satisfies

$$R_1^n = O(\Delta t), \quad R_2^n = f^n - I_h f^n, \quad R_3^n = \delta_t \eta^{n-\frac{1}{2}}. \tag{5.167}$$

From Theorem 60 we obtain

$$\begin{aligned} \|e^n\|^2 &\leq \|e^0\|^2 + \Delta t^\gamma A(e^0, e^0) + C\Delta t \sum_{k=1}^{n} \|R^k\|^2 \\ &\leq C(\Delta t + h^{r+1-\alpha}). \end{aligned} \tag{5.168}$$

Using $\|u_h^n - u(t_n)\| \leq \|\eta^n\| + \|e^n\|$ yields the desired result. The proof is ended. □

The convergence rates for methods (5.156), (5.157), and(5.158) can be similarly derived, which are of orders $O(\Delta t + h^{r+1-\alpha})$, $O(\Delta t^{1+\gamma} + h^{r+1-\alpha})$, and $O(\Delta t + h^{r+1-\alpha})$, respectively in the L^2 sense.

Bibliography

[1] A.H. Bhrawy and A.S. Alofi. The operational matrix of fractional integration for shifted Chebyshev polynomials. *Appl. Math. Lett.*, 26: 25–31, 2013.

[2] A.H. Bhrawy, E.H. Doha, D. Baleanu, and S.S. Ezz-Eldien. A spectral tau algorithm based on Jacobi operational matrix for numerical solution of time fractional diffusion-wave equations. *J. Comput. Phys.*, 2014. DOI: 10.1016/j.jcp.2014.03.039.

[3] S.C. Brenner and L.R. Scott. *The Mathematical Theory of Finite Element Methods*. Springer-Verlag, New York, 2008.

[4] H. Brunner, H.D. Han, and D.S. Yin. Artificial boundary conditions and finite difference approximations for a time-fractional diffusion-wave equation on a two-dimensional unbounded spatial domain. *J. Comput. Phys.*, 276(1): 541–562, 2014.

[5] H. Brunner, L. Ling, and M. Yamamoto. Numerical simulations of 2D fractional subdiffusion problems. *J. Comput. Phys.*, 229(18): 6613–6622, 2010.

[6] W.P. Bu, Y.F. Tang, Y.C. Wu, and J.Y. Yang. Crank–Nicolson ADI Galerkin finite element method for two-dimensional fractional FitzHugh–Nagumo monodomain model. *Appl. Math. Comput.*, 2014. DOI: 10.1016/j.amc.2014.09.034.

[7] W.P. Bu, Y.F. Tang, Y.C. Wu, and J.Y. Yang. Finite difference/finite element method for two-dimensional space and time fractional Bloch–Torrey equations. *J. Comput. Phys.*, 2014. DOI:10.1016/j.jcp.2014.06.031.

[8] W.P. Bu, Y.F. Tang, and J.Y. Yang. Galerkin finite element method for two-dimensional Riesz space fractional diffusion equations. *J. Comput. Phys.*, 276: 26–38, 2014.

[9] C. Canuto, M.Y. Hussaini, A. Quarteroni, and T.A. Zang. *Spectral Methods: Fundamentals in Single Domains*. Springer-Verlag, Berlin, 2006.

[10] J.X. Cao and C.P. Li. Finite difference scheme for the time-space fractional diffusion equations. *Cent. Eur. J. Phys.*, 11(10): 1440–1456, 2013.

[11] J.X. Cao, C.P. Li, and Y.Q. Chen. Compact difference method for solving the fractional reaction-subdiffusion equation with Neumann boundary value condition. *Int. J. Comput. Math.*, 92(1): 167–180, 2015.

[12] J.X. Cao, C.P. Li, and Y.Q. Chen. High-order approximation to Caputo derivatives and Caputo-type advection-diffusion equations (II). *Fract. Calc. Appl. Anal.*, in press.

[13] J.Y. Cao and C.J. Xu. A high order schema for the numerical solution of the fractional ordinary differential equations. *J. Comput. Phys.*, 238: 154–168, 2013.

[14] C. Çelik and M. Duman. Crank–Nicolson method for the fractional diffusion equation with the Riesz fractional derivative. *J. Comput. Phys.*, 231: 1743–1750, 2012.

[15] A. Chen and C.P. Li. Numerical algorithm for fractional calculus based on Chebyshev polynomial approximation. *J. Shanghai Univ. (Nat. Sci.)*, 18(1): 48–53, 2012.

[16] C.M. Chen, F. Liu, and V. Anh. A Fourier method and an extrapolation technique for Stokes' first problem for a heated generalized second grade fluid with fractional derivative. *J. Comput. Appl. Math.*, 223: 777–789, 2009.

[17] C.M. Chen, F. Liu, and K. Burrage. Finite difference methods and a fourier analysis for the fractional reaction–subdiffusion equation. *Appl. Math. Comput.*, 198: 754–769, 2008.

[18] C.M. Chen, F. Liu, I. Turner, and V. Anh. A Fourier method for the fractional diffusion equation describing sub-diffusion. *J. Comput. Phys.*, 227: 886–897, 2007.

[19] M. R. Cui. Compact finite difference method for the fractional diffusion equation. *J. Comput. Phys.*, 228: 7792–7804, 2009.

[20] G.G. Dahlquist. A special stability problem for linear multistep methods. *BIT Numer. Math.*, 3: 27–43, 1963.

[21] P.J. Davis and I. Polonsky. *Numerical Interpolation, Differentiation, and Integration, Handbook of Mathematical Functions, M. Abramowitz and I.A. Stegun, eds.* Dover Publications, Inc., New York, 1972.

[22] W.H. Deng. Short memory principle and a predictor-corrector approach for fractional differential equations. *J. Comput. Appl. Math.*, 206: 174–188, 2007.

[23] W.H. Deng and J.S. Hesthaven. Local discontinuous Galerkin Methods for Fractional Diffusion Equations. *M2AN*, 47: 1845–1864, 2013.

[24] Z.Q. Deng, V.P. Singh, and L. Bengtsson. Numerical solution of fractional advection-dispersion equation. *J. Hydraul. Eng.*, 130: 422–431, 2004.

[25] K. Diethelm. An algorithm for the numerical solution of differential equations of fractional order. *Electron. Trans. Numer. Anal.*, 5: 1–6, 1997.

[26] K. Diethelm. Generalized compound quadrature formulae for finite-part integrals. *IMA J. Numer. Anal.*, 17: 479–493, 1997.

[27] K. Diethelm. Predictor-corrector strategies for single and multi-term fractional differential equations. In *Proceedings of the 5th Hellenic-European Conference on Computer Mathematics and Its Applications*. LEA Press, Athens, pp.117–122, 2001.

[28] K. Diethelm, J.M. Ford, N.J. Ford, and M. Weilbeer. Pitfalls in fast numerical solvers for fractional differential equations. *J. Comput. Appl. Math.*, 186: 482–503, 2006.

[29] K. Diethelm and N.J. Ford. Analysis of Fractional Differential Equations. *J. Math. Anal. Appl.*, 265: 229–248, 2002.

[30] K. Diethelm and N.J. Ford. Multi-order fractional differential equations and their numerical solution. *Appl. Math. Comput.*, 154: 621–640, 2004.

[31] K. Diethelm, N.J. Ford, and A.D. Freed. A predictor-corrector approach for the numerical solution of fractional differential equations. *Nonlinear Dynam.*, 29: 3–22, 2002.

[32] K. Diethelm, N.J. Ford, and A.D. Freed. Detailed error analysis for a fractional Adams method. *Numer. Algor.*, 36: 31–52, 2004.

[33] K. Diethelm and A.D. Freed. On the solution of nonlinear fractional-order differential equations used in the modeling of viscoplasticity. In *Scientific Computing in Chemical Engineering II: Computational Fluid Dynamics, Reaction Engineering, and Molecular Properties*. Springer, Heidelberg, pp. 217–224, 1999.

[34] K. Diethelm and G. Walz. Numerical solution of fractional order differential equations by extrapolation. *Numer. Algor.*, 16: 231–253, 1997.

[35] H.F. Ding and C.P. Li. Mixed spline function method for reaction–subdiffusion equations. *J. Comput. Phys.*, 242: 103–123, 2013.

[36] H.F. Ding and C.P. Li. Numerical algorithms for the fractional diffusion-wave equation with reaction term. *Abstr. Appl. Anal.*, 2013: 493406, 2013.

[37] H.F. Ding and C.P. Li. High-order Algorithms for Riesz Derivative and Their Applications (III). Submitted, 2014.

[38] H.F. Ding, C.P. Li, and Y.Q. Chen. High-order Algorithms for Riesz Derivative and Their Applications (I). *Abstr. Appl. Anal.*, 2014: 653797, 2014.

[39] H.F. Ding, C.P. Li, and Y.Q. Chen. High-order Algorithms for Riesz Derivative and Their Applications (II). *J. Comput. Phys.*, 2014. DOI: 10.1016/j.jcp.2014.06.007.

[40] H.F. Ding and Y.X. Zhang. New numerical methods for the Riesz space fractional partial differential equations. *Comput. Math. Appl.*, 63: 1135–1146, 2012.

[41] Z.Q. Ding, A.G. Xiao, and M. Li. Weighted finite difference methods for a class of space fractional partial differential equations with variable coefficients. *J. Comput. Appl. Math.*, 233: 1905–1914, 2010.

[42] J. Dixon. On the order of the error in discretization methods for weakly singular second kind non-smooth solutions. *BIT*, 25: 623–634, 1985.

[43] E.H. Doha, A.H. Bhrawy, and S.S. Ezz-Eldien. A chebyshev spectral method based on operational matrix for initial and boundary value problems of fractional order. *Comput. Math. Appl.*, 62: 2364–2373, 2011.

[44] E.H. Doha, A.H. Bhrawy, and S.S. Ezz-Eldien. Efficient chebyshev spectral methods for solving multi-term fractional orders differential equations. *Appl. Math. Model.*, 35: 5662–5672, 2011.

[45] E.H. Doha, A.H. Bhrawy, and S.S. Ezz-Eldien. A new Jacobi operational matrix: an application for solving fractional differential equations. *Appl. Math. Model.*, 36: 4931–4943, 2012.

[46] M.L. Du, Z.H. Wang, and H.Y. Hu. Measuring memory with the order of fractional derivative. *Sci. Rep.*, 3: 3431, 2013.

[47] R. Du, W.R. Cao, and Z.Z. Sun. A compact difference scheme for the fractional diffusion-wave equation. *Appl. Math. Model.*, 34: 2998–3007, 2010.

[48] D. Elliott. Three algorithms for Hadamard finite-part integrals and fractional derivatives. *J. Comput. Appl. Math.*, 62: 267–283, 1995.

[49] V.J. Ervin and J.P. Roop. Variational formulation for the stationary fractional advection dispersion equation. *Numer. Meth. P. D. E.*, 22: 558–576, 2006.

[50] V.J. Ervin and J.P. Roop. Variational solution of fractional advection dispersion equations on bounded domains in R^d. *Numer. Meth. P. D. E.*, 23: 256–281, 2007.

[51] S. Esmaeili and M. Shamsi. A pseudo-spectral scheme for the approximate solution of a family of fractional differential equations. *Commun. Nonlinear Sci. Numer. Simulat.*, 16: 3646–3654, 2011.

[52] N.J. Ford, J.Y. Xiao, and Y.B. Yan. A finite element method for time fractional partial differential equations. *Fract. Calc. Appl. Anal.*, 14(3): 454–474, 2011.

[53] L. Galeone and R Garrappa. On multistep methods for differential equations of fractional order. *Mediterr. J. Math.*, 3: 565–580, 2006.

[54] L. Galeone and R. Garrappa. Fractional Adams-Moulton methods. *Math. Comput. Simulat.*, 79: 1358–1367, 2008.

[55] L. Galeone and R. Garrappa. Explicit methods for fractional differential equations and their stability properties. *J. Comput. Appl. Math.*, 228: 548–560, 2009.

[56] G.H. Gao and Z.Z. Sun. A compact finite difference scheme for the fractional subdiffusion equations. *J. Comput. Phys.*, 230: 586–595, 2001.

[57] Z. Gao and X. Z. Liao. Discretization algorithm for fractional order integral by Haar wavelet approximation. *Appl. Math. Comput.*, 218: 1917–1926, 2011.

[58] R. Garrappa. On some explicit Adams multistep methods for fractional differential equations. *J. Comput. Appl. Math.*, 229: 392–399, 2009.

[59] R. Garrappa. On linear stability of predictor-corrector algorithms for fractional differential equations. *Int. J. Comput. Math.*, 87: 2281–2290, 2010.

[60] R. Gorenflo and E.A. Abdel-Rehim. Convergence of the Grünwald-Letnikov scheme for time-fractional diffusion. *J. Comput. Appl. Math.*, 205: 871–881, 2007.

[61] T. Hasegawa and H. Sugiura. An approximation method for high-order fractional derivatives of algebraically singular functions. *Comput. Math. Appl.*, 62: 930–937, 2011.

[62] X.L. Hu and L.M. Zhang. Implicit compact difference schemes for the fractional cable equation. *Appl. Math. Model.*, 36: 4027–4043, 2012.

[63] J. Huang, Y. Tang, L. Vázquez, and J. Yang. Two finite difference schemes for time fractional diffusion-wave equation. *Numer. Algor.*, 64: 707–720, 2013.

[64] J.F. Huang, N.N. Nie, and Y.F. Tang. A second order finite difference-spectral method for space fractional diffusion equation. *Sci. China Math.*, 57: 1303–1317, 2014.

[65] Y.J. Jiang and J.T. Ma. High-order finite element methods for time-fractional partial differential equations. *J. Comput. Appl. Math.*, 235: 3285–3290, 2011.

[66] B. Jin, R. Lazarov, and Z. Zhou. Error estimates for a semidiscrete finite element method for fractional order parabolic equations. *SIAM J. Numer. Anal.*, 51: 445–466, 2013.

[67] B. Jin, R. Lazarov, and Z. Zhou. On two schemes for fractional diffusion and diffusion-wave equations. *arXiv:1404.3800*, 2014.

[68] A.A. Kilbas, H.M. Srivastava, and J.J. Trujillo. *Theory and Applications of Fractional Differential Equations*. Elsevier, Amersterdam, 2006.

[69] P. Kumar and Om P. Agrawal. An approximate method for numerical solution of fractional differential equations. *Signal Proc.*, 86: 2602–2610, 2006.

[70] M. Lakestani, M. Dehghan, and S. Irandoust-pakchin. The construction of operational matrix of fractional derivatives using B-spline functions. *Commun. Nonlinear Sci. Numer. Simulat.*, 17: 1149–1162, 2012.

[71] T.A.M. Langlands and B.I. Henry. The accuracy and stability of an implicit solution method for the fractional diffusion equation. *J. Comput. Phys.*, 205: 719–736, 2005.

[72] G.W. Leibniz. *Leibnizens Mathematische Schriften.* Olms, Hildesheim, 1962.

[73] C.P. Li, A. Chen, and J.J. Ye. Numerical approaches to fractional calculus and fractional ordinary differential equation. *J. Comput. Phys.*, 230: 3352–3368, 2011.

[74] C.P. Li, Y.Q. Chen, and J. Kurths. Introduction: Fractional Calculus and its Applications. *Phil. Trans. R. Soc. A*, 371(1990): 20130037, 2013.

[75] C.P. Li, Y.Q. Chen, B.M. Vinagre, and I. Podlubny. Introduction. *Int. J. Bifurcation Chaos*, 22(4): 1202002, 2012.

[76] C.P. Li, X.H. Dao, and P. Guo. Fractional derivatives in complex planes. *Nonlinear Anal.: TMA*, 71: 1857–1869, 2009.

[77] C.P. Li and W.H. Deng. Remarks on fractional derivatives. *Appl. Math. Comput.*, 187: 777–784, 2007.

[78] C.P. Li and H.F. Ding. Higher order finite difference method for the reaction and anomalous diffusion equation. *Appl. Math. Model.*, 38: 3802–3821, 2014.

[79] C.P. Li and H.F. Li. High-order approximation to Caputo derivatives and Caputo-type advection-diffusion equations (III). Preprinted, 2014.

[80] C.P. Li and F. Mainardi. Editorial. *The European Physical Journal-Special Topics*, 193: 1–4, 2011.

[81] C.P. Li, D.L. Qian, and Y.Q. Chen. On Riemann-Liouville and Caputo Derivatives. *Discrete Dyn. Nat. Soc.*, 2011: 562494, 2011.

[82] C.P. Li and C.X. Tao. On the fractional Adams method. *Comput. Math. Appl.*, 58: 1573–1588, 2009.

[83] C.P. Li and Y.H. Wang. Numerical algorithm based on Adomian decomposition for fractional differential equations. *Comput. Math. Appl.*, 57(10): 1672–1681, 2009.

[84] C.P. Li, R.F. Wu, and H.F. Ding. High-order approximation to Caputo derivatives and Caputo-type advection-diffusion equations (I). *Commun. Appl. Ind. Math.*, in press, 2014.

[85] C.P. Li, Y.J. Wu, and R.S. Ye eds. *Recent Advances in Applied Nonlinear Dynamics with Numerical Analysis.* World Scientific, Singapore, 2013.

[86] C.P. Li and F.H. Zeng. Finite difference methods for fractional differential equations. *Int. J. Bifurcat. Chaos*, 22: 1230014, 2012.

[87] C.P. Li and F.H. Zeng. The finite difference methods for fractional ordinary differential equations. *Numer. Func. Anal. Opt.*, 34: 149–179, 2013.

[88] C.P. Li and F.H. Zeng. Finite element methods for frctional differential equations, in *Recent Advances in Applied Nonlinear Dynamics with Numerical Analysis*, C.P. Li, Y.J. Wu, and R.S. Ye eds. *World Scientific, Singapore*, pp. 49–68, 2013.

[89] C.P. Li, F.H. Zeng, and F.W. Liu. Spectral approximations to the fractional integral and derivative. *Fract. Calc. Appl. Anal.*, 15: 383–406, 2012.

[90] C.P. Li, F.R. Zhang, J. Kurths, and F.H. Zeng. Equivalent system for a multiple-rational-order fractional differential system. *Phil. Trans. R. Soc. A*, 371(1990): 20120156, 2013.

[91] C.P. Li and Z.G. Zhao. Introduction to fractional integrability and differentiability. *Eur. Phys. J.-Special Topics.*, 193: 5–26, 2011.

[92] C.P. Li, Z.G. Zhao, and Y.Q. Chen. Numerical approximation of nonlinear fractional differential equations with subdiffusion and superdiffusion. *Comput. Math. Appl.*, 62: 855–875, 2011.

[93] X.C. Li, M.Y. Xu, and X.Y. Jiang. Homotopy perturbation method to time-fractional diffusion equation with a moving boundary condition. *Appl. Math. Comput.*, 208(2): 434–439, 2009.

[94] X.J. Li and C.J. Xu. Existence and uniqueness of the weak Solution of the space-time fractional diffusion equation and a spectral method approximation. *Commun. Comput. Phys.*, 8(5): 1016–1051, 2010.

[95] R. Lin and F. Liu. Fractional high order methods for the nonlinear fractional ordinary differential equation. *Nonlinear Anal.: TMA*, 66: 856–869, 2007.

[96] Y.M. Lin and C.J. Xu. Finite difference/spectral approximations for the time-fractional diffusion equation. *J. Comput. Phys.*, 225: 1533–1552, 2007.

[97] F. Liu, V. Anh, and I. Turner. Numerical solution of the space fractional Fokker-Planck equation. *J. Comput. Appl. Math.*, 166: 209–219, 2004.

[98] F. Liu, S. Shen, V. Anh, and I. Turner. Analysis of a discrete non Markovian random walk approximation for the time fractional diffusion equation. *ANZIAM J.*, 46: C488–C504, 2005.

[99] F. Liu, Q.Q. Yang, and I. Turner. Two new implicit numerical methods for the fractional cable equation. *J. Comput. Nonlinear Dyn.*, 6: 1–7, 2011.

[100] F. Liu, P. Zhuang, and K. Burrage. Numerical methods and analysis for a class of fractional advection-dispersion models. *Comput. Math. Appl.*, 64: 2990–3007, 2012.

[101] Q. Liu, F.H. Zeng, and C.P. Li. Finite difference method for time-space-fractional Schrödinger equation. *Int. J. Comput. Math.*, 2014. DOI: 10.1080/00207160.2014.945440.

[102] C. Lubich. Fractional linear multistep methods for Abel–Volterra integral equations of the second kind. *Math. Comp.*, 45: 463–469, 1985.

[103] C. Lubich. A Stability analysis of convolution quadratures for Abel–Volterra integral equations. *IMA J. Numer. Anal.*, 6: 87–101, 1986.

[104] C. Lubich. Discretized fractional calculus. *SIAM J. Math. Anal.*, 17: 704–719, 1986.

[105] V.E. Lynch, B.A. Carreras, D. del Castillo-Negrete, K.M. Ferreira-Mejias, and H.R. Hicks. Numerical methods for the solution of partial differential equations of fractional order. *J. Comput. Phys.*, 192: 406–421, 2003.

[106] J. Tenreiro Machado. Numerical calculation of the left and right fractional derivatives. *J. Comput. Phys.*, 2014. DOI: 10.1016/j.jcp.2014.05.029.

[107] R.L. Magin, O. Abdullah, D. Baleanu, and X.J. Zhou. Anomalous diffusion expressed through fractional order differential operators in the bloch–torrey equation. *J. Magn. Reson.*, 190(2): 255–270, 2008.

[108] B.B. Mandelbrot. *Fractals: Form, Chance and Dimension*. W.H. Freeman and Company, San Francisco, 1977.

[109] W. McLean and K. Mustapha. Convergence analysis of a discontinuous Galerkin method for a sub-diffusion equation. *Numer. Algor.*, 52(1): 69–88, 2009.

[110] M.M. Meerschaert, J. Mortensen, and H.P. Scheffler. Vector Grünwald formula for fractional derivatives. *Fract. Calc. Appl. Anal.*, 7: 2004, 2004.

[111] M.M. Meerschaert and C. Tadjeran. Finite difference approximations for fractional advection-dispersion flow equations. *J. Comput. Appl. Math.*, 172: 65–77, 2004.

[112] M.M. Meerschaert and C. Tadjeran. Finite difference approximations for two-sided space-fractional partial differential equations. *Appl. Numer. Math.*, 56: 80–90, 2006.

[113] R. Metzler and J. Klafter. The random walk's guide to anomalous diffusion: a fractional dynamics approach. *Phys. Rep.*, 339: 1–77, 2000.

[114] R. Metzler and J. Klafter. The restaurant at the end of the random walk: recent developments in the description of anomalous transport by fractional dynamics. *J. Phys. A: Math. Gen.*, 37: R161–R208, 2004.

[115] K.S. Miller and B. Ross. *An Introduction to the Fractional Calculus and Fractional Differential Equations*. Wiley, New York, 1993.

[116] E. D. Nezza, G. Palatucci, and E. Valdinoci. Hitchhiker's guide to the fractional Sobolev spaces. *Bull. Sci. Math.*, 136: 521–573, 2012.

[117] Z. Odibat and S. Momani. An algorithm for the numerical solution of differential equations of fractional order. *J. Appl. Math. Informatics*, 26: 15–27, 2008.

[118] K.B. Oldham and J. Spanier. *The Fractional Calculus*. Academic Press, New York, 1974.

[119] M.D. Ortigueira. Riesz potential operators and inverses via fractional centred derivatives. *Int. J. Math. Math. Sci.*, 2006: 48391, 2006.

[120] J. Padovan. Computational algorithms for FE formulations involving fractional operators. *Computat. Mech.*, 2: 271–287, 1987.

[121] R.E.A.C. Paley and N. Wiener. *Fourier Transforms in the Complex Domain*. American Mathematical Society, 1934.

[122] G.F. Pang, W. Chen, and Z.J. Fu. Space-fractional advection-dispersion equations by the Kansa method. *J. Comput. Phys.*, 2014. DOI:10.1016/j.jcp.2014.07.020.

[123] D.W. Peaceman and H.H. Rachford. The numerical solution of parabolic and elliptic differential equations. *J. Soc. Indust. Appl. Math.*, 3: 28–41, 1955.

[124] I. Podlubny. *Fractional Differential Equations*. Academic Press, New York, 1999.

[125] I. Podlubny. Matrix approach to discrete fractional calculus. *Fract. Calc. Appl. Anal.*, 3: 359–386, 2000.

[126] I. Podlubny, A. Chechkin, T. Skovranek, Y.Q. Chen, and B.M. Vinagre Jara. Matrix approach to discrete fractional calculus II: partial fractional differential equations. *J. Comput. Phys.*, 228: 3137–3153, 2009.

[127] A. Quarteroni, R. Sacco, and F. Saleri. *Numerical Mathematics, 2nd Edition*. Springer-Verlag, New York, 2007.

[128] M. Rehman and R.A. Khan. The Legendre wavelet method for solving fractional differential equations. *Commun. Nonlinear Sci. Numer. Simulat.*, 16: 4163–4173, 2011.

[129] M. Rehman and R.A. Khan. A numerical method for solving boundary value problems for fractional differential equations. *Appl. Math. Model.*, 36: 894–907, 2012.

[130] J.C. Ren, Z.Z. Sun, and X. Zhao. Compact difference scheme for the fractional sub-diffusion equation with Neumann boundary conditions. *J. Comput. Phys.*, 232: 456–467, 2013.

[131] J.P. Roop. *Variational Solution of the Fractional Advection Dispersion Equation.* Clemson University, PhD thesis, 2004.

[132] A. Saadatmandi and M. Dehghan. A new operational matrix for solving fractional-order differential equations. *Comput. Math. Appl.*, 59: 1326–1336, 2010.

[133] A. Saadatmandi and M. Dehghan. A Legendre collocation method for fractional integro-differential equations. *J. Vibrat. Control*, 17: 2050–2058, 2011.

[134] S.G. Samko, A.A. Kilbas, and O.I. Marichev. *Fractional Integrals and Derivatives: Theory and Applications.* Gordon & Breach Science Publishers, Switzerland, 1993.

[135] J. Shen, T. Tang, and L.L. Wang. *Spectral Methods: Algorithms, Analysis and Applications.* Springer-Verlag, Heidelberg, Berlin, 2011.

[136] S. Shen and F. Liu. Error analysis of an explicit finite difference approximation for the space fractional diffusion equation with insulated ends. *ANZIAM J.*, 46: C871–C887, 2005.

[137] E. Sousa. Finite difference approximations for a fractional advection diffusion problem. *J. Comput. Phys.*, 228: 4038–4054, 2009.

[138] E. Sousa. Numerical approximations for fractional diffusion equations via splines. *Comput. Math. Appl.*, 62: 938–944, 2011.

[139] E. Sousa. A second order explicit finite difference method for the fractional advection-diffusion equation. *Comput. Math. Appl.*, 64: 3141–3152, 2012.

[140] E. Sousa. How to approximate the fractional derivative of order $1 < \alpha \leq 2$. *Int. J. Bifurcat. Chaos*, 22: 1250075, 2012.

[141] L.J. Su, W.Q. Wang, and Z.X. Yang. Finite difference approximations for the fractional advection-diffusion equation. *Phys. Lett. A*, 373: 4405–4408, 2009.

[142] H. Sugiuraa and T. Hasegawa. Quadrature rule for Abel's equations: uniformly approximating fractional derivatives. *J. Comput. Appl. Math.*, 223: 459–468, 2009.

[143] H.G. Sun, W. Chen, C.P. Li, and Y.Q. Chen. Finite difference schemes for variable-order time fractional diffusion equation. *Int. J. Bifurcation Chaos*, 22(4):1250085, 2012.

[144] Z.Z. Sun and X.N. Wu. A fully discrete difference scheme for a diffusion-wave system. *Appl. Numer. Math.*, 56: 193–209, 2006.

[145] C. Tadjeran, M.M. Meerschaert, and H.P. Scheffler. A second-order accurate numerical approximation for the fractional diffusion equation. *J. Comput. Phys.*, 213: 205–213, 2006.

[146] W.Y. Tian, H. Zhou, and W.H. Deng. A class of second order difference approximations for solving space fractional diffusion equations. *arXiv:1201.5949*, 2012.

[147] C. Trinks and P. Ruge. Treatment of dynamic systems with fractional derivatives without evaluating memory-integrals. *Comput. Mech.*, 29: 471–476, 2002.

[148] V.K. Tuan and R. Gorenflo. Extrapolation to the limit for numerical fractional differentiation. *Z. angew. Math. Mech.*, 75(8): 646–648, 1995.

[149] D. Valério and J. Costa. Variable-order fractional derivatives and their numerical approximations. *Signal Proc.*, 91: 470–483, 2011.

[150] H. Wang and N. Du. Fast alternating-direction finite difference methods for three-dimensional space-fractional diffusion equations. *J. Comput. Phys.*, 258: 305–318, 2014.

[151] H. Wang, K.X. Wang, and T. Sircar. A direct $O(N\log^2 N)$ finite difference method for fractional diffusion equations. *J. Comput. Phys.*, 229: 8095–8104, 2010.

[152] H. Wang and D.P. Yang. Wellposedness of variable-coefficient conservative fractional elliptic differential equations. *SIAM J. Numer. Anal.*, 51(2): 1088–1107, 2013.

[153] T. Wei and Z.Q. Zhang. Stable numerical solution to a Cauchy problem for a time fractional diffusion equation. *Eng. Anal. Bound. Elem.*, 40: 128–137, 2014.

[154] C.H. Wu and L.Z. Lu. Implicit numerical approximation scheme for the fractional Fokker-Planck equation. *Appl. Math. Comput.*, 216: 1945–1955, 2010.

[155] R.F. Wu, H.F. Ding, and C.P. Li. Determination of coefficients of high-order schemes for Riemann-Liouville derivative. *Sci. World J.*, 2014: 402373, 2014.

[156] J.Y. Yang, J.F. Huang, D.M. Liang, and Y.F. Tang. Numerical solution of fractional diffusion-wave equation based on fractional multistep method. *Appl. Math. Model.*, 38: 3652–3661, 2014.

[157] Q. Yang, F. Liu, and I. Turner. Numerical methods for fractional partial differential equations with Riesz space fractional derivatives. *Appl. Math. Model.*, 34: 200–218, 2010.

[158] B. Yu, X.Y. Jiang, and H.Y. Xu. A novel compact numerical method for solving the two-dimensional non-linear fractional reaction-subdiffusion equation. *Numer. Algor.*, 2014. DOI: 10.1007/s11075-014-9877-1.

[159] L.X. Yuan and Om P. Agrawal. A numerical scheme for dynamic systems containing fractional derivatives. *J. Vibrat. Acoustics*, 124: 321–324, 2002.

[160] S.B. Yuste. Weighted average finite difference methods for fractional diffusion equations. *J. Comput. Phys.*, 216: 264–274, 2006.

[161] S.B. Yuste and L. Acedo. An explicit finite difference method and a new von neumann-type stability analysis for fractional diffusion equations. *SIAM J. Numer. Anal.*, 42: 1862–1874, 2005.

[162] M. Zayernouri and G.E. Karniadakis. Exponentially accurate spectral and spectral element methods for fractional ODEs. *J. Comput. Phys.*, 257: 460–480, 2014.

[163] F.H. Zeng. Second-order stable finite difference schemes for the time-fractional diffusion-wave equation. *J. Sci. Comput.*, 2014. DOI: 10.1007/s10915-014-9966-2.

[164] F.H. Zeng and C.P. Li. Numerical approach to the Caputo derivative of the unknown function. *Central Eur. J. Phys.*, 11: 1433–1439, 2013.

[165] F.H. Zeng and C.P. Li. Fractional differential matrices with applications. *arXiv:1404.4429*, 2014.

[166] F.H. Zeng and C.P. Li. A new Crank–Nicolson finite element method for the time-fractional subdiffusion equation. Preprinted, 2014.

[167] F.H. Zeng, C.P. Li, and F. Liu. High-order explicit-implicit numerical methods for nonlinear anomalous diffusion equations. *Eur. Phys. J.-Special Topics*, 222(8): 1885–1900, 2013.

[168] F.H. Zeng, C.P. Li, F. Liu, and I. Turner. The use of finite difference/element approaches for solving the time-fractional subdiffusion equation. *SIAM J. Sci. Comput.*, 35(6): A2976–A3000, 2013.

[169] F.H. Zeng, C.P. Li, F. Liu, and I. Turner. Numerical methods for time-fractional subdiffusion equation with second-order accuracy. *SIAM J. Sci. Comput.*, 37(1): A55–A78, 2015.

[170] F.H. Zeng, F. Liu, C.P. Li, K. Burrage, I. Turner, and V. Anh. A Crank–Nicolson ADI spectral method for a two-dimensional Riesz space fractional nonlinear reaction-diffusion equation. *SIAM J. Numer. Anal.*, 52(6): 2599–2622, 2014.

[171] H. Zhang, F. Liu, and V. Anh. Galerkin finite element approximation of symmetric space-fractional partial differential equations. *Appl. Math. Comput.*, 217: 2534–2545, 2010.

[172] Y.N. Zhang, Z.Z. Sun, and H.L. Liao. Finite difference methods for the time fractional diffusion equation on non-uniform meshes. *J. Comput. Phys.*, 265: 195–210, 2014.

[173] Y.N. Zhang, Z.Z. Sun, and H.W. Wu. Error estimates of Crank–Nicolson-type difference schemes for the subdiffusion equation. *SIAM J. Numer. Anal.*, 49: 2302–2322, 2011.

[174] Y.N. Zhang, Z.Z. Sun, and X. Zhao. Compact alternating direction implicit scheme for the two-dimensional fractional diffusion-wave equation. *SIAM J. Numer. Anal.*, 50(3): 1535–1555, 2012.

[175] J.J. Zhao, J.Y. Xiao, and Y. Xu. A finite element method for the multiterm time-space Riesz fractional advection-diffusion equations in finite domain. *Abstr. Appl. Anal.*, 2013: 868035, 2013.

[176] X. Zhao and Z.Z. Sun. A box-type scheme for fractional sub-diffusion equation with Neumann boundary conditions. *J. Comput. Phys.*, 230: 6061–6074, 2011.

[177] X. Zhao, Z.Z Sun, and G.E. Karniadakis. Second-order approximations for variable order fractional derivatives: algorithms and applications. *J. Comput. Phys.*, 2014. DOI: 10.1016/j.jcp.2014.08.015.

[178] Z.G. Zhao and C.P. Li. A numerical approach to the generalized nonlinear fractional Fokker-Planck equation. *Comput. Math. Appl.*, 64(10): 3075–3089, 2012.

[179] Z.G. Zhao and C.P. Li. Fractional difference/finite element approximations for the time-space fractional telegraph equation. *Appl. Math. Comput.*, 219(6): 2975–2988, 2012.

[180] Y.Y. Zheng. A discontinuous finite element method for a type of fractional Cauchy Problem, in *Recent Advances in Applied Nonlinear Dynamics with Numerical Analysis*, C.P. Li, Y.J. Wu, and R.S. Ye eds. *World Scientific, Singapore*, pp. 105–120, 2013.

[181] Y.Y. Zheng, C.P. Li, and Z.G. Zhao. A fully discrete discontinuous Galerkin method for nonlinear fractional Fokker-Planck equation. *Math. Probl. Eng.*, 2010: 279038, 2010.

[182] Y.Y. Zheng, C.P. Li, and Z.G. Zhao. A note on the finite element method for the space-fractional advection diffusion equation. *Comput. Math. Appl.*, 59(5): 1718–1726, 2010.

[183] P. Zhuang, F. Liu, V. Anh, and I. Turner. New solution and analytical techniques of the implicit numerical method for the anomalous subdiffusion equation. *SIAM J. Numer. Anal.*, 46: 1079–1095, 2008.

Index

Printed and bound by CPI Group (UK) Ltd, Croydon, CR0 4YY
24/10/2024
01778283-0007